Laboratory Manual for INTRODUCTORY CHEMISTRY

fifth edition

Laboratory Manual for INTRODUCTORY CHEMISTRY

Otto W. Nitz • University of Wisconsin - Stout
Martin G. Ondrus • University of Wisconsin - Stout
Tammy Melton • St. Norbert College

WCB Wm. C. Brown Publishers
Dubuque, Iowa • Melbourne, Australia • Oxford, England

Book Team

Editor *Craig S. Marty*
Developmental Editor *Elizabeth M. Sievers*
Production Editor *Karen L. Nickolas*

Wm. C. Brown Publishers
A Division of Wm. C. Brown Communications, Inc.

Vice President and General Manager *Beverly Kolz*
Vice President, Publisher *Kevin Kane*
Vice President, Publisher *Earl McPeek*
Vice President, Director of Sales and Marketing *Virginia S. Moffat*
Marketing Manager *Christopher T. Johnson*
Advertising Manager *Janelle Keeffer*
Director of Production *Colleen A. Yonda*
Publishing Services Manager *Karen J. Slaght*

Wm. C. Brown Communications, Inc.

President and Chief Executive Officer *G. Franklin Lewis*
Corporate Senior Vice President, President of WCB Manufacturing *Roger Meyer*
Corporate Senior Vice President and Chief Financial Officer *Robert Chesterman*

Cover image: Longcore Maciel Studio © Wm. C. Brown Publishers

Copyedited by Sue Dillon

Interior design by Kay D. Fulton

A Times Mirror Company

ISBN 0–697–12482–7

Some of the laboratory experiments included in this text may be hazardous if materials are handled improperly or if procedures are conducted incorrectly. Safety precautions are necessary when you are working with chemicals, glass test tubes, hot water baths, sharp instruments, and the like, or for any procedures that generally require caution. Your school may have set regulations regarding safety procedures that your instructor will explain to you. Should you have any problems with materials or procedures, please ask your instructor for help.

Printed in the United States of America by Wm. C. Brown Communications, Inc., 2460 Kerper Boulevard, Dubuque, IA 52001

10 9 8 7 6 5 4 3 2 1

CONTENTS

PREFACE

TO THE STUDENT AND INSTRUCTOR

The fifth edition of *Laboratory Manual for Introductory Chemistry* retains many of the practical laboratory exercises that Otto Nitz developed when he published the first edition of this manual over 35 years ago. Experiments such as Impurities in Natural Water, Phosphate Contamination in Water, Percent by Volume of Alcohol in Wine, and Corrosion provide the students with pertinent applications of solid chemical principles. This manual has a wide selection of concepts for a one-semester course, but, with the inclusion of a series of qualitative analysis exercises, it provides the coverage and variety necessary for two semesters of General Chemistry laboratory.

These experiments have been used for years with a great deal of success and positive feedback from many students and faculty. Many of the exercises are quantitative, and often an unknown is included to challenge the student. A discussion section is included with each experiment to provide the student with immediately accessible information and to supplement the textbook. Access to balances having 0.001 g sensitivity is recommended for quantitative experiments.

Experiment descriptions are sufficiently detailed to allow students to carry out the procedure with minimal direction from the instructor. As with all laboratory work, the directions should he perused prior to the beginning of the laboratory period. The student should try to visualize the steps involved in the procedure before ever doing them. The report sheets allow the data, results, and conclusions to be summarized in a logical fashion to facilitate understanding by the student and grading by the instructor.

The experiments include discussion questions from which the instructor may pick and choose as desired. For some labs, the amount of work associated with doing all the review questions is quite extensive, and the number of questions assigned should be limited. For other experiments, the student could reasonably be expected to complete all of the review questions along with the laboratory work. Some of the discussion questions require that students search literature other than their textbook. Students have access to outstanding resources at college and university libraries. They should learn that information from literature sources in addition to the collection of data in the laboratory is an important part of scientific investigations.

Some of the equipment needed for these experiments (such as the gas-law apparatus in Experiment 25) is not commercially available and must be constructed on-site. If there is a question concerning the details of a piece of equipment, the authors would gladly provide additional information.

The authors have tried to compile a wide variety of practical and workable laboratory exercises in keeping with the "applied chemistry" emphasis found in earlier editions authored by Dr. Nitz. We will be open to questions, concerns, and suggestions from all who use this manual. Every effort will be made to use those suggestions to improve future editions.

Otto W. Nitz
Martin G. Ondrus
Tammy Melton

LABORATORY SAFETY

The following rules and regulations have been established to minimize hazardous conditions for everyone in the laboratory. They will not restrict your learning and will help to provide a safe environment in which you can learn. Read these rules carefully. You will be expected to follow them when you work in the laboratory.

1. Never work in the laboratory alone.
2. Conduct only authorized experiments and activities.
3. Wear eye protection at all times.
4. Know the location of fire extinguishers, eyewash fountains, safety showers, solid and liquid waste containers, and fume hoods. Know how to use these safety features.
5. Tie back long hair to keep it from falling into chemicals or a flame.
6. Leave sweaters, coats, purses, backpacks, and extra books in a designated area, not on the laboratory bench.
7. Always wear shoes in the laboratory.
8. Do not eat, drink, or smoke in the laboratory.
9. Treat all chemicals as if they were poisons; they could be dangerous if absorbed through the skin, inhaled, or tasted. Read all labels and directions carefully.
10. Wash chemical spills from hands, face, and eyes immediately and thoroughly using copious quantities of water.
11. Wipe up chemical spills immediately. This includes reagent shelves, reagent bottles, balance tables and shelves, and balances.
12. Leave the reagent bottles on the designated reagent shelf. When obtaining solid reagents, put a small amount on a weighing paper or in an appropriate container. Do not put spoons, spatulas, stirring rods, and so on into bottles or containers to obtain material; this causes contamination of reagents.
13. Never weigh granular reagents directly on a balance pan. Always use a suitable container or weighing paper.
14. Always replace the stopper or cap on a reagent bottle after removing a sample.
15. Never return unused chemicals (solids or liquids) to the reagent bottles or containers from which they were originally dispensed.
16. Dispose of waste chemicals in the manner prescribed by the instructor. Use hazardous waste disposal containers and toxic metal disposal containers as required.

EXPERIMENT 1

Laboratory Techniques

Purpose: To become familiar with common laboratory equipment; to learn useful techniques.

Materials: 6mm soft-glass tubing, crucible, and crucible cover. Copper, aluminum, and iron wire. Solutions (0.2 N) of lead nitrate and potassium chromate.

Safety Precautions: Use care with hot glass. It cools slowly and readily burns fingers.

Hazardous Waste Disposal: Lead chromate prepared and filtered in part D should be disposed of in a laboratory container designated for waste heavy metals.

INTRODUCTION

"What is all that stuff in my lab drawer, and how do I use it?"

This activity is designed to be used on the day of lab check-in to provide an introduction to some of the equipment found in the laboratory and in the lab drawer. The tools and techniques introduced here will be useful throughout the remainder of the semester. Students are encouraged to learn the location of regularly-used lab items such as ring stands, clamps, iron rings, Bunsen burners, balances, waste containers, purified laboratory water (distilled, deionized, or reverse-osmosis), fire extinguishers, fume hoods, eyewash fountains, and even the broom and dustpan. It is appropriate to fill the wash bottle located in the equipment drawer with laboratory water so that a source of purified water is always available at the student's desk for use in an experiment or for rinsing glassware.

PROCEDURE

A. Use of the Bunsen Burner

The Bunsen burner mixes air with the fuel (usually methane, CH_4, which is commonly called "natural gas"). Mixed in correct proportions with air, the methane burns completely. The reaction is:

$$\begin{array}{ccccccc} CH_4 & + & 2O_2 & \longrightarrow & CO_2 & + & 2H_2O \\ \text{methane} & + & \text{oxygen} & \longrightarrow & \text{carbon dioxide} & + & \text{water} \end{array}$$

Air, which supplies oxygen for the reaction, is drawn into the Bunsen burner through openings at the base and mixed with the fuel in the barrel of the burner. When the fuel-air mixture is properly adjusted the flame produced is blue, with a lighter blue cone in the center. The flame is nonluminous (produces very little light) and burns with a rustling sound.

If an insufficient amount of air is mixed with the fuel, particles of unburned carbon will be produced in the flame and heated until they emit light (become luminous). The flame will produce soot and smoke. Under these conditions the temperature of the flame is much lower than that of the properly adjusted blue flame.

Experiment with your Bunsen burner until you obtain a nonluminous flame with an inner cone. The light blue inner cone marks the region where combustion starts. Its center, which consists of unburned gases, is relatively cold. The hottest point of the flame is above the tip of this inner cone.

If the burner fails to light, burns above the barrel, or blows out, the gases may be coming out of the barrel faster than they can burn. The remedy is to reduce the rate of flow at the gas jet or change the proportion of air to permit faster burning. Sometimes the gas will start burning at the base and the barrel will become very hot. The burner must be turned off and relighted after the mixture is changed. Foreign matter in the gas jet in the base of the burner or on the "gallery" that surrounds the top of the barrel will prevent the burner from working properly.

1. *Temperature zones in a nonluminous (blue) flame.*

   The burner should now be adjusted to produce a blue flame with a distinct inner cone. Obtain a wire gauze and hold it horizontally across the flame about half an inch above the burner (use the tongs). Observe the region where the flame is hot enough to heat the wire gauze red-hot. Then raise the wire gauze about half an inch and repeat the observation. Continue to move the wire gauze up a half an inch at a time until it no longer gets red-hot. On the report sheet, sketch the nonluminous flame, showing the inner cone, and label the hottest area and the cool area.

   With a pair of tongs, hold a piece of iron wire (B & S #15) in the hottest part of the flame. Repeat using copper and aluminum wires. Record whether each melts. From the melting points of these metals (see table) estimate the approximate flame temperature. Record your results on the report sheet.

Metal	Melting Point
iron	1535°C
copper	1083°C
aluminum	660°C
tin	231°C

   Solder is used to connect copper wires. Why do you think solder is made of tin instead of iron?

2. *Temperature zones in a luminous (yellow) flame.*

   Adjust the burner to obtain a luminous flame (yellow flame like that of a candle) and repeat the preceding observations.

3. *Heating a crucible.*

   Place an empty, uncovered crucible on a wire triangle supported by an iron ring on a ring stand (Figure 1.1). Adjust the height of the ring so that the bottom of the crucible is above the top part of a nonluminous burner flame. Grasp the base of the burner in your hand, and *slowly* raise the operating burner so that the top of the burner moves closer and closer to the bottom of the crucible. This should be done slowly enough to observe changes in the intensity of the red glow at the bottom of the crucible produced by the flame. Try to determine the region of the flame that causes the bottom of the crucible to glow most brightly. If heating something in a crucible, how would the flame be positioned for strongest heating?

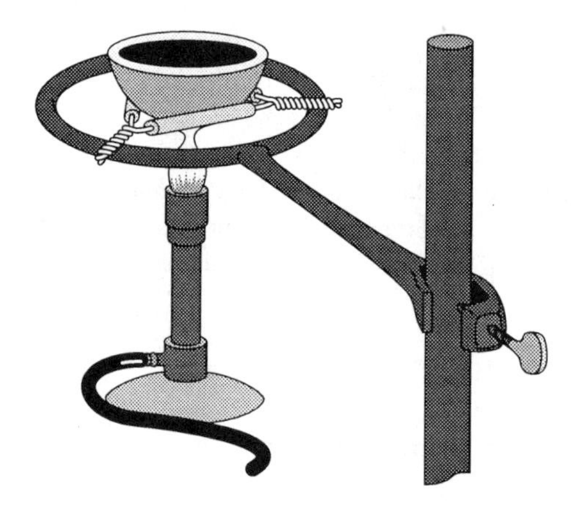

FIGURE 1.1
Heating crucible.

B. Glass Bending

Glass tubing can be cut by making a single mark with a sharp file at the desired place. This starts a crack that can be opened by placing the thumbs on the opposite side of the tube from the crack and bending very gently while pulling. The sharp edges become rounded.

The heat of a Bunsen burner will soften glass sufficiently for bending, but the bend must be a curve rather than a sharp angle. Too sharp a bend introduces strain and causes dangerous weak spots (Figure 1.2). In order to get a curve, a section of glass about two inches long must be heated, and this requires a flame spreader or "wing top."

Put the wing top on the burner and make sure the flame burns evenly. This may require removing the wing top and making the opening uniform with the handle of a file. When the flame burns properly, hold the tube in the flame above the light blue section that marks the unburned gases and rotate it to heat all sides evenly. When the glass is flexible, *remove it from the flame*, bend it gently, and hold it until it stiffens. Lay it on an ceramic square to cool, not on the stone table top. When it is cool enough to handle, fire polish both ends. *Take care! It is very easy to burn your fingers on hot glass.*

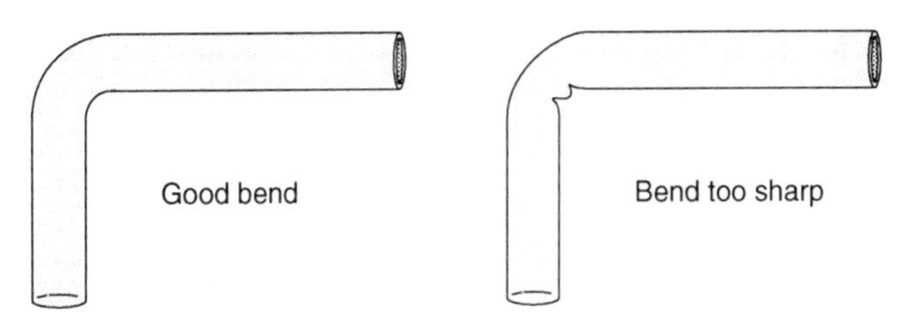

FIGURE 1.2
Examples of glass bends.

Obtain two pieces of glass tubing, one about six inches long and another about one foot long. Make a 90 degree bend about the center of the shorter piece. Make a 90 degree bend about two inches from one end of the longer one. Fire polish all raw ends. Before the close of the period have both pieces checked by the instructor and store them in your drawer for future use.

C. Volume Measurement

A graduated cylinder is commonly used for measuring volumes. The surface of water in a glass tube is never flat and tends to curve up the sides slightly. The curved surface is called the meniscus, and in a transparent liquid the volume is always measured at its lowest point; that is, at the center of the tube (the bottom of the meniscus). If the graduated cylinder is made of a polymer (plastic), the surface of the liquid is nearly flat (Figure 1.3). In either case, the cylinder must be held at eye level when reading the calibration marks. Too high or too low an eye position can result in errors, as shown in the diagram. With a little practice, cylinders calibrated in 1 mL divisions can be read to the nearest 0.1 mL by estimating between calibration marks. Many measurements require this degree of precision. The meniscus is best seen if you look toward a bright background such as a window.

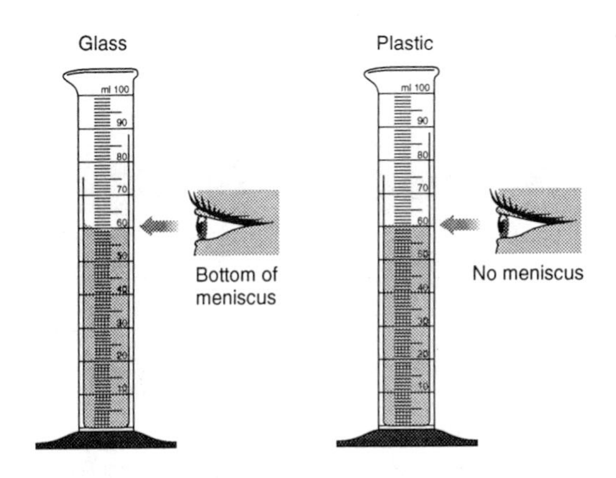

FIGURE 1.3
Volume measurements with glass and plastic graduated cylinders.

Glass graduated cylinders usually have the letters TC printed on the cylinder barrel. This means the cylinder is calibrated "to contain" whatever volume is read from the barrel markings. When the liquid is poured from the graduated cylinder, slightly less (a nearly negligible amount) than the amount contained will be delivered because a thin film of liquid remains adhering to the glass surface. Plastic graduated cylinders are usually calibrated TC/TD ("to contain" and "to deliver") because water solutions have almost no tendency to adhere to a plastic surface. This means that the amount read in the cylinder can be completely delivered from the cylinder.

Measure the volume of a small test tube from your lab drawer by filling the test tube with water and then pouring that water into an empty graduated cylinder. Record the volume on the report sheet.

D. Filtration

Solids are separated from liquids by filtration. Place a small funnel in a ring on the stand with a beaker under it. The long edge of the funnel tube should touch the wall of the beaker. This allows liquid to move from funnel to beaker without splashing.

Prepare the filter paper by folding it in half and again in quarters. Open it to form a cone and fit it to the funnel. Wet the paper with a few drops of water from a wash bottle to cause the paper to stick to the funnel. The rim of the paper should not extend above the top of the funnel. The funnel should never be filled with liquid above the top edge of the paper (Figure 1.4).

With your graduated cylinder measure exactly 10.0 mL of lead nitrate solution and pour it into a small beaker. Rinse the cylinder and measure exactly 10.0 mL of potassium

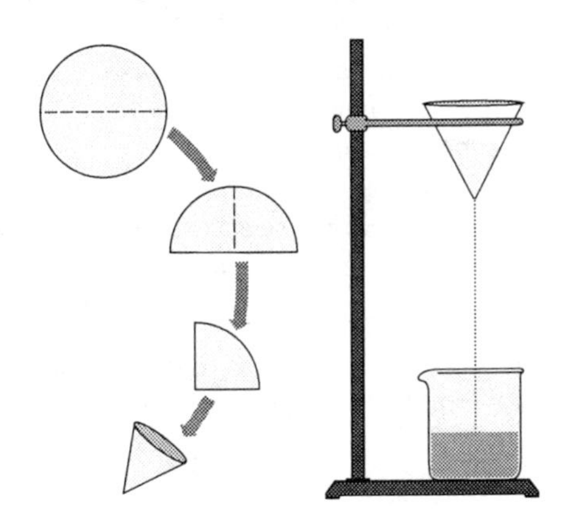

FIGURE 1.4
Filter paper folding and filtration apparatus.

chromate solution. Pour it into the beaker and thoroughly mix the two solutions. Pour the mixture carefully into the funnel and filter out the solid (lead chromate). The liquid that runs through is the filtrate, and it consists of potassium nitrate dissolved in water. Although colored, it should be clear; that is, free of any solid. A word equation describes the reaction:

$$\text{lead nitrate} + \text{potassium chromate} \longrightarrow \text{lead chromate} + \text{potassium nitrate}$$

Get the instructor's OK before the end of the period.

The yellow lead chromate on the filter paper (the **precipitate**) should be scraped into a heavy metal waste container using a spatula. The filter paper may then be disposed of in a wastepaper container, and the filtered liquid (the **filtrate**) may be poured down a sink drain.

E. Weighing

Several kinds of balances may be used in chemistry laboratories. Their selection depends on the precision required. They are

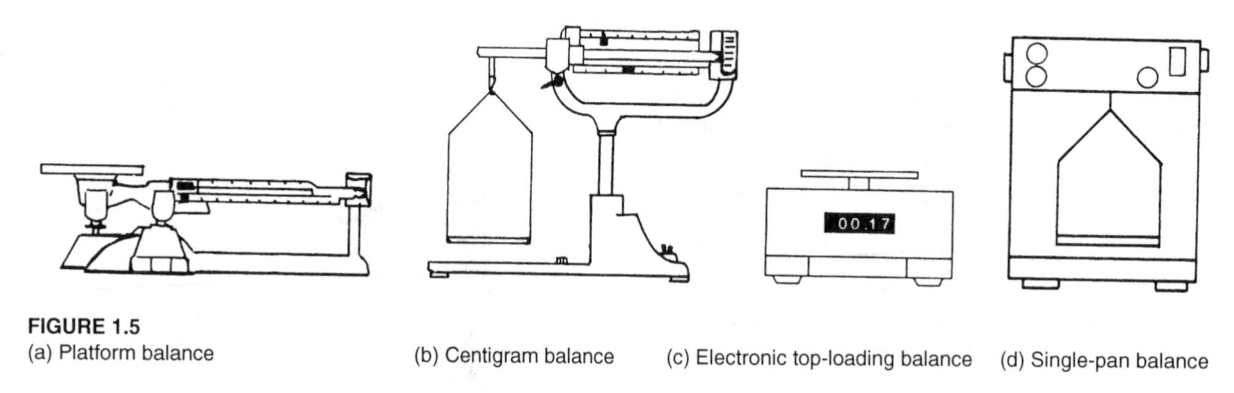

FIGURE 1.5
(a) Platform balance (b) Centigram balance (c) Electronic top-loading balance (d) Single-pan balance

illustrated in Figure 1.5.

The platform balance weighs to the nearest 0.1 gram. It is used when loads up to 1600 g are weighed and high precision is not needed. The centigram balance handles loads up to about 310 g and weighs to the nearest 0.01 g. Top-loading electronic and single pan balances are limited to loads of several hundred grams that will vary with the make and model of the balance. Most electronic or top-loading balances used in general chemistry laboratories weigh to the nearest milligram (0.001 g). Some weigh to 1/10 of a milligram (0.0001 g).

Balances that weigh in the same range often differ greatly in detail, and directions for the use of specific balances will be given by the instructor (Figure 1.5).

Certain instructions, however, are common to the use of all balances. You should learn them.

1. Make sure that the balance is level. Most instruments with sensitivities of 0.01 g, 0.001 g, or 0.0001 g have leveling bubbles.
2. Make sure that the balance is adjusted to zero when empty.
3. If the balance does not have a glass case, protect it from air currents when weighing.
4. Never put chemicals directly on a balance pan. Use a weighing paper or a vessel of some kind.
5. Lock the beam when finished (if a lock is provided on some mechanical balances) or leave a weight on the beam so that it cannot swing in air currents.
6. Never overload the balance.
7. If a chemical is spilled anywhere on the balance, clean it up at once.
8. Many electronic balances used in laboratories cost over $1000. They are often the most costly and sensitive piece of equipment in the general laboratory. Treat them with the care due to any finely tuned, high-quality product.

Weigh a crucible (*making sure it has cooled to room temperature*) and cover separately using the most precise balance available. Record the weights on the report sheet and add them. Then weigh the two parts together and record your results. Close agreement between the two figures indicates good weighing technique. Get the instructor's OK when you have finished.

F. Rough Measurements

For many purposes in chemistry, the most precise measurement possible must be used, and there is no substitute for painstaking care. This is usually the case when the measurements are to be used in a calculation. However, in many procedures, amounts are not at all critical and are given merely to avoid waste of materials. Here the extra time needed for precise measurements would be wasted. You should learn to judge the need.

In many procedures, about 5 mL of solution is needed. This amount can be roughly estimated by the width of your thumb on a 30-mL test tube. The drawer equipment will usually include a spatula for estimating small amounts of solids. The spatula *should not* be used to remove solids from stock bottles on the reagent shelf because chemicals on the spatula could contaminate the entire contents of the stock bottle. However, the spatula is useful for transferring rough amounts of granular solid from a beaker, watch glass, weighing paper, or similar container to another container.

Experiment 1
Laboratory Techniques

Name____________________
Date____________________

REPORT SHEET

A. Bunsen Burner

1. Nonluminous (blue) flame

Flame sketch

Wire	Temperature Estimate Melting Point	Behavior in Flame
Iron		
Copper		
Aluminum		

Estimated flame temperature ____________

Why do you think solder is made of tin instead of iron?

2. Luminous (yellow) flame

Flame sketch

Wire	Temperature Estimate Melting Point	Behavior in Flame
Iron		
Copper		
Aluminum		

Estimated flame temperature ____________

3. How would the crucible be positioned in a burner flame to achieve the most intense heating?

B. Glass Bending Instructor's OK

C. Volume Measurement

Volume of water poured from test tube ____________________

D. Filtration Instructor's OK

E. Weighing

Type(s) of balance(s) used in your laboratory ____________________________________

Capacity ____________________ Sensitivity ____________________

weight of crucible =

weight of cover = ____________________

Sum of above weights =

Weight of crucible and cover when weighed together ____________________

Instructor's OK ____________________

EXPERIMENT 2

Measurement and Density

Purpose: To become familiar with measurements in the metric system; to measure density, an intrinsic property of a substance.

Materials: Wooden blocks; bars of aluminum, tin, zinc, lead, copper, iron, and magnesium (approximately 1.5 cm x 5 cm); water; a solution of ethanol in water; a solution of sodium chloride in water; graduated 10-mL pipets and pipetting bulbs; irregular solids such as nuts and bolts, rocks, plastic and glass pieces.

Safety Precautions: Never draw liquids into a pipet by mouth. Always use a pipetting bulb.

Hazardous waste disposal: None

INTRODUCTION

All measurements are made by comparison with some selected standard. A system of measurement consists of a series of standards (e.g., standards of length, weight, volume, time, etc.). By comparing an object to accepted standards we can obtain measurements that are easily communicated to others. It is essential, of course, to designate which standard we have used (to "name the unit") in recording the measurement.

As you learn to use tools in the laboratory, you will measure several properties of substances. A measurement has two parts: a numerical quantity and the proper unit. Both parts are necessary for a correct measurement; one part without the other is incorrect. If a nurse records "kg" for your weight on your medical chart, you would conclude that the record is incorrect because he or she forgot to include the numerical value of the weight. Likewise the record would be incomplete if only the number "95" were recorded because anyone reading the chart would be unsure if the number represents pounds, kilograms, or some other weight unit. Never assume that someone else understands the appropriate unit of measurement for a numerical quantity. It is always the responsibility of the person who makes a measurement to communicate the correct numerical quantity *and* unit.

When working with numerical quantities, you must consider *precision, accuracy,* and *error*. **Accuracy** is the closeness with which an experimental result approaches the true value. **Precision** is the closeness with which several measurements of the same quantity made with the same measuring tool agree. **Error** is the difference between the true value and the experimental measurement.

You will work with measured values and calculated values as you complete this activity and others during the semester. Not only must the numerical quantity be recorded correctly, but the unit must be recorded correctly as well. You are also expected to record measured numbers to reflect the precision limit of the measuring tool. This is usually the smallest numerical value that can be estimated with the measuring device. For example, the precision limit of a centigram balance is ±0.01 g. Therefore, any weight measurement made on a centigram balance must have two digits after the decimal point, even if both are zeros (that is, 25.00 g, not 25 g). Conversely, a recorded number such as 34.56 mL indicates that the measurement was made with an instrument having a precision limit of ±0.01 mL. The precision limit is sometimes referred to as the uncertainty of the measurement.

When measured values are used to calculate a relationship between parameters (for example, density), the calculated value must reflect the uncertainty of the measurements. This is the time when significant figure rules are applied to the numerical quantities of measurements. If you are unfamiliar with these rules, find them in your textbook or in the Appendix at the end of this laboratory manual and review them at this time.

Accurate measurements are obtained only when the measuring tool is operating properly. Several sources of error lead to inaccurate measurements. Incorrect tool calibration is one. Most inaccuracies, however, result from incorrect use of the tool. A common source of error, especially with balances, is lack of zeroing before a measurement is made. Since several people will be using the same balance in the lab, be sure to check the zero adjustment before each measurement. Always leave the balance clean and undamaged for others.

Mass, volume, and length are examples of *extrinsic* properties; that is, those that depend on the amount of a substance examined. *Intrinsic* properties, on the other hand, depend on what the substance is, not on the amount present. Color, density, and specific heat are examples of intrinsic properties. They are frequently used to identify substances.

In this experiment mass and volume measurements will be combined to determine the densities of various substances. The density is the mass of a definite volume of the substance. It may be expressed in several ways; for instance, as tons per cubic yard, pounds per cubic foot, pounds per gallon, ounces per cubic inch, etc. In science, metric units expressed as grams per cubic centimeter or grams per milliliter are used. (One cubic centimeter is equal by definition to one milliliter.)

Most methods of measuring density require that both the weight (mass) and the volume of the sample be determined. The density, then, is the mass divided by the volume. The accuracy of the calculated result cannot be better than that of the least accurate measurement used. It is important to record all measurements carefully, including the exact number of digits actually determined.

PROCEDURE

In Experiment 1, we performed simple volume and mass measurements. Those, along with the measurement of length, will again be used with increased emphasis on precision and accuracy. These measurements will also be combined to determine density.

A. Measurement of Weight

1. *Centigram balance.* In general chemistry laboratories, centigram balances get a great deal of use because they are dependable, rugged, and relatively inexpensive (costing between $100 and $200). Learn how to use them well. The smallest graduation or division on this type of balance is 0.01 g, which is 1 centigram (1 cg). Therefore, a sample can be weighed with an uncertainty or precision of ± 0.01 g.

   To determine the reproducibility from one centigram balance to another, the mass of a coin (nickel, dime, or quarter) will be measured three times. Begin by adjusting the zero. Then place the coin on the balance, and record the mass on the report sheet to the nearest 0.01 g. Reweigh the coin on two other centigram balances in the laboratory, making sure that each is zeroed first.
2. *Milligram balance.* Milligram electronic balances (costing between $1,000 and $3,000) vary in capacity and appearance but allow the operator to adjust the zero and obtain measurements more quickly than with mechanical centigram balances. Use a milligram balance to determine the mass of the same coin just measured on the centigram balance. Remember to first zero the balance and to record the mass to the nearest 0.001 g.

B. Measurement of Volume

1. *Graduated cylinder.* In Experiment 1, the slight difference between reading glass and plastic graduated cylinders was discussed. The volume markings on various brands and sizes of graduated cylinders may vary—division marks may represent different volume increments from one cylinder to another. Examine the markings on each of the graduated cylinders in the lab drawer to determine the volume represented by each of the smallest divisions. Draw the small divisions between three of the large divisions in the sketches on the report sheet. Label the large divisions. Record the volume represented by the smallest division and the precision limit for each graduated cylinder. Complete the following activity:

   Measure the volume of a 250-mL beaker when it is filled to the top by using the large graduated cylinder. To do this, fill the 250-mL beaker with water. Pour water from the beaker into the large graduated cylinder and determine the volume transferred by reading the liquid level in the graduated cylinder. Do not overfill the cylinder so that the water level is higher than the top mark. Record on the report sheet the volume of water transferred and then discard the water. Repeat the procedure to obtain the total volume of the beaker. Add the measured volumes and record this value as the total volume of the beaker.

2. *Pipet.* Pipets are designed to measure the volume of liquid delivered to another container (and are calibrated "TD" rather than "TC" like most graduated cylinders). Obtain a graduated 10-mL pipet and examine the markings on the pipet to determine the volume represented by each of the smallest divisions. Draw the smallest divisions between three large divisions in the sketch on the report sheet. Compare these markings with the markings on a graduated cylinder.

C. Measurement of Length

Another fundamental measurement is distance (length). Measure the length and width of this page in centimeters, and record each value on the report sheet to the nearest 0.1 cm. Repeat the measurement in inches also recording to the nearest 0.1 inch.

Divide the number of centimeters by the number of inches for both the length and the width measurements. Record the answer as the experimental conversion factor. Compare your result with the accepted conversion factor of 2.54 cm/in. If your calculated conversion factor differs by more than ± 0.03 cm/in. from the accepted value, repeat the measurements more carefully and perform the calculation with the new measurements. Record the difference between the measured and the accepted conversion factors in the "error" space on the report sheet.

D. Combination of Measurements

Density is a property of matter that can be used to identify many substances because different substances often have different densities. Knowing the density of a substance can, for example, allow one to predict whether the substance will float or sink in water (or any other liquid). A substance with a density greater than that of the water or other liquid will sink in the liquid; a substance with density less than that of the liquid will float.

Density is calculated by dividing the mass (weight) of a sample of material by the volume (space) this mass occupies. In the activities that follow, the mass and volume of several materials will be measured followed by calculation of the density. In most density measurements for solids the sample is weighed, but several methods of determining the volume are available:

Geometric calculation from measured dimensions. If the shape is regular, the dimensions are measured and the volume is calculated.

Displacement. A volume of water sufficient to immerse the sample is carefully measured in a graduated cylinder, the object is put into the water, the new water level is read, and the difference is determined.

Archimedes' principle. The object is weighed in air and again when immersed in a liquid, and the volume is calculated from the weight change caused by buoyancy of the object in the liquid. This method is only applicable to solids having density greater than that of the liquid so that they will submerge completely.

If the sample is a liquid, it may be weighed, the volume measured, and the density calculated as with solids. By an alternate method, the density may be measured directly with a hydrometer.

1. *Regular wooden solid.* Obtain one of the wood blocks from the reagent shelf and record the identification number of the block on the report sheet. Weigh the block to the nearest 0.01 g and record this value. Carefully determine the dimensions of the block in centimeters, recording the values to the nearest 0.1 cm. Calculate the volume of the block by multiplying length times width times height. Then calculate the density of the piece of wood in grams per cubic centimeter.
2. *Metal.* Obtain one metal bar, not necessarily of simple geometric shape, and weigh it as accurately as the balance permits. Record the weight. Measure the volume by liquid displacement and Archimedes' principle using the following directions, and calculate the density using as many significant figures as occur in the measurement having the fewest significant figures. The bar may be zinc, tin, copper, aluminum, lead, iron, or magnesium.

   Displacement. Measure into a 25-mL graduated cylinder some definite volume of water (15.0 mL is suitable). Read the volume measurement to the nearest 0.1 mL. Tie a thread to the bar, lower the bar carefully into the water, and read the new water level to the nearest 0.1 mL. Make sure that the bar is completely immersed, that there are no adhering air bubbles, and that the water level does not rise above the calibration marks. Calculate the density using the correct number of significant figures.

   Find out which metal you have, look up its density in the *Handbook of Chemistry and Physics*, compare it with your value, and calculate the percent error as follows: Determine the difference between the measured and the correct value and divide this difference by the correct value. This gives the fraction of error of the correct value. When this fraction is multiplied by 100%, the percent error is obtained. **Note:** If the percent error is greater than 5%, the instructor may ask you to repeat the measurement.

   Archimedes' principle. Tie a minimum length of thread to the bar and suspend the bar from the arm of the balance. Fasten the platform over the pan of the balance so that it does not interfere with the balance swing. Put a 250-mL beaker on the platform and adjust its position so that the bar hangs inside it but does not touch the sides or bottom. Add enough water to completely immerse the bar. See Figure 2.1.

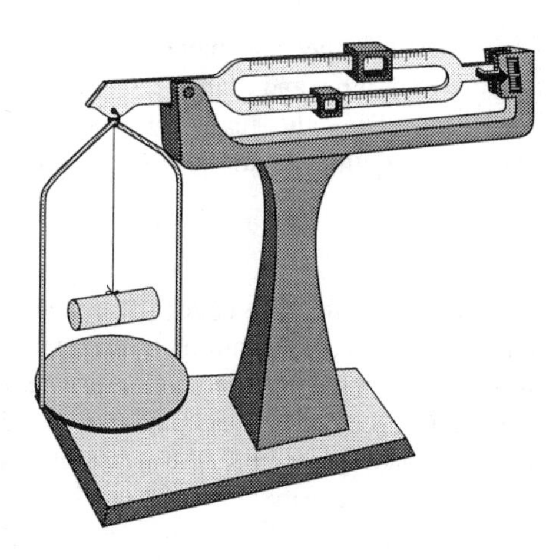

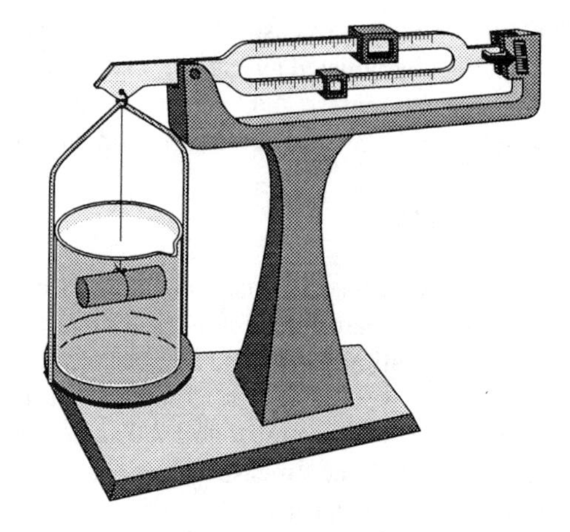

FIGURE 2.1
Suspend bar from arm of balance and weigh. Submerge bar in water and reweigh.

Weigh the immersed bar as accurately as the balance permits and record the weight. According to the principle of Archimedes, an object immersed in a fluid loses weight equal to the weight of the fluid displaced. Because the density of water is essentially 1.00 g/mL, the weight of water displaced in grams is the same as the volume of the object in milliliters. Calculate the density of the bar in the space provided on the report sheet. Compare the experimental value to the handbook value, and calculate the percent error.

3. *Density of a Liquid.* Measurement of volume with a graduated cylinder. Weigh the small (25-mL) graduated cylinder using a centigram balance. Make sure the cylinder is completely dry before weighing it. Record this value. Pour about 20 mL of laboratory water into the graduated cylinder. It is not necessary to obtain exactly 20.0 mL, but it is necessary to record the exact volume of liquid measured to the best precision possible with the graduated cylinder used. Record the volume reading. Place the graduated cylinder on the centigram balance and determine the mass of the cylinder plus its liquid contents.

   Calculate the weight of the water in the cylinder by subtracting the weight of the empty graduated cylinder from the weight of the cylinder plus liquid. The density of the liquid can now be calculated. The density of water at room temperature is 1.00 g/mL. Compare your measured value with the true value by calculating the percent error.

   Measurement of volume with a 10-mL pipet. Weigh a stoppered Erlenmeyer flask with a milligram balance. Draw a sample of laboratory water into a 10-mL pipet (using a rubber pipetting bulb) so that the liquid level is above the top (0.00 mL) calibration mark. Slip your finger over the upper end of the pipet, and by gently manipulating your finger, allow the liquid level to drop until the bottom of the meniscus just reaches the 0.00-mL calibration line. Touch the end of the pipet to the vessel containing the extra water to remove adhering drops, then transfer 10.00 mL to the weighed flask. Again touch the tip to the surface of the liquid in the flask but do not blow out the last bit of liquid inside. The instructor will demonstrate the use of the pipet. Stopper and weigh the flask and contents as accurately as possible and calculate the density of the water or other liquid.

E. Unknowns

Obtain two unknowns from the dispensing area: one solid and one liquid. The solid may be regular or irregular in shape and may be a metal, wood, rock, or plastic and may or may not be higher in density than water. The liquid may be an organic solvent such as ethanol or acetone, or it may be water or a water solution.

For each unknown, record the unknown number in the space provided on the report sheet. Note that the report sheet consists only of a large blank space for each unknown. Decide on an appropriate method to determine the density of each unknown. Identify the method used on the report sheet in the space provided. Record all measured values in a neat, logical fashion with appropriate units, and show the setup on each calculation. Identify the density for each unknown by positioning it so that it can be easily seen on the sheet and by underlining it. **Note:** You are not expected to calculate percent error for the unknowns because the identity of the sample is "unknown" to the student.

Experiment 2

Measurement and Density

Name________________________

Date________________________

REVIEW QUESTIONS

1. Express the density of (a) aluminum and (b) water in each of the following units:

   g/cm^3 lb/ft^3 g/L lb/gallon

2. An astronaut's camera was weighed on earth using a double pan balance and also a spring scale. It weighed 2.00 kilograms by each method. The camera was taken to the moon, where the gravity is about 1/6 that of earth, and the weighings were repeated. Did the results still agree? Justify your answer.

3. Is the weight of an object exactly the same all over the earth? Explain.

4. We say that an object is weightless in space. Is this true? Is the object massless? Explain.

5. Is a satellite that is orbiting the earth two hundred miles above the surface beyond the attraction of earth's gravity? Explain.

6. Explain how a satellite can maintain a fixed position above the earth.

7. Show how Archimedes' principle applies to a human body floating in a swimming pool.

   Estimate the overall density of the human body.

8. Would more of the same floating object be submerged in Lake Michigan or in Great Salt Lake? Explain.

9. Explain the relationship of the displacement of a ship to its weight.

10. Why are the displacement and Archimedes' principle methods for measuring the volume of a solid not useful for a wooden sample?

11. The numerical quantity of a weight measurement can be both precise and inaccurate. Give an example of this type of measurement.

12. Which is more precise, a gram balance or a milligram balance? Explain.

13. Convert 55.5 inches to centimeters. Use the accepted value for the conversion factor.

14. A student calculated a conversion factor of 2.63 cm/in. from some measured values of length. Calculate the percent error of the conversion factor.

15. Would it be preferable to use a 25-mL or a 100-mL graduated cylinder to measure 11.5 mL of a liquid? Explain.

Experiment 2
Measurement and Density

Name________________________________

Date________________________________

REPORT SHEET

A. Measurement of Weight

Coin Weighed	Centigram Balances			
	1	2	3	Milligram Balance
	g	g	g	g

What is the uncertainty of measurement in weights obtained using the:

a. Centigram balances? ____________________

b. Milligram balances? ____________________

Normally the same balance should be used for measurements throughout an experiment. Why is this a wise practice to follow?

B. Measurement of Volume

1. Graduated cylinder

____________________-mL cylinder

volume per division: __________

precision limit: __________

____________________-mL cylinder

volume per division: __________

precision limit: __________

a. Volume of 250-mL beaker = __________ + __________ + __________ = __________

2. 10-mL pipet

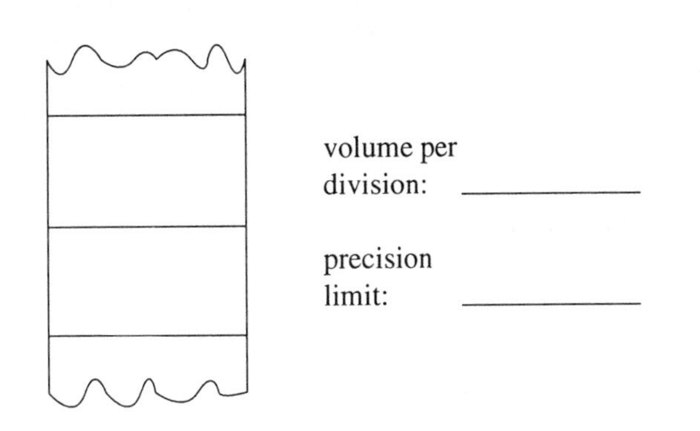

Compare and contrast the markings on the pipet with the markings on the graduated cylinders.

C. Measurement of Distance

Measurement	Page length		Page width	
	Trial 1	Trial 2 (if necessary)	Trial 1	Trial 2 (if necessary)
Metric units				
English units				
Experimental Conversion Factor				
Accepted Conversion Factor	2.54 cm/in.	2.54 cm/in.	2.54 cm/in.	2.54 cm/in.
Error				

D. Combination of Measurements—Density

1. Regular wooden solid (identification number: ______________________)

length = ______________________

width = ______________________

height = ______________________

weight = ______________________

volume = ______________________

Density = $\frac{\text{weight}}{\text{volume}}$ = ______________________

2. Metal

Number on metal ___________ Kind of metal ___________ Mass of metal ___________

Displacement

Final volume =

Initial volume =

____

Volume of metal =

Density calculation

(watch significant figures)

Handbook value for density of the metal = ______________________

Percent error = ______________________

Archimedes' principle

Weight in air =

Weight in water =

____

Loss of weight =

Density calculation

(watch significant figures)

Handbook value for density of the metal = ______________________

Percent error = ______________________

3. Density of a liquid (water)

Measurement of volume with graduated cylinder

		Trial 2 (if necessary)
Weight of graduated cylinder + water =	______________	______________
Weight of empty graduated cylinder =	______________	______________
Weight of liquid =	______________	______________
Volume of liquid =	______________	______________
Density of liquid =	______________	______________
Accepted density of water =	______________	______________
Percent error =	______________	______________

Measurement of volume with pipet

		Trial 2 (if necessary)
Weight of Erlenmeyer + water =	______________	______________
Weight of stoppered Erlenmeyer =	______________	______________
Weight of liquid =	______________	______________
Volume of liquid =	______________	______________
Density of liquid =	______________	______________
Accepted density of water =	______________	______________
Percent error =	______________	______________

E. Unknowns

1. Liquid (unknown number ______________)

Show all figures (including correct units) and all calculations in a logical fashion in the space provided.

2. Solid (unknown number ______________)

Show all figures (including correct units) and all calculations in a logical fashion in the space provided.

EXPERIMENT 3

Physical and Chemical Properties

Purpose: To observe some physical and chemical properties of a group of substances; to identify an unknown on the basis of its properties.

Materials: Biphenyl, acetone, sodium chloride, hexane, cyclohexane, phenyl salicylate, ethanol (95%), $AgNO_3$ (0.2 N), capillary melting-point tubes, hot plates, boiling chips, small rubber bands made by cutting 1/16-inch segments from rubber tubing.

Safety Precautions: Be particularly cautious with combustible solids and liquids. Keep them away from Bunsen burner flames. Use small samples when testing combustibility.

Hazardous Waste Disposal: Place unused liquids (with the exception of silver nitrate) in the "waste organic solvents" container. Silver nitrate contains a heavy metal and should be poured into the "heavy metals" disposal container. Phenyl salicylate and biphenyl should be disposed of in the "waste organic solids" container.

INTRODUCTION

A substance is recognized by its properties, and by its use of them, it can often be separated from other substances. The properties of a substance also determine its use. If a property can be observed without changing the composition of a substance, it is a **physical property.** Melting point, boiling point, color, and density are examples of physical properties. **Chemical properties** are observed when a substance reacts to produce one or more different substances in a chemical change. Such changes can usually be recognized by the evolution of a gas, formation of a precipitate, the evolution or absorption of heat, evolution of light, or the production of an electric current.

When a solid changes to a liquid at its melting point, a **physical change** occurs. Physical changes involve a change in "form" without a change in chemical identity. Likewise, a liquid changing to a gas at its boiling point or evaporation of the liquid at a lower temperature is a physical change because the form (but not the identity) changes. Chemical properties usually, but not always, are manifested by **chemical changes** producing new substances. Thus, the *tendency* for a substance to burn (combustibility) is a *chemical property* and the *process* of burning is a *chemical change*. The inability of some substances to burn is also a chemical property, but no chemical change is associated with lack of burning.

In this experiment you will observe physical and chemical properties and use some of them to identify an unknown substance. Several of the substances used as examples may be familiar to you:

Biphenyl: used in organic synthesis
Sodium chloride: table salt
Phenyl salicylate: used in making certain plastics

Hexane: a component of some gasolines
Cyclohexane: an organic solvent
Acetone: an organic solvent

The unknown you identify will be one of these substances.

PROCEDURE

A. Properties of Some Known Substances

Obtain from the storeroom a small sample of each of the substances just listed. Make the following tests and record your results on the report sheet. Identify each property observed as chemical or physical.

1. *Direct observation.* Examine each sample and record the physical state (solid, liquid, or gas), color, and odor if detectable.
2. *Combustibility.* Because the liquids used in this experiment are highly combustible, the instructor will demonstrate flammability or combustibility of these substances. A few drops of each liquid will be tested for combustibility **by the instructor.** For liquids that have vapors heavier than air, the **instructor** will pour the vapor from a 250-mL beaker into a 150-mL beaker and then demonstrate the combustibility of the vapor.

   Students should test combustibility on solid samples only. Place a match-head size portion of the solid on the end of a Monel metal spatula. Heat the spatula gently about an inch from the sample. If the sample seems to melt readily, plan to make a regular melting point test (described in the next paragraph). Now move the substance on the spatula directly into the flame and note whether it will burn.
3. *Melting point.* Fill a 150-mL beaker about half full of water, mount it on a ring stand as shown (Figure 3.1), and put a small ring around it to reduce danger of spillage. Obtain a melting point tube and a 1/8 inch piece of rubber tubing. Press the open end of the glass tube repeatedly against the sample of a piece of paper, turn the tube closed end down, and tap it on the desk top or rub it gently with a file. The vibration causes the crystals to fall to the closed end. When about 1/4 inch of the sample has been transferred to the tube, slip the rubber band over your thermometer and move it up until it is 2–3 inches above the bulb and lower part of the sample tube immersed in water. The thermometer is most easily held by the clamp if a rubber stopper is slid onto the thermometer near the top of the thermometer. Cutting a slit vertically through one side of the stopper will allow the stopper to be moved more easily on the thermometer.

   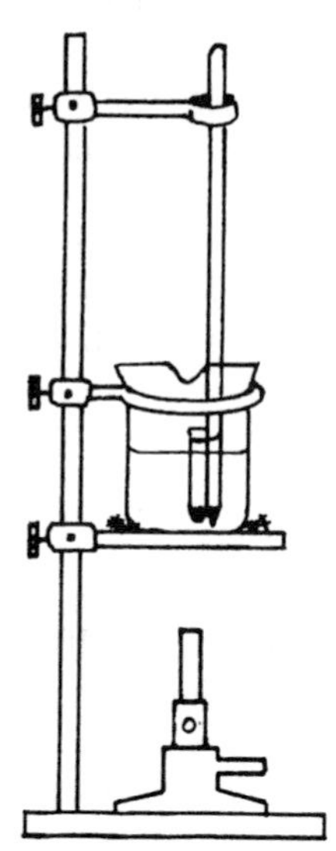

   FIGURE 3.1
   Melting point apparatus.

   Heat the water gently, stirring actively with a glass rod. The rubber tube should not touch the water. Bring the temperature up slowly, watching the sample and the temperature while stirring. Record the temperature at which the sample melts. If the water bath is heated too rapidly, the reading may be indefinite. Replace the water with cool water while preparing another sample. Then heat at 2–3 degrees per minute to obtain an exact melting point. Record your results.

   Repeat the experiment with each sample that seemed to melt on the spatula in the flame.
4. *Boiling point.* Determine the boiling points of the liquid samples using the apparatus diagrammed in Figure 3.2. Fit a large test tube with a two-hole stopper holding a thermometer. **Note:** As with the melting point apparatus, slitting vertically down the stopper into one of the two holes will allow the thermometer to slide much more easily and greatly reduce the chance of breaking a thermometer. Mount the tube vertically on a ring stand with a clamp. Put about 10 mL of one of the liquid samples into the tube, add a boiling chip, and adjust the thermometer so that the bulb is about 2 inches above the liquid. Heat in a beaker of boiling water **with a hotplate** until the liquid boils and liquid drips off of the bulb. Do not allow the sample to boil rapidly. Record your result.

Repeat the test with the other liquids, using a clean test tube each time. Wipe the thermometer and stopper well between samples.

5. *Density of liquid substance.* Determine the density of the liquids using the graduated cylinder method from Experiment 2. Measure about 10 mL of the liquid being tested in a small, preweighed graduated cylinder. Record the measured volume of the liquid, the weight of the liquid, and its calculated density on the report sheet.
6. *Solubility in water and ethanol.* Put about 10 mL of distilled water and a sample of one of the substances being tested into a small test tube. Use a match-head size portion of a solid or about 5 drops of a liquid. Shake and note if dissolving occurs. Record your observation. If the sample dissolves, proceed directly to Test 6 using the solution you have. Repeat with each sample. Repeat the entire series of solubility tests, using ethanol instead of water as the solvent. Record your observations.
7. *Treatment with silver nitrate solution.* Select the substances that are water-soluble for this test. To the aqueous (water) solution of the substance add about 5 mL of silver nitrate solution. **Caution!** Silver nitrate solution will stain clothing or skin black if it is spilled. Observe and record your results.

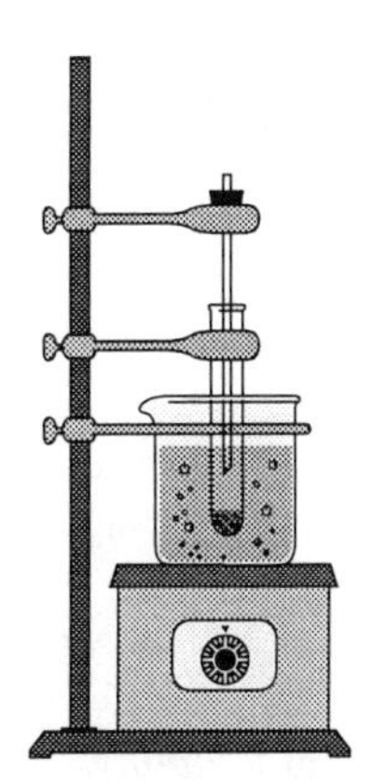

FIGURE 3.2
Boiling point apparatus.

B. Identification of an Unknown Substance.

Obtain an unknown substance (one of the samples you tested in Part A) and record its number. Identify this unknown by observing as many of its properties as necessary. If the unknown is a liquid, do not measure combustibility (we know all three liquids burn) or melting point (for obvious reasons). If the unknown is a solid, do not measure density or boiling point. Record your data.

Experiment 3
Physical and Chemical Properties

Name______________________________

Date______________________________

REVIEW QUESTIONS

1. Are the physical properties of a substance changed if that substance undergoes a chemical change? Explain. Are the chemical properties changed if the substance undergoes a physical change? Explain.

2. Define the term *solubility*.

3. Look up the meanings of these terms: *solute, solvent, concentration of solution.*

4. True solutions are clear and transparent, although they may be colored. What conclusions can you draw as to the size of the dissolved particles?

5. Properties may be classified as extrinsic or intrinsic. Look up the meaning of the words (see the introduction to Experiment 2) and state whether the properties observed in the experiment are extrinsic or intrinsic.

6. From references find out what the effect on the melting point would have been had the sample not been pure.

7. From references find out how the sample might have been purified if the melting point had indicated it to be impure.

8. What dangers to operator or equipment exist in taking melting points?

9. Using references, find out whether the effect of temperature changes on solubility is the same for all substances.

10. What is the material that stains hands or clothing dark when silver nitrate is spilled on them, and how is it formed?

Experiment 3

Physical and Chemical Properties

Name________________________________

Date________________________________

REPORT SHEET

A. Properties of Some Known Substances

Property	Biphenyl	Sodium Chloride	Hexane	Phenyl Salicylate	Cyclohexane	Acetone	Physical or Chemical
1. Physical state							
Color							
Odor							
2. Combustibility							
3. Melting Point							
4. Boiling Point							
5. Density Volume Liquid Mass Liquid Density							
6. Solubility in H_2O							
Solubility in Ethanol							
7. Silver Nitrate							

Record any additional notes, observations, and calculations on properties of the known substances in the space provided below.

B. Identification of an Unknown Substance

Number of sample ______________________

Property	Unknown
1. Physical state	
Color	
Odor	
2. Combustibility	
3. Melting Point	
4. Boiling Point	
5. Density Volume Liquid Mass Liquid Density	
6. Solubility in H_2O	
Solubility in Ethanol	
7. Silver Nitrate	

Identity of unknown ______________________

What observations led you to your conclusion?

EXPERIMENT 4

Water of Hydration

Purpose: To calculate the percent water by mass in several potential unknowns; to dehydrate a solid sample and identify it by percent by mass of water.

Materials: $CaSO_4 \cdot H_2O$, $CuSO_4 \cdot 5H_2O$, $NiCl_2 \cdot 6H_2O$, $BaCl_2 \cdot 2H_2O$, $MgSO_4 \cdot 7H_2O$ (the magnesium sulfate crystals must be CLEAR, TRANSPARENT crystals, the white crystals are 6-hydrate and lower).

Apparatus: Bunsen burner, crucible and cover, wire triangle, ring stand, tongs.

Safety Precautions: Objects remain hot long after redness disappears—touch the crucible, cover, ring, and wire triangle cautiously. The solid material may tend to spatter as it is dehydrated. Wear your safety glasses at all times during this experiment.

Hazardous Waste Disposal: All the solids may be dissolved in water and rinsed down the sink with plenty of water.

INTRODUCTION

Many solids, especially inorganic salts, occur naturally as *hydrates*. This means that H_2O molecules are incorporated into the crystalline lattice in specific ways; that is, they are chemically combined. The materials may or may not look or feel "wet." They may spontaneously tend to lose water molecules in dry air, *(efflorescence)*. Other hydrates absorb water molecules from humid air, are classified as *hygroscopic*, and are useful as *desiccants* (drying agents).

Some hydrates exist in equilibrium with the moisture in air, absorbing or releasing water molecules depending on the relative humidity. An example is cobalt(II) chloride, which is used in novelty items to predict the weather (often attached to thermometers). $CoCl_2 \cdot 6H_2O$ is red, while the *anhydrous* $CoCl_2$ is blue. When the relative humidity is high, $CoCl_2$ tends to absorb water and the weather strip turns red, which the legend indicates signals rain (humidity *must* be high for it to rain). When the relative humidity is low, as on fair days, $CoCl_2 \cdot 6H_2O$ loses water molecules and the weather strip turns blue. The process is reversible

$$\underset{\text{(red)}}{CoCl_2 \cdot 6H_2O} \rightleftharpoons \underset{\text{(blue)}}{CoCl_2 + 6H_2O}$$

Note that hydrates are not solids that are simply wet; most materials will absorb some water onto their surfaces. Hydrates, however, contain water molecules that are chemically combined to the cations and anions in the salt, and are present in *specific molar ratios*.

The common notation for hydrates is a raised dot (·) between the formula of the salt and the number of moles of water molecules per mole of salt. Thus

$$CuSO_4 \cdot 5H_2O \text{ can also be written } CuSO_4(H_2O)_5$$

Hydrates are named by adding the number of water molecules to the end of the standard name of the salt. The Greek prefixes mono, di, tri, tetra, penta, etc. are used to indicate the number of moles of water. Thus

$CuSO_4 \cdot 5H_2O$ "copper (II) sulfate pentahydrate"
$CaCl_2 \cdot 2H_2O$ "calcium chloride dihydrate"

Water of hydration can be removed usually by simple heating to just above 100° C, the boiling point of water. In this experiment, you will identify an unknown salt from a list of potential unknowns by calculating the amount of water lost after heating.

PROCEDURE

1. Clean a porcelain crucible and cover. Do not worry about stains. Support the crucible and cover (slightly ajar) on a clay triangle on a ring stand. Heat with a Bunsen burner, first gently, and then to redness for about 3 minutes. Allow to cool *completely*.
2. Weigh the crucible and cover to the nearest 0.001 g. Do NOT handle the crucible with your hands. Moisture and oils from your skin will adhere to the porcelain and artificially increase the weight. Use crucible tongs or a piece of paper.
3. Select an unknown and record its code. Transfer about one gram of the unknown to the crucible and reweigh. By difference, you now know the weight of the sample.
4. Support the crucible and cover (slightly ajar to allow steam to escape) on the clay triangle and heat gently for 5 minutes. Best results are obtained with gentle heat. Do not fry or melt the solid as this can lead to decomposition of the salt itself. You may wish to lift the cover occasionally to observe any changes in the material. When heating is finished, place the cover completely on the crucible (this prevents reabsorption of water) and allow the crucible to cool *completely*. Describe any changes in appearance on your report sheet.
5. Weigh the crucible and *anhydrous* salt. Record.
6. Return the crucible to the clay triangle and heat gently for 5 minutes. Cool and weigh. Record the weight. If the weight after the second heating agrees with the first to the nearest 0.001 g, then no further heating is necessary. If the second heating appeared to drive off more water, heat for a third time, cool and re-weigh. Consult a TA or instructor if you think a fourth heating is necessary.
7. Scrape out the excess anhydrous salt into an evaporating dish or small beaker and save for later. It is not necessary to clean the crucible completely. Repeat steps 2–6 (with the same unknown!) for a second trial. Average your values for the % H_2O.
8. After the second trial, again scrape out the excess anhydrous salt into the container. Add a few drops of water. Describe any change on your report sheet. Completely dissolve any remaining solid and rinse down the sink. Wash the crucible and cover, and return.

CALCULATIONS

Calculate the % water by mass in your unknown. First calculate the weight of the hydrated salt. Then calculate the weight of the anhydrous salt (last weighing). The difference between these two numbers is the weight of the water lost. Then

$$\%\ \text{water} = \frac{\text{weight of water lost}}{\text{weight of hydrated salt}} \times 100\%$$

On your report sheet, calculate the % water by mass in each of the possible unknowns. Show your work on a separate piece of paper. Then identify your unknown by making the closest match. If you used other data in your identification (e.g., color), cite that evidence as well.

Experiment 4
Water of Hydration

Name______________________________

Date______________________________

REVIEW QUESTIONS

1. How does the appearance of your unknown change when it is dehydrated? Rehydrated? Why do you think this happens?

2. Assuming your identification is correct, calculate your % error. Be sure to indicate whether it is a positive or negative error. Speculate on the possible sources of your error, paying attention to both the amount and direction of the error.

3. Write the formula of your unknown. Give its correct chemical name.

4. Two of the unknowns are also commonly known as "gypsum" and "Epsom salts." Identify these.

5. A student takes 1.082 g of hydrated nickel(II) sulfate and heats it in a crucible. The anhydrous material weighed 0.596 g. What is the complete formula for this hydrated salt? Show your work.

6. A student made calculations for the preparation of 1 liter of 4M barium chloride solution, but neglected to account for the 2 moles of water per formula unit when he weighed out the solid. Was the actual molarity of the solution he prepared too high or too low?

Experiment 4
Water of Hydration

Name______________________________

Date______________________________

REPORT SHEET

Unknown code____________________

	Trial 1	Trial 2	
Mass of crucible + cover	____________	____________	g
Mass of crucible + hydrated salt	____________	____________	g
Mass of crucible + anhydrous salt (weight I)	____________	____________	g
Mass of crucible + anhydrous salt (weight II)	____________	____________	g
Mass of crucible + anhydrous salt (weight III)	____________	____________	g

Observations:

Calculations

Mass of hydrated salt	____________	____________	g
Mass of anhydrous salt	____________	____________	g
Mass of H_2O lost	____________	____________	g
% H_2O in unknown	____________	____________	%

AVERAGE ____________ %

IDENTITY ____________

Possible unknowns % H_2O (Show your calculations)

$BaCl_2 \cdot 2H_2O$

$CaSO_4 \cdot 2H_2O$

$CuSO_4 \cdot 5H_2O$

$NiCl_2 \cdot 6H_2O$

$MgSO_4 \cdot 7H_2O$

EXPERIMENT 5

Two Laws of Chemical Combination

Purpose: To study two fundamental laws of chemical combination: the law of definite proportions and the law of multiple proportions.

Materials: Copper gauze or wire (B & S #26 or smaller), powdered sulfur.

Safety Precautions: The reaction between copper and sulfur must be run in a fume hood. Be sure the fume hood is operating.

Hazardous Waste Disposal: All reaction products and unused starting materials may be disposed of in a waste basket.

INTRODUCTION

Literally millions of different compounds have been identified by chemists, and all of them are formed when a relatively small number of elements combine in unique ways. At first you might assume that the combination of elements to form molecules is a random process and that making sense of it all is impossible. Just the opposite is true. The combination of elements to form compounds occurs in an orderly manner according to several natural laws of combination.

Scientists gathered information about compounds in a systematic way when they began to study the relationships between elements in compounds during the latter part of the 18th century. The most useful tool in the measurement of the quantities of materials at that time was the balance. Chemists then used carefully determined mass measurements of combining elements to calculate mass ratios between elements for specific compounds. It was found that *for a given compound, the mass ratio between elements was always the same*, no matter how or where the compound originated. Water molecules in the rain have the same oxygen-to-hydrogen mass ratio (8.00 g oxygen/1.00 g hydrogen) as do water molecules in your body. You can find more examples of mass ratios in your textbook. These examples illustrate the natural **law of definite proportions,** or the law of constant composition. As a consequence of the law of definite proportions, we can define a compound as a combination of two or more elements in fixed proportion by mass.

Another law governing the formation of compounds is the law of multiple proportions. We know from the law of definite proportions that the mass ratio between two elements in a compound is always the same. The **law of multiple proportions** tells us that when two elements combine to form two different compounds, the two compounds will have different mass ratios. In fact, the mass ratios of the two elements in the different compounds are related to one another by a simple whole number ratio. For example, in nitrous oxide (a compound of nitrogen and oxygen) the ratio of nitrogen to oxygen is 0.57 g N/1.00 g O, and in nitric oxide (another compound of nitrogen and oxygen) the ratio of nitrogen to oxygen is 1.14 g N/1.00 g O. The two mass ratios, 0.57 and 1.14, are related to each other by the simple whole number ratio of 1 to 2.

Using the results of this experiment, we will attempt to verify both of the laws governing the formation of compounds. Two different sulfides of copper are possible, one of which you will prepare. The weights of the two elements involved will be measured; their mass ratios will be determined and then compared with those obtained by other students. The mass ratio will be used to determine a mole ratio from which the empirical formula will be developed. Possible sources of error include inaccurate weighing, difficulty in getting the last bit of each substance to react, and the possibility that secondary reactions may occur.

A second sulfide of copper can be prepared by indirect methods. Because the procedure is rather long and difficult, data concerning the amounts of each element used in such an experiment is given to you. The mass ratio is calculated and the empirical formula determined as before. Then the amounts of copper for the same amount of sulfur in each compound (the mass ratios) are compared. The comparison gives you evidence of the generalization that is called the law of multiple proportions.

Calculation of the mass ratio is a simple mathematical procedure. For example, if you know that 1.208 g of oxygen (O) combines with 0.151 g of hydrogen (H), you can calculate the O/H ratio or the H/O ratio using these weight measurements. For the O/H ratio, divide the weight of oxygen by the weight of hydrogen. The H/O ratio is calculated in reverse.

$$\frac{\mathrm{O}}{\mathrm{H}} = \frac{1.208 \text{ g oxygen}}{0.151 \text{ g hydrogen}} = \frac{8.00 \text{ g oxygen}}{1.00 \text{ g hydrogen}} = 8.00 \text{ g O}/1.00 \text{ g H}$$

$$\frac{\mathrm{H}}{\mathrm{O}} = \frac{0.151 \text{ g hydrogen}}{1.208 \text{ g oxygen}} = \frac{1.00 \text{ g hydrogen}}{8.00 \text{ g oxygen}} = 0.125 \text{ g H}/1.00 \text{ g O}$$

Usually the smaller of the two numerical values is chosen to be the denominator in the ratio. Since the two weight measurements can be arranged to obtain two ratios, you must be careful to define the ratio being calculated.

Formulas represent the number of atoms of elements in a molecule of a compound. To determine the empirical formula of a compound starting with weight measurements, you must first convert the weight measurements of elements in the compound to a value that expresses quantity of atoms in the compound. The quantity unit we use is the mole, since we know that there are 6.02×10^{23} atoms in a mole of an element and that a mole of an element has a mass equivalent to the atomic weight of that element in gram units. For example, 1.208 g O and 0.151 g H can each be converted to mole units using the technique of dimensional analysis.

$$1.208 \text{ g O} \times \frac{1 \text{ mol O}}{16.0 \text{ g O}} = 0.0755 \text{ mol O}$$

$$0.151 \text{ g H} \times \frac{1 \text{ mol H}}{1.00 \text{ g H}} = 0.151 \text{ mol H}$$

Once the mass of each element in the compound is converted to moles, the mole ratio between elements can be calculated. Usually the smaller mole value is chosen for the denominator in this ratio.

$$\frac{\mathrm{H}}{\mathrm{O}} = \frac{0.151 \text{ mol H}}{0.0755 \text{ mol O}} = \frac{2.00 \text{ mol H}}{1.00 \text{ mol O}}$$

This mole ratio is then translated to a chemical formula: OH_2 or more commonly H_2O. Your textbook gives many more problems of this type and you may want to review some of these at this time.

PROCEDURE

A. Synthesis of Copper(I) Sulfide

Clean a porcelain crucible, support it on a triangle mounted on a ring stand, and heat strongly for about a minute. When it is completely cool, weigh it as accurately as your balance permits (preferable to the nearest milligram). Record all your weighing. Two sets of blanks are provided in case the experiment must be repeated.

Obtain a piece of clean copper wire or gauze roughly weighing one gram, coil or fold it compactly, put it in the crucible, and weigh accurately again. The difference in weights is the weight of your sample. Add enough powdered sulfur to half-fill the crucible. The exact amount is not important because the excess burns away. Put the lid on the crucible, mount it on a ring stand as shown in Figure 5.1, and heat carefully **under the fume hood. Be sure the hood blower is operating.** Keep the burner in your hand but hold it sideways so that your **hand is not directly under the crucible.** Burning molten sulfur could drop onto your hand or into the burner. Move the flame around the crucible, making sure that the top and all sides of it are strongly heated. Continue to heat for five minutes after the sulfur vapor stops burning around the cover. Allow to cool to room temperature before removing the cover. This reduces the effect of air oxidation. When the crucible is cool, weigh it accurately and record the weight. The product copper(I) sulfide is quite brittle and should not be handled.

To ensure that all of the metal has reacted, a second portion of sulfur is added and the sample is heated as before. When cool, it is again accurately weighed, and the second weighing is recorded. The two weighings of the product should agree unless some copper remained unreacted the first time or some excess sulfur is present. If they do not agree, a third heating and weighing may be necessary. **Calculate your results before destroying the sample.** If your results are in error, you may still be able to save the experiment.

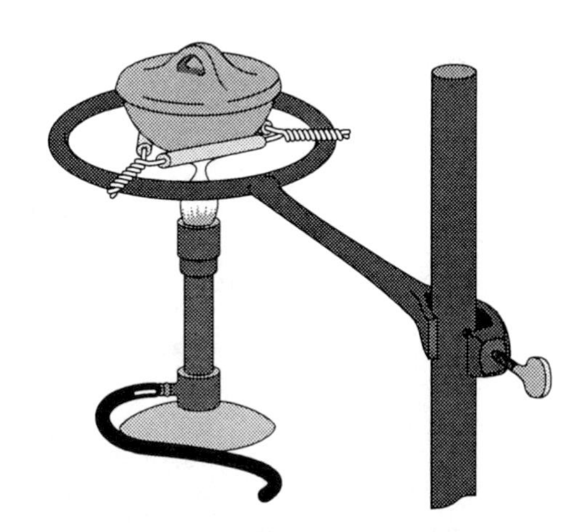

FIGURE 5.1
Heating covered crucible containing copper and sulfur.

B. Calculation of Mass Ratio

From your data, calculate the weights of copper and sulfur in your compound and the mass ratio of the two elements present.

1. Determine the weight of copper reacted by subtraction.

$$\text{wt(crucible + cover + copper)} - \text{wt(crucible + cover)} = \text{wt(copper)}$$

2. Determine the weight of sulfur reacted by subtraction.

$$\text{wt(crucible + cover + product)} - \text{wt(crucible + cover + copper)} = \text{wt(sulfur)}$$

3. Calculate the ratio of mass of copper to mass of sulfur reacted by dividing the weight of copper by the weight of sulfur.

$$\frac{\text{Cu}}{\text{S}} = \frac{\text{wt Cu reacted}}{\text{wt S reacted}}$$

4. Collect and tabulate the mass ratios of three other members of the class. Calculate an average mass ratio. Use the data to justify the law of definite proportions.

C. Empirical Formula Calculation

1. Determine the weights of copper and sulfur as before.
2. Using your data, calculate the number of moles of each element present. If necessary, *see the sample calculations in the introduction.*
3. Reduce the numbers of moles to a whole number ratio, and round off to the nearest whole number ratio.
4. Translate the Cu/S mole ratio to the empirical formula of the compound.

D. Calculation of Formula of a Second Sulfide of Copper: Copper(II) Sulfide

Since preparation of the second sulfide of copper is rather impractical, weight data is given. A sample of copper metal weighing 1.562 g was dissolved in nitric acid, the excess acid was neutralized, and the copper was precipitated with hydrogen sulfide. After the solid copper sulfide was filtered off and dried, it was found to weigh 2.350 g. Calculate its copper/sulfur mole ratio and its empirical formula as you did in Part A.

E. Comparison of the Two Sulfides of Copper

Compare the mass ratios of copper to sulfur found in Parts A and B for each compound. From your data write a statement of the law of multiple proportions as it applies to the two sulfides of copper.

Experiment 5

Two Laws of Chemical Combination

Name____

Date____

REVIEW QUESTIONS

1. Would the percentage of metal in the compound in Part A have been too high, too low, or unaffected if
   (a) some unburned sulfur were left in crucible?

   (b) some copper in a strand of wire did not react?

   (c) some Cu_2O were formed?

   (d) some CuO were formed?

   (e) some particles of product were lost before the final weighing?

2. Why did the use of excess sulfur make no difference? What gaseous product was formed from this excess?

3. Is the formation of the two substances Cu_2S and CuS an exception to the law of definite proportions? Explain.

4. Look up the formulas of the following pairs of compounds: water and hydrogen peroxide; ammonia and hydrazine; benzene and methane. Show how the law of multiple proportions applies to each pair.

5. Look up the formulas of benzene and acetylene. Calculate the percent composition of each. Since these are different compounds with very different properties, it may seem that this is an exception to the law of definite proportions. Explain why it is not and tell what unusual property of the carbon atom makes these two compounds possible.

6. Write equations for the following chemical changes, all of which relate to this experiment:
(a) metallic copper combines with powdered sulfur

(b) sulfur burns in air

(c) copper metal dissolves in nitric acid

(d) the gas H_2S is bubbled through a solution containing Cu(II) ions

7. Calculate the sulfur/iron mass ratio when 0.304 g of iron reacts with sulfur to form 0.783 g of an iron-sulfur compound.

8. The mass ratio between the elements in the compound calcium carbide is 6.68 g Ca/1.00 g C. How many grams of carbon would react with 1.00 g of calcium?

9. Calculate the empirical formula of a compound known to have the mass ratio 4.67 g nitrogen/1.00 g hydrogen.

10. The initial heating of the empty crucible removes dirt and other materials that might add error to the experimental results. If a student did not heat the crucible initially but instead weighed it with contaminants that were later lost during the heating of the copper with sulfur, how would this affect the experimental results?

Experiment 5
Two Laws of Chemical Combination

Name______________________________
Date_______________________________

REPORT SHEET

A. Synthesis of Copper(I) Sulfide

Descriptive data

Physical appearance of copper ______________________________

Physical appearance of sulfur ______________________________

Physical appearance of copper(I) sulfide ________________________

What evidence leads you to infer that a chemical reaction has taken place?

Numerical Data

	Trial 1	Trial 2 (if necessary)
Mass of crucible after heating and cooling	________	________
Mass of crucible + copper	________	________
Mass of crucible + copper(I) sulfide (First heating)	________	________
Mass of crucible + copper(I) sulfide (Second heating)	________	________

B. Calculation of Mass Ratio (Use space provided to show calculations)

Mass of copper reacted	________	________
Mass of sulfur reacted	________	________
Cu/S mass ratio	________	________
Mass ratios from other students	________	________
	________	________
	________	________
Average mass ratio		________

Are these Cu/S mass ratios consistent with the law of definite proportions? Explain.

C. Empirical Formula Calculation

Mass of copper reacted ________

Mass of sulfur reacted ________

Moles of copper reacted ______________________

Moles of sulfur reacted ______________________

Cu/S mole ratio ______________________

Empirical formula ______________________

D. Calculation of Formula of a Second Sulfide of Copper: Copper(II) Sulfide

Mass of copper reacted ______________________

Mass of sulfur reacted ______________________

Cu/S mass ratio ______________________

Moles of copper reacted ______________________

Moles of sulfur reacted ______________________

Cu/S mole ratio ______________________

Empirical formula ______________________

E. Comparison of the Two Sulfides of Copper

Record the grams of copper combined with 1.00 gram of sulfur for each compound:

Copper(I) sulfide ______________________

Copper(II) sulfide ______________________

How do these weights of copper per gram of sulfur compare?

Are these Cu/S mass ratios consistent with the law of multiple proportions? Explain.

EXPERIMENT 6

Bohr Model of the Hydrogen Atom

Purpose: To calculate the energy changes possible for the first ten orbits of the hydrogen atom; to calculate the corresponding wavelengths; to view an actual hydrogen spectrum; to correlate the visible lines to the calculated wavelengths.

Materials: None

Apparatus: Tube of hydrogen gas, emission source, spectroscope with wavelength calibrations (if spectroscope is unavailable, your instructor may provide you with a printed emission spectrum of hydrogen).

Each student should bring a calculator to lab.

Safety Precautions: Follow all directions concerning the spectroscope to avoid electrical hazards. If in the laboratory, you may be required to wear your safety glasses.

Hazardous Waste Disposal: None

INTRODUCTION

Chemists and physicists adopt a variety of ways to describe how nuclei and electrons interact in atoms. The particular *model* chosen depends on which characteristic is of primary importance to the discussion. A model is a description or mental picture of something that represents a physical phenomenon from a specific viewpoint. We don't usually consider whether or not a model is *true*, but a model is considered successful if it readily explains some behavior and can accurately predict future behavior.

About 1913 Niels Bohr developed a model to describe the movement of electrons in atoms. He chose the hydrogen atom because it is the simplest case of one proton and one electron. Since the 1880s, emission and absorption spectra of several gaseous elements had been observed. Bohr's model was developed to explain how the spectra are generated.

When hydrogen gas is subjected to a high energy source, say an electric current, a pinkish glow is visible (you have seen the same type of effect in neon signs). If this *emission* is passed through a prism, it is found that the light is actually made up of four distinct wavelengths that can be observed and classified separately (also called a bright line spectrum).

Recall that "white light" is made up of all visible wavelengths. If this light is passed through a prism, a *continuous spectrum* is produced, made up of all wavelengths of light overlapping in such a way that it appears to be continuous. This is what happens when we see a rainbow—the white light of the sun passes through tiny water droplets suspended in air, each acting like a prism to separate out the various energy wavelengths into an ordered spectrum.

Why does hydrogen only emit energy in four unique, discrete (separate) wavelengths (in the visible range)? This is what Bohr set out to explain. In his explanation, the *Bohr Model*, the electrons are described as being in orbit about the nucleus at special, specific distances. Moreover, he said, the *energy* of the electron is a function of the *distance* it is away from nucleus. Electrons closest to the nucleus are the lowest in energy; i.e., the most stable. Electrons farther away are in higher energy states (less stable). These states are called *orbits* or *levels* and are given the symbol n (n = 1,2,3, . . .).

An electron can exist in the first orbit (n = 1) or in the second orbit (n = 2), but it *cannot orbit anywhere in between.* You can stand on a staircase on the fifth step or the sixth step, but you cannot stand unsupported in space between the two (at least not for any reasonable length of time).

This is the basis of *quantum theory.* We say the energy of the electrons is *quantized.* An electron might exist in the n = 1 level and, therefore, have a certain specific energy associated with the n = 1 level. Or the electron might orbit in the n = 2 level and have a higher energy, but there is no way the electron can have any energy in between these two values. The quantity of energy is predetermined by the level the electron lives in and no other energies are allowed.

Bohr worked out the calculations for the orbital energies. The energy of an electron in the *nth* level is given by

$$E_n = \frac{-B}{n^2} \quad (n = 1, 2, 3, \ldots)$$

where B is the Bohr constant (not surprising), which has a different value for each element. For the hydrogen atom, in useful units, B = 1312 kJ/mol. Note that there is a negative sign in the equation. Bohr assigned (somewhat arbitrarily) $E_\infty = 0$ so that the energies of the levels decrease down the n values with n = 1 being the most negative number; i.e., lowest number and, therefore, most stable.

To illustrate, the energy of the n = 3 level of the hydrogen atom is found by

$$E_3 = \frac{-1312}{3^2} \text{kJ/mol} = -145.78 \text{ kJ/mol}$$

Now you can move from the third step up to the sixth step. By analogy, this is how Bohr envisioned the movement of electrons between orbits. On the staircase you must expend energy—against gravity—to move from the third step to the sixth step. Likewise, a hydrogen electron must absorb energy to move from n = 3 to n = 6 since it is moving to a higher energy position (less stable). Now suppose you jump down from the sixth step to the third step—energy is released in the form of your body's momentum against the step. When the electron moves "down" from n = 6 to n = 3, energy is released also; the analogy is so good we sometimes say the electron "falls" from n = 6 to n = 3. How much energy is released in this process? Exactly the same amount of energy that had to be absorbed to move it "up" from n = 3 to n = 6. Well, how much is that? Exactly the difference in energy between the two levels in question.

To illustrate, for an electron moving from E_6 down to E_3, we have seen previously that E_3 = −145.78 kJ/mol. And

$$E_6 = \frac{-1312}{6^2} \text{ kJ/mol} = -36.44 \text{ kJ/mol}$$

Therefore, the energy difference, ΔE is

$$\Delta E = E_{\text{hi}} - E_{\text{lo}} = E_6 - E_3 = -36.44 - (-145.78) = +109.34 \text{ kJ/mol}$$

Notice that the answer is positive. This tells us that 109.34 kilojoules of energy are *released* for each mole of hydrogen atoms. This energy can conveniently be expressed as a wavelength via

$$\lambda = \frac{1.196 \times 10^5}{\Delta E} \cdot \frac{\text{kJ} \cdot \text{nm}}{\text{mol}}$$

Note that if the energy is put in the equation in units of kJ/mol, those units will cancel and the resulting wavelength will be expressed in nanometers (nm).

In this exercise, you will perform many of these calculations in lab in order to gain experience in the Bohr theory. Then you will view a hydrogen emission spectrum in lab, record your observations, and correlate them with your calculations.

Experiment 6

Bohr Model of the Hydrogen Atom

Name________________________________

Date_________________________________

REVIEW QUESTIONS

1. Explain why a series of transitions is specified by where the electron ends up, not by where it starts.

2. In what part of the electromagnetic spectrum is the series with $n_{lo} = 3$? Its name?

3. How much energy must be expended to completely remove an electron in the third orbit from a hydrogen atom? (This is called the ionization energy.)

4. Given that $\Delta E = \dfrac{hc}{\lambda}$, show by conversion factors that

$$\lambda = \frac{1.196 \times 10^5}{\Delta E} \text{ kJ} \cdot \text{nm/mol}$$

(**Hint:** This is most easily demonstrated by *choosing* a value for ΔE, say 93.30 kJ/mol, and using conversion factors to arrive at the wavelength 1282 nm.)

5. If the Bohr constant for He^+ is 5248.16 kJ/mol, what is the energy and wavelength for a transition from $n = 4$ to $n = 3$? Could you see this?

6. What kind of energy lies just above the visible? Just below?

7. Do a little research: where and how was helium discovered?

Experiment 6

Bohr Model of the Hydrogen Atom

Name____

Date____

REPORT SHEET

The equations you will need:

$$E_n = \frac{-1312}{n^2} \text{ kJ/mol} \qquad \Delta E = E_{\text{hi}} - E_{\text{lo}}$$

$$\lambda = \frac{1.196 \times 10^5}{\Delta E} \text{ kJ} \cdot \text{nm/mol}$$

First, calculate each of the first ten energies.

E_1 = ___________________________ kJ/mol

E_2 = ___________________________ kJ/mol

E_3 = ___________________________ kJ/mol

E_4 = ___________________________ kJ/mol

E_5 = ___________________________ kJ/mol

E_6 = ___________________________ kJ/mol

E_7 = ___________________________ kJ/mol

E_8 = ___________________________ kJ/mol

E_9 = ___________________________ kJ/mol

E_{10} = ___________________________ kJ/mol

For convenience, you may round off each to the nearest 0.01.

Using these values, you can fill in the table on the next page. Calculate the ΔE's for all the possible transitions for orbits $n = 10$ down. Round off the energies to the nearest 0.01. Place this value in the upper portion of the square. Convert the energy to wavelength in nanometers and place this value in the lower portion of the square. Round the wavelength to four significant figures. You will find it convenient to do all the ΔE's first, then do the conversions to nanometers. The transition from $n = 5$ to $n = 3$ has been done for you.

Now view the hydrogen emission spectrum provided in the lab. Use the scale in the viewer to estimate the wavelengths. If the scale is marked off in angstroms, Å, convert to units of nanometers. Since 10 Å = 1 nm, simply divide by ten to get nanometers; e.g. 4100 Å = 410 nm.

Draw in lines where you see them. Use colored pens or pencils for a nice effect or simply note in words the colors you see. Include the wavelength values. (Some people are unable to see the most energetic line, but most people should be able to see four lines.)

E_{lo} \ E_{hi}	$n = 2$	$n = 3$	$n = 4$	$n = 5$	$n = 6$	$n = 7$	$n = 8$	$n = 9$	$n = 10$	$n = \infty$
$n = 1$										
$n = 2$										
$n = 3$				93.30 1282						
$n = 4$										
$n = 5$										
$n = 6$										
$n = 7$										
$n = 8$										
$n = 9$										
$n = 10$										

Label each line A, B, C, D. Using your chart of transitions, state the transition that accounts for each line.

A: B:

C: D:

This is the *visible spectrum* of hydrogen emission. It is called the ______________ series and its common element is that all transitions ____.

EXPERIMENT 8

Calorimetry and Hess's Law

Purpose: To perform several measurements of calorimetry and to verify Hess's Law.

Materials: 1.0M HCl, 1.0M NaOH, solid NaOH

Apparatus: Coffee cup calorimeter, thermometer, 100-mL graduated cylinder, stopwatch or wristwatch that measures seconds, 25- or 50-mL Erlenmeyer flask with stopper

Safety Precautions: One-molar solutions of hydrochloric acid and sodium hydroxide are corrosive. If you spill any on your skin, wash it off immediately. Sodium hydroxide pellets should not be touched or handled. Weigh them in glass containers and keep from air. The thermometers used in this experiment are fragile and expensive. Please exercise caution—especially do not set thermometer on the bench top and allow it to roll off. Safety glasses are required at all times.

Hazardous Waste Disposal: All the solutions may be washed down the drain with plenty of water.

INTRODUCTION

Calorimetry is the study of heat flow. It is not simply temperature, but the total change in temperature for a specific mass of a specific material. That is

$$q = \text{S.H.} \times m \times \Delta t$$

where q is *heat* (also sometimes written ΔH), S.H. stands for *specific heat*, m is *mass*, and $\Delta t = t_{\text{final}} - t_{\text{initial}}$. If the temperature increases, Δt will be positive and, therefore, $q_{\text{calorimeter}}$ will be positive. However,

$$q_{\text{reaction}} = -q_{\text{calorimeter}}.$$

The heat that we measure in a calorimeter is equal in magnitude but opposite in sign to that of the heat of the reaction, which is what we are usually interested in. This is because we are not really measuring the heat that came out of the reaction, but the heat that went into the calorimeter. For practical purposes, whenever heat flows out of a reaction (it gets hotter), the reaction is *exothermic* (out-heat) and the sign of q is *negative*. Conversely, whenever heat flows into a reaction (it gets colder), the reaction is said to be *endothermic* (in-heat) and the sign of q is *positive*. All of the reactions in this experiment are exothermic.

A calorimeter is any device that is used to measure heat flow. Its main characteristic is that it must be well insulated so that no heat is lost to the surroundings. A simple calorimeter, which will be employed in this experiment, is the "coffee cup" calorimeter. It uses a styrofoam cup (sometimes two cups stacked together) with a styrofoam or cardboard cover. It works well enough to allow reasonable results.

More elaborate types of equipment include devices modeled on the familiar Thermos bottle. A "bomb" calorimeter is the best sealed and most accurate equipment, but it is relatively expensive and somewhat tedious to operate. (Despite its name, a "bomb" calorimeter does not explode.)

Using calorimetry, it is possible to test the validity of *Hess's Law*: the heat flow in a reaction that is the sum of two other reactions is equal to the sum of the heat flows in those two reactions. For example, in this experiment, you will measure the heat of *neutralization* for aqueous solutions of a strong acid + a strong base, the heat of *solution* for dissolving the strong base (solid), and the heat of *reaction* for doing both simultaneously. Hess's Law says that the overall is the sum of the parts.

$$HCl_{(aq)} + NaOH_{(aq)} \rightarrow H_2O_{(l)} + NaCl_{(aq)} \qquad \Delta H_N$$

$$NaOH_{(S)} \rightarrow NaOH_{(aq)} \qquad \Delta H_S$$

$$HCl_{(aq)} + NaOH_{(S)} \rightarrow H_2O_{(l)} + NaCl_{(aq)} \qquad \Delta H_R$$

All three ΔH's will be measured to confirm Hess's Law: $\Delta H_N + \Delta H_S = \Delta H_R$.

PROCEDURE

A. Heat of Neutralization ΔH_N

1. For this experiment you will work in pairs. The pair will collect one set of data, but each person will make his or her own graphs and submit independent results. Each pair will need one time measuring device. Any watch with a sweep second hand or any electronic or digital watch that measures seconds is fine. A limited number of stopwatches will be available in the lab for pairs who do not have the appropriate timepieces.
2. Each pair will need to obtain two coffee cup calorimeters and one thermometer. The calorimeter should include a cover with a hole in it large enough to insert the thermometer.
3. Measure 50.0 mL of 1.0 M HCl in a graduated cylinder and transfer completely to one calorimeter. Rinse and dry the graduate before measuring 50.0 mL of 1.0 M NaOH, and transfer the solution completely to the other calorimeter. Since both solutions have been stored at the same room temperature, it is reasonable to assume that they are both at the same temperature. (You may check this briefly but be sure to wipe off the thermometer first to avoid transferring any solution from one cup to the other.)
4. Place the thermometer in the cup with the HCl and allow about one minute to equilibrate. Using your timing device, record the temperature for three minutes at 1-minute intervals (±0.1°C). On the fourth minute pour the NaOH solution into the HCl solution quickly and mix thoroughly. Avoid splashing or losing any solution. Replace the cover and the thermometer and record temperature readings for the fifth, sixth, and seventh minutes. Dump the solution down the sink (it is only salt water now) and rinse and dry both calorimeters.

B. Heat of Solution ΔH_S

1. Into a stoppered 25- or 50-mL Erlenmeyer flask weigh about 2.00 g (±0.01 g) of solid NaOH pellets. This material is very hygroscopic (absorbs moisture from the air) so keep the bottle closed except when you are actually using it. Keep the stopper on the Erlenmeyer flask until it is ready for use. Record the mass of NaOH actually weighed out.
2. Measure 50.0 mL of distilled water into one of the calorimeters, cover it, and insert the thermometer. Wait a minute or so for equilibration and then start measuring and recording temperatures at 1-minute intervals for 3 minutes. On the fourth minute dump all the NaOH pellets quickly into the water, stir gently to facilitate dissolving and re-cover. Measure and record the temperature for the fifth, sixth, and seventh minutes. You may find it helpful to swirl the water gently with the thermometer to ensure that all the NaOH is dissolved.
3. Pour the resulting solution down the drain and flush with copious amounts of water. Rinse and dry the calorimeter.

C. Heat of Reaction ΔH_R

1. Again, weigh about 2.00 g of NaOH (±0.01 g) into the stoppered Erlenmeyer flask. Record the weight used.
2. Pour 55 mL of 1.0 M HCl into a clean, dry graduated cylinder and add distilled water to make a total of 100.0 mL of solution. (The slight excess of acid will ensure complete reaction with the solid.)

3. Pour the HCl solution completely into a calorimeter, cover and insert the thermometer. Allow about a minute to equilibrate. Measure and record the temperature at 1-minute intervals for 3 minutes. On the fourth minute, add the NaOH pellets quickly to the HCl solution, mix gently and re-cover. You may need to stir carefully with the thermometer until all the NaOH is dissolved. Record the temperature for the fifth, sixth, and seventh minutes.
4. Dump the solution down the drain. Rinse and dry the calorimeter.
5. Time permitting, perform second trials of each of the three reactions.
6. Clean and dry any borrowed equipment and return.

CALCULATIONS

The calculations for this experiment can be somewhat intimidating, but if you take everything one step at a time, you will find it is not so very difficult.

First, Δt for each trial must be obtained graphically. Refer to notes in the appendix of this manual for correct graphing techniques. You may put two graphs on one page if you wish.

For each graph, you will need to extrapolate (draw the best-fitting straight line) to the time of mixing (fourth minute). Plot temperature (°C) vs. time (minutes). Two examples are sketched for you below. If the temperatures go up with time, extrapolate straight back to obtain the greatest Δt. The trials with the solid NaOH are the most likely to increase in temperature over time, as a result of the slowness of dissolving.

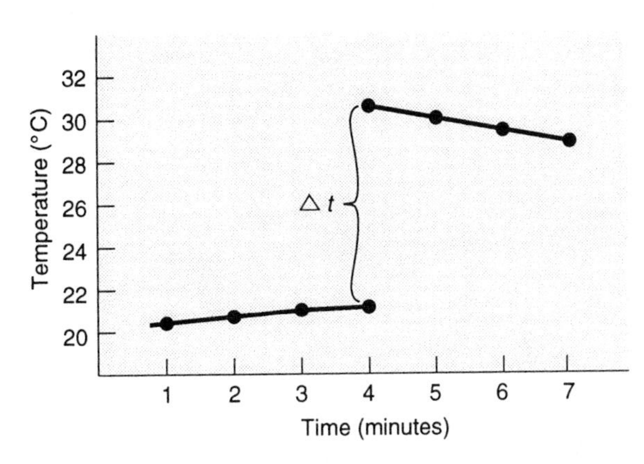

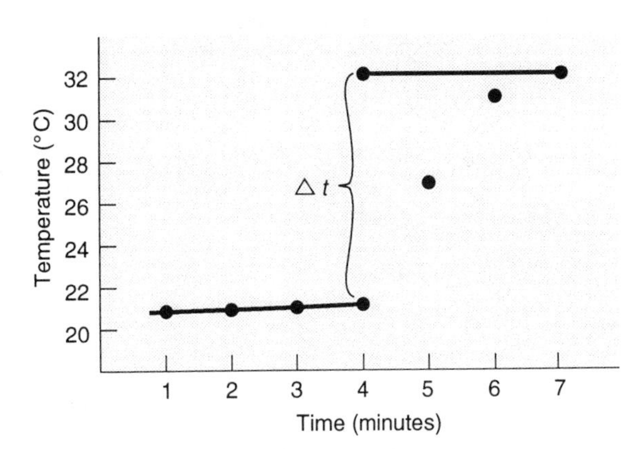

Δt is the difference in temperature between where the upper and lower extrapolated lines intercept the 4-minute line. (Note that the temperature scale should not start at zero.)

Record the Δt (to nearest 0.1°C) on the report form. Now recall that

$$q = \text{S.H.} \times m \times \Delta t$$

S. H. is "specific heat," which is a constant for a particular substance.

Part A: S.H. = 4.02 J/g°C (0.5 M NaCl solution)
Part B: S.H. = 3.93 J/g°C (1.0 M NaOH solution)
Part C: S.H. = 4.02 J/g°C (0.5 M NaCl solution)

(Since the resulting solutions in parts A and C are the same, the constants are the same. Notice that the specific heats are not too different from that for pure water, 4.184 J/g°C.)

With S.H. and Δt identified, all that is left is the mass. The solutions were measured out by volume, not by mass, but if the densities are known, the mass can be calculated. The density for parts A and C is the same (1.02 g/mL) and the calculation is identical. Both solutions were exactly 100.0 mL, so

$$(100.0 \text{ ml})(1.02 \text{ g/mL}) = 102 \text{ g},$$

which is the mass in both cases (A and C).

What about part B? Well, you used 50.0 mL of pure water and about 2 g of NaOH, say, for example, 2.05 g of NaOH. The density of pure water is 1.00 g/mL, so the water weighed 50.0 g. Now just add the weight of the NaOH you used; in this example then the total mass is 52.05 g.

Show on your report sheet the calculations (including units) for $q_{calorimeter}$ and enter the value under Trial 1. Enter the value for Trial 2, but you need not show this calculation.

Recalling that $q_{reaction}$ is equal in magnitude but opposite in sign to $q_{calorimeter}$, fill in the values for $q_{reaction}$.

Before you can compare the three results, you must set them on the same scale. The amounts of reactants were chosen somewhat randomly, but Hess's Law refers to summations on an equimolar basis. Therefore, you know heat, but you need to know *heat per mole*. Based on the chemical equations, this implies Part A: heat per mole of H_2O formed; Part B: heat per mole of solid NaOH; Part C: heat per mole of solid NaOH.

Parts B and C are the same, just change grams to moles. Again, using the example of 2.05 g of NaOH, with the molecular weight of 40.00 g/mol

$$2.05 \text{ g} \times \frac{1 \text{ mole}}{40.00 \text{ g}} = 0.05125 \text{ mol NaOH.}$$

Of course, you will use your own weights of NaOH, which may differ from Trial 1 to Trial 2.

How many moles of water are formed? The chemical equation indicates that one mole of HCl reacts with one mole of NaOH to form one mole of H_2O and one mole of NaCl (1:1:1:1) HCl and NaOH are present in equal amounts so we can use either one.

$$50.0 \text{ mL} = 0.050 \text{ L}$$

$$0.050 \text{ L HCl} \times \frac{1.0 \text{ mole HCl}}{\text{L HCl}} = 0.050 \text{ mol HCl}$$

Therefore, 0.050 mole of H_2O is formed in Part A.

To get ΔH, the heat per mole, remember that "per" usually means divide, so divide the heat by the number of moles (don't lose the negative sign in the process). Show the calculation for Trial 1.

Average the values for the two trials. Now you are ready to test Hess's Law. Do they add up? Comment briefly on your results on your report sheet. Turn in your graphs with your report sheet.

Experiment 8

Calorimetry and Hess's Law

Name______________________________

Date______________________________

REVIEW QUESTIONS

1. Estimate your error in this experiment. Calculate the percent error for what the sum should be and what you measured for Part C. Discuss significant sources of error.

2. Why is styrofoam a fairly good insulator?

3. What is a calorie? A Calorie?

4. It is found that 252 J of heat must be absorbed to raise the temperature of 50.0 g of nickel from 20.0°C to 31.4°C. What is the specific heat of nickel?

5. When the first two of our chemical equations are added together to get the third, what species "cancels out"? Demonstrate (see Introduction).

6. What is the relationship between a calorie and a joule? Which one is an SI unit? Why?

7. What is the origin of the unit "Joule"?

8. Why do you think heat is released when NaOH dissolves?

Experiment 8

Calorimetry and Hess's Law

Name________________________________

Date________________________________

REPORT SHEET

A. Heat of Neutralization ΔH_N

Trial 1

______	______	______	(mix)	______	______	______
1	2	3	4	5	6	7

Δt from graph ________ °C

Trial 2

______	______	______	(mix)	______	______	______
1	2	3	4	5	6	7

Δt from graph ________ °C

Calculations	Trial 1		Trial 2	
q_{cal}:	________	J	________	J
$q_{reaction}$:	________	J	________	J
moles of H_2O:	________	mol	________	mol
ΔH per mole:	________	J/mol	________	J/mol

average: ________ J/mol

B. Heat of Solution ΔH_S

Trial 1

______	______	______	(mix)	______	______	______
1	2	3	4	5	6	7

weight of NaOH ________ g

Δt from graph ________ °C

Trial 2

______	______	______	(mix)	______	______	______
1	2	3	4	5	6	7

weight of NaOH ________ g

Δt from graph ________ °C

Calculations	Trial 1		Trial 2	
q_{cal}:	________	J	________	J
$q_{reaction}$:	________	J	________	J
moles of NaOH:	________	mol	________	mol
ΔH per mole:	________	J/mol	________	J/mol

average: ________ J/mol

C. Heat of Reaction ΔH_R

	1	2	3	4	5	6	7
Trial 1	________	________	________	(mix)	________	________	________

weight of NaOH ________ g Δ*t* from graph ________ °C

	1	2	3	4	5	6	7
Trial 2	________	________	________	(mix)	________	________	________

weight of NaOH ________ g Δ*t* from graph ________ °C

Calculations	Trial 1		Trial 2	
q_{cal}:	________	J	________	J
$q_{reaction}$:	________	J	________	J
moles of NaOH:	________	mol	________	mol
ΔH per mole:	________	J/mol	________	J/mol

average: ________ J/mol

Conclusions on results. Explain.

EXPERIMENT 9

Degree of Ionization

Purpose: To observe the differences in degree of ionization of a number of substances as indicated by the electrical conductivity of their solutions. This experiment is designed to be a demonstration performed by the laboratory instructor. However, most of it could be run by small groups of students if an adequate number of safe conductivity apparatus were available.

Materials: 0.1 N solutions of HCl, HNO_3, H_2SO_4, NaOH, NH_4OH, $HC_2H_3O_2$, NaCl, $CuSO_4$, and $CaCl_2$; 0.1 M sucrose solution, glacial acetic acid, benzene, concentrated H_2SO_4; solid NaCl and KSCN.

Apparatus: Conductivity apparatus shown in diagram. The HCl generator should be set up in a hood with that part of the demonstration being done in the hood.

Waste Disposal: Benzene should be disposed of in the nonhalogenated organic solvent waste container. The HCl generator may be carefully rinsed with tap water, with the rinse liquid being run down the sink drain with ample water.

INTRODUCTION

When some elements react to form compounds, electrons are shared more or less equally between them, forming molecules that exist in solid or in solution. Such particles are uncharged, their solutions do not conduct an electric current, and they are nonelectrolytes.

In other reactions, electrons are transferred more or less completely from one atom to another, forming ions that exist in solid or in solution. Since ions carry charges and can migrate in solution, these solutions are conductors and are called strong electrolytes. The strong electrolytes include the strong acids (HCl, HNO_3, and H_2SO_4), the strong bases (NaOH, KOH), and most of the soluble salts.

Some compounds are partly ionic and partly molecular. Their solutions have limited numbers of free ions and are rather poor electrical conductors. They are weak electrolytes, and they include the weak acids ($HC_2H_3O_2$, H_2CO_3, and most other acids), the weak bases such as NH_4OH, and some salts such as $HgCl_2$ and $Pb(C_2H_3O_2)_2$.

In an aqueous solution water molecules attach themselves to ions in a crystal by polar forces and pull them free or utilize the attraction of unshared electron pairs to break up polar molecules.

$$\underset{\displaystyle\text{H}}{\text{H—}\underset{|}{\text{O}}} + \text{H—Cl} \rightarrow \text{H—}\underset{\substack{|\\ \text{H}}}{\text{O}}\text{—H}^+ + \text{Cl}^-$$

Organic solvents such as benzene, toluene, acetone, and alcohol, especially those that are nonpolar, have little or no dissociating influence on solutes.

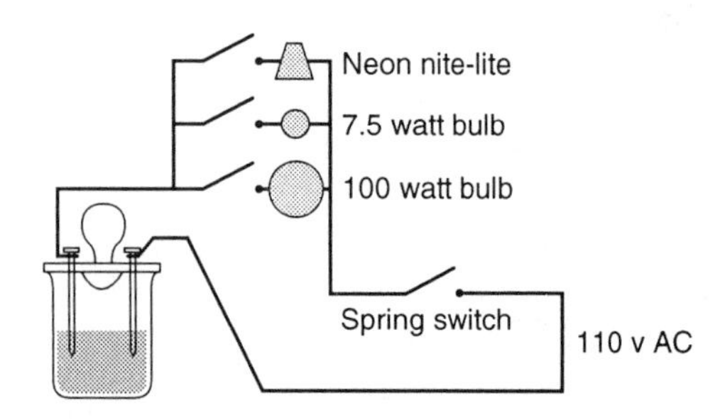

Demonstration Conductivity Apparatus.

The degree of ionization can be indicated by the electrical conductivity of a solution. The more ions a solution contains, the better conductor it becomes, and its conductivity can be shown by the brightness of the lamp in the circuit (demonstration apparatus). The neon nite-light requires very little current and may light even if the larger bulbs do not. This apparatus should be used for demonstration only. For safety's sake, the circuit should be controlled by a spring-loaded, push-button switch.

PROCEDURE

Conductivity Demonstration

Each of the substances or solutions listed on the report sheet is to be tested. Put about 50 mL of the solution into a 150-mL beaker and lower the electrodes of the conductivity apparatus into it. Switch in the largest light bulb first, then close the spring-loaded switch. If necessary try the other two bulbs in succession. After each test, switch off the current, wipe the electrodes, and rinse them with distilled water. Thoroughly clean the beaker between tests.

Mix 0.1 N solutions of NH_4OH and $HC_2H_3O_2$ in equal proportions to prepare the $NH_4C_2H_3O_2$ solution.

A solid salt is melted and tested in order to determine whether the process of melting liberates ions as well as the process of solution. Put about 25 g of solid KSCN into a beaker and test its conductivity. Then heat until molten and make the conductivity test again with dry electrodes.

In order to investigate the role of the solvent in ionization, first test the conductivity of pure benzene, C_6H_6. Be sure to use a dry beaker and dry electrodes.

Prepare a HCl generator by fitting a gas bottle with a two-hole rubber stopper carrying a right angle glass bend and a thistle tube that reaches to the bottom of the bottle. Connect another right angle glass bend to the outlet tube by a short section of rubber tubing. Put 15 g NaCl into the gas bottle, insert the stopper, and put the delivery tube into the beaker containing the benzene. Cover the beaker with a piece of cardboard and run the delivery tube through it. Take the apparatus to the fume hood and add 20 mL of concentrated H_2SO_4 to the generator through the thistle tube. Allow the gaseous HCl to bubble through the benzene for a few minutes in order to form a solution. Test the conductivity of this solution of HCl/C_6H_6.

Add about 25 mL of water to the solution and stir well with a glass rod. Water is more dense than benzene and immiscible with it, so it underlies benzene. The HCl distributes itself between the two solvents according to its solubility in each. With the current turned on, slowly lower the electrodes to the bottom of the mixture and note the result.

Experiment 9

Degree of Ionization

Name____________________________________

Date_____________________________________

REVIEW QUESTIONS

1. Does the area of the electrode immersed in the solution and the distance between the electrodes have any bearing on the conductivity?

2. Look up the meaning of the term *specific conductivity.*

3. Why is a solid salt a nonconductor?

4. A 1 M $HC_2H_3O_2$ solution is about 0.42% ionized at 25° C and a 0.1 M $HC_2H_3O_2$ solution is 1.33% ionized. Which is the better conductor? Explain.

5. A student wanted to make a table lamp with a transparent plastic column and proposed conducting current to the bulb through two channels bored through the plastic and filled with salt solution. Criticize the idea.

6. Write the equation showing the role water plays in the ionization of HNO_3 and label the conjugate acids and bases. Do the same with $HC_2H_3O_2$. Explain in terms of the Bronsted theory why HNO_3 is a strong acid and $HC_2H_3O_2$ is a weak one.

7. Give the Bronsted definitions of an acid and a base.

8. What are the Lewis definitions of an acid and a base? Give examples of each.

Experiment 9

Degree of Ionization

Name______________________

Date______________________

REPORT SHEET

Sample	Strong Electrolyte, Weak Electrolyte, or Nonelectrolyte?	Ionization Equation for Electrolytes (Show a Reversible Reaction for Weak Electrolytes)
Distilled H_2O		
Tap		
0.1 N H_2SO_4		
0.1 N HCl		
0.1 N NaOH		
0.1 N HNO_3		
0.1 N NH_4OH		
0.1 N $HC_2H_3O_2$		
$NH_4C_2H_3O_2$ (Mix of NH_4OH + $HC_2H_3O_2$)		
$HC_2H_3O_2$ (glacial)		
0.1 N NaCl		
0.1 N $CuSO_4$		
0.1 N $CaCl_2$		
0.1 M Sucrose		
Molten KSCN		
Benzene (C_6H_6)		
HCl in C_6H_6		
HCl in H_2O		

Is distilled water a complete nonelectrolyte?______________________

Explain:______________________

____

List four ions that may be responsible for conductivity in tap water.

1. ____________ 2. ____________ 3. ____________ 4. ____________

Use the Bronsted theory to explain why an HCl solution in H_2O is a better conductor than an $HC_2H_3O_2$ solution.

____

____

Why is dilute $HC_2H_3O_2$ a better conductor than glacial $HC_2H_3O_2$? ______

______

Is the salt $NH_4C_2H_3O_2$ a stronger or a weaker electrolyte than NH_4OH or $HC_2H_3O_2$? ______

Explain. ______

What ions account for the conductivity? ______

What general statement can be made about the conductivity of salt solutions?

______

______

Why is molten KSCN a conductor? ______

______

The reaction between H_2SO_4 and solid NaCl is an ion combination involving only one hydrogen ion in H_2SO_4. Write the equation for it: ______

______

Explain the difference in conductivity behavior between the HCl-H_2O solution and the HCl-C_6H_6 solution:

______

______

Show by means of an electronic equation how water promotes ionization of HCl.

______

Define:

1. Ionization ______

   ______

   ______

2. Strong acid ______

   ______

   ______

3. Weak acid ______

   ______

   ______

4. Hydronium ion ______

   ______

   ______

EXPERIMENT 10

Periodic Properties of the Elements

Purpose: To note how certain properties of elements tend toward a periodic similarity when the elements are arranged in order of increasing atomic number; to note that elements (or ions of those elements) in the same vertical column (family) of the periodic table have similar properties; to note that elements (or ions of those elements) in different families of the periodic table have fewer similarities and greater differences.

Materials: Display samples of as many elements as possible; photographs and descriptions of elements; tables of properties; student textbook; *Handbook of Chemistry and Physics*; graph paper; solid samples of Li, Na, and K metals for instructor performed demonstration; Mg, Ca, and Al metals for student experiments; NaCl, KCl, $MgCl_2$, $CaCl_2$, $AlCl_3$, and "unknown" salts; and steel wool. Reagents include 6 M HCl, 0.2 M NaCl, 0.2 M NaBr, 0.2 M NaI, 0.2 M Na_3PO_4, 0.2 M Na_2CO_3, 0.05 M $AgNO_3$, 6 M HNO_3, 1 M NaOH, and 6 M NaOH.

Safety Precautions: The alkali and alkaline earth metals are highly reactive. The instructor should use appropriate care in demonstrating lithium, sodium, and potassium. Students should also use care in allowing calcium and magnesium to react with water or aqueous solutions.

Hazardous Waste Disposal: The only heavy metal used in this experiment is silver. The contents of all test tubes to which $AgNO_3$ has been added should be disposed of in the heavy metal waste container.

INTRODUCTION

Early in the nineteenth century people noted similarities among various elements and tried to find a pattern of relationship among them. In 1869 Dimitri Mendeleev of Russia and Lothar Meyer of Germany independently arranged the elements in order of increasing atomic weight and noted that similarities appeared at intervals. Nothing was known of atomic structure at that time, and no reasons could be given for the periodic similarities. In 1913 Henry Moseley, an English scientist, determined the nuclear charges (atomic numbers) of the elements and pointed out that the fundamental order of arrangement of the elements should be based on increasing atomic numbers. In the next few years, theories of atomic structure founded on the work of Rutherford and Bohr, and advanced by many scientists, were able to relate repeated similarities of behavior with structural similarities. The modern periodic law can be stated as follows: When the elements are arranged in order of increasing atomic number, similarities of properties occur periodically.

The periodic table is a central tool in helping us to understand regularities in the behavior of elements and compounds. It is a very organized system that allows us to study and understand the behavior of *groups* of elements rather than *individual* elements. This streamlines and makes our study more efficient and, at times, helps us to make predictions about the physical or chemical properties of elements that are unsafe, unavailable, or too expensive to work with in the laboratory.

One way to learn how a tool works is to take it apart and look at the structure and function of the component parts. This lab activity has been designed to help you examine the structure and function of some of the subsections (component parts) of the periodic table. The subsections you will be examining are based on vertical groupings of elements found in the modern periodic table. You will be given the opportunity to examine the physical and chemical characteristics of elements and their compounds within selected groups and to compare characteristics among groups. In lecture, these vertical groupings will be related to atomic theory and the outer electron configuration of electrons in atoms. As you proceed with the lab activities you should keep the following questions in mind.

Do elements in a vertical column (called a group or family) have similar physical and chemical properties?

Do elements in different groups have different physical and chemical properties?

Vertical columns in the periodic table are usually labeled with a Roman numeral and a letter. The first column, for instance, is labeled IA, the second IIA, and so on. Sometimes groups are labeled differently, so you must examine the periodic table carefully. While there may be disagreement about how to label a group, this does not change the observable regularity of physical and chemical properties of elements in that group. The labeling system used for this activity is consistent with that in the periodic table found as an appendix of this lab manual and in the tables in most general chemistry textbooks.

This activity does not involve numerical data and calculations. It is qualitative in nature, which means that you must rely on your skills of observation, inference, and written communication.

PROCEDURE

A. Physical Properties of Selected Elements

Complete the table on the first page of the report sheet. Record the state of the elements at normal temperature and pressure conditions, their metallic characteristics, and their color. This information is available from a variety of sources. There may be samples of some of the elements in the lab for you to examine. Physical properties of elements are given in resources such as the *Handbook of Chemistry and Physics* and your textbook. Some periodic tables also list properties of elements. Write the accepted symbol for each element in the table.

B. Chemical Reactions of Metals

1. **Group IA: Li, Na, K (alkali metals) (demonstration).** These elements are very reactive and are not found in nature as the free elemental form. The following reaction equation represents the typical violent reaction between an alkali metal and water.

$$2\,Na + 2\,H_2O \longrightarrow 2\,NaOH + H_2$$

   The instructor will perform and demonstrate the reactions of the alkali metals with water. Record your observations on the report sheet. Using this physical evidence, make inferences about the chemical behavior of alkali metals.
2. **Group IIA: Mg and Ca (alkaline-earth metals).** *CAUTION! DO NOT HOLD TEST TUBES WITH REACTING SUBSTANCES IN YOUR HAND.* Put a small piece of Mg and a small piece of Ca into separate test tubes. Add 5 mL of cold distilled water to each of the test tubes. Observe and record any physical changes that indicate that a chemical reaction has or has not occurred.

   Add 5 drops of 6 M HCl to the test tube containing Mg plus H_2O. Record your observations.
3. **Group IIIA: Al.** Clean a small piece of aluminum metal with steel wool. (Do this on a piece of paper so you do not scratch the lab bench.) Put the burnished aluminum into a clean test tube and add 5 mL of cold distilled water. Record your observations.

   Add 5 drops of 6 M HCl to the test tube contents. If no change is observed, add an additional 5 mL of 6 M HCl to the test tube. Record your observations.

C. Chemical Reactions of Compounds Containing Ions

1. **Groups IA, IIA, and IIIA metal ions.** Put 4 mL of distilled water into each of five clean test tubes. Put about one-half spatula of sodium chloride (NaCl) into the first test tube. In succeeding test tubes add about the same amount of potassium chloride (KCl), magnesium chloride ($MgCl_2$), calcium chloride ($CaCl_2$), and aluminum chloride ($AlCl_3$). Mix each thoroughly and record your observations.

If any of the tubes contain solid material, decant (pour off) the liquid into another clean test tube to use for the next step and discard the solid. Now add 1 mL of 1 M NaOH to each of the test tubes containing solutions of metal chlorides. Mix each thoroughly and note any changes. Then add 1 mL of 6 M NaOH to the contents of each test tube. Mix thoroughly. Note any evidence that a reaction has taken place.

2. **Nonmetal Ions: halides (group VIIA ions), phosphates, and carbonates.** Put 3 mL of each of the following solutions into separate clean test tubes. Note that chlorine, bromine, and iodine belong to group VIIA which is the family of elements known as the **halogens.** Negatively charged ions (*anions*) of the halogens are known as **halide ions.**

   0.2 M NaCl, source of chloride ion (Cl^-)
   0.2 M NaBr, source of bromide ion (Br^-)
   0.2 M NaI, source of iodide ion (I^-)
   0.2 M Na_3PO_4, source of phosphate ion (PO_4^{3-})
   0.2 M Na_2CO_3, source of carbonate ion (CO_3^{2-})

   Each of the solutions is to be treated exactly the same. Add 1 mL of 0.05 M $AgNO_3$ (silver nitrate) solution to each. Mix thoroughly. Allow these mixtures to stand for about 5 minutes, then make and record your observations. Then slowly add 1-mL increments of 6 M HNO_3 to each test tube with mixing until a distinct change is noted in the phosphate and carbonate solutions. Observe the contents of the test tubes as you add nitric acid.

3. **Salt containing a group IA, IIA, or IIIA metal ion (cation) and a halide, carbonate, or phosphate negative ion (*anion*).** Obtain a sample of an unknown solid and record the number of this salt on the report sheet. Measure 3 mL of ion-free water into two separate test tubes and add 1/2 spatula of the unknown salt to each test tube. Mix well to dissolve the solid, and decant the solution into another test tube only if some remains undissolved. Add 1 mL of 1 M NaOH to the first test tube. Mix thoroughly and note whether a precipitate forms. If a precipitate forms, add 1 mL of 6 M NaOH to the contents of the tube. Mix thoroughly. Note any evidence that a reaction has taken place. Use this evidence to indicate (on the report sheet) to which of the three groups the cation in this salt belongs.

   To the second test tube add 1 mL of 0.05 M $AgNO_3$ (silver nitrate) solution. Mix thoroughly. Allow the mixture to stand for about 5 minutes, then make and record your observations. Then slowly add two or three 1-mL increments of 6 M HNO_3 to the test tube with mixing until a distinct change (or no change) is noted. Record your observations and identify the anion as being either carbonate, phosphate, or a halide.

D. Plots of Properties of the Elements

In this section of the experiment, two plots are drawn using data obtained by physicists and chemists. The first property plotted, **first ionization energy,** is the energy required to remove one electron from an atom of the element in the gaseous state. It is measured in electron volts (eV). The second property to be plotted, **atomic radius,** is one-half the shortest distance between nuclei in the pure element. It is measured in Angstrom units (Å).

1. **Ionization energy.** On the upper section of the graph paper plot the first ionization energy of **each element (for which a value is given in Table 1)** on the vertical axis against increasing atomic number on the horizontal axis. Locate the point representing the ionization energy of **each element** and connect successive points with a line. Write the symbol of each element that represents a maximum in the curve above the point for that element. If any atomic value is unknown, leave a gap in the line. Connect with a light line, or line of a different color, the noble gases; He, Ne, Ar, Kr, and Xe. In a similar manner connect the alkali metals; Li, Na, K, Rb, and Cs and the halogens; F, Cl, Br, and I.
2. **Atomic radius.** On the lower half of the graph sheet, plot in a similar manner the atomic radius of each element on the vertical axis against the atomic number on the horizontal axis. Connect each successive point by means of a line segment. Write the symbol of each element that represents a maximum in the curve above the point for that element. With a light line, or a different color, connect the alkali metals, Li, Na, K, Rb, and Cs; the alkaline-earth metals, Be, Mg, Ca, Sr, and Ba; and the noble gases, He, Ne, Ar, Kr, and Xe. Note the characteristic pattern.

Experiment 10

Periodic Properties of the Elements

Name______________________________

Date______________________________

REVIEW QUESTIONS

1. In Group IIB, zinc is in the solid state and mercury is in the liquid state at room temperature. What is the trend for melting points in Group IIB as atomic number increases?

2. Using Group IVA as a guide, make a statement about change in metallic properties of elements in the same group as their atomic weights increase.

3. How are the melting points of elements in Group VIIA related to increasing atomic weight?

4. Which of the following elements is most reactive to water? (Circle.)

   Al Na Mg

5. Using your observations of Group IA metals, how would you expect the element Rb to react with water?

6. Is there a relationship between ionization energy and reactivity to water? (Look at your experimental data for reactivity and at the ionization energy table.)

7. List the metals Li, Mg, K, and Ca in order of increasing reactivity with water.

8. Identify the gas, by name, that is produced when metallic sodium reacts with water.

9. Compare the atomic radii of Zr and Hf. Using your textbook as a reference, find an explanation for the close similarity.

10. Plot the ionization energies for removing the first, second, third, and fourth electrons for Na. On the same graph paper, plot similar information for Mg and Al. Use your textbook or similar references.

TABLE 1 First Ionization Energies and Atomic Radii (van der Waals')

Atomic Number	Symbol	Ionization Engery eV	Atomic Radii Å
1	H	13.6	1.2
2	He	24.6	1.18
3	Li	5.39	1.55
4	Be	9.32	1.12
5	B	8.30	0.98
6	C	11.3	0.91
7	N	14.5	0.92
8	O	13.6	—
9	F	17.4	1.35
10	Ne	21.6	1.60
11	Na	5.14	1.96
12	Mg	7.64	1.60
13	Al	5.98	1.43
14	Si	8.15	1.32
15	P	11.0	1.28
16	S	10.4	1.27
17	Cl	13.0	1.40
18	Ar	15.8	1.92
19	K	4.34	2.35
20	Ca	6.11	1.97
21	Se	6.56	1.44
22	Ti	6.83	1.47
23	V	6.74	1.34
24	Cr	6.67	1.27
25	Mn	7.43	1.26
26	Fe	7.90	1.26
27	Co	7.86	1.25
28	Ni	7.63	1.24
29	Cu	7.72	1.28
30	Zn	9.39	1.38
31	Ga	6.00	1.41
32	Ge	7.88	1.37
33	As	9.81	1.39
34	Se	9.75	1.40
35	Br	11.8	1.65
36	Kr	14.00	1.97
37	Rb	4.18	2.48
38	Sr	5.69	2.15
39	Y	6.5	1.80
40	Zr	6.95	1.60
41	Nb	6.77	1.46
42	Mo	7.10	1.39
43	Te	7.28	1.36
44	Ru	7.36	1.34
45	Rh	7.46	1.34
46	Pd	8.33	1.37
47	Ag	7.57	1.44
48	Cd	8.99	1.54
49	In	5.79	1.66
50	Sn	7.34	1.62
51	Sb	8.64	1.59
52	Te	9.01	1.60
53	I	10.4	1.77
54	Xe	12.1	2.17
55	Cs	3.89	2.67
56	Ba	5.21	2.22
57	La	5.61	1.87
58	Ce	6.9	1.81
59	Pr	5.7	1.82
60	Nd	6.3	1.82
61	Pm	5.8	—
62	Sm	5.6	1.66
63	Eu	5.7	2.04
64	Gd	6.2	1.79
65	Tb	6.7	1.77
66	Dy	6.8	1.77
67	Ho	6-6.9	1.76
68	Er	6-6.9	1.75
69	Tm	6-6.9	1.74
70	Yb	6.2	1.92
71	Lu	5.0	1.74
72	Hf	5.5	1.58
73	Ta	7.88	1.46
74	W	7.98	1.39
75	Re	7.87	1.37
76	Os	8.7	1.35
77	Ir	9.	1.36
78	Pt	9.0	1.38
79	Au	9.22	1.44
80	Hg	10.4	1.57
81	Tl	6.11	1.48
82	Pb	7.42	1.75
83	Bi	7.29	1.70
84	Po	8.43	1.76
85	At	—	—
86	Rn	10.8	—
87	Fr	—	—
88	Ra	5.28	—
89	Ac	6.6	—
90	Th	6.95	—
91	Pa	—	—
92	U	6.1	—

Experiment 10 Periodic Properties of the Elements

First Ionization Energy (Electron Volts)

30
20
10

Atomic Radius (Angstrom Units)

3.0
2.0
1.0
0

0 10 20 30 40 50 60 70 80 90 100

Atomic Number

Experiment 10
Periodic Properties of the Elements

Name________________________________

Date________________________________

REPORT SHEET

A. Physical Properties of Selected Elements

Group	Element	State (solid, liquid, gas)	Metal, Nonmetal, Semimetal	Color	Symbol
IA	Lithium				
	Sodium				
	Potassium				
IIA	Beryllium				
	Magnesium				
	Calcium				
IB	Copper				
IIB	Zinc				
	Cadmium				
	Mercury				
IIIA	Aluminum				
IVA	Carbon				
	Silicon				
	Tin				
	Lead				
VA	Nitrogen				
	Phosphorus				
	Arsenic				
	Antimony				
	Bismuth				
VIA	Oxygen				
	Sulfur				
VIIA	Bromine				
	Iodine				

B. Chemical Reactions of Metals

1. Group IA: Li, Na, K (alkali metals) (Demonstration).

Element	Behavior in cold water
Lithium	
Sodium	
Potassium	

Do the elements in Group IA react with water in a similar way? If so, describe this similarity. If not, describe the dissimilarity.

Which of the elements observed in Group IA appears to be:

a. Most reactive with water?____

b. Least reactive with water?____

For the alkali metals, state the trend for reactivity with water as atomic number increases.

2. Group IIA: Mg and Ca (alkaline-earth metals)

Element	Behavior in cold water	Behavior with 5 drops 6 M HCl added
Magnesium		
Calcium		XXXXXXXXXXXXXX

Which of the elements observed in Group IIA appears to be more reactive? Explain on the basis of your observations.

3. Group IIIA: Al

Element	Behavior		
	In cold water	With 5 drops 6 M HCl added	With 5 mL 6 M HCl added
Aluminum			

From the information collected thus far, arrange the three periodic groups of metals studied in order of increasing reactivity with cold water (least reactive listed first).

____________ < ____________ < ____________

C. Chemical Reactions of Compounds Containing Ions

1. Groups IA, IIA, and IIIA metal ions.

Compound	Behavior		
	In water	With 1 M NaOH added	With 6 M NaOH added
NaCl			
KCl			
$MgCl_2$			
$CaCl_2$			
$AlCl_3$			

In the table, put brackets in the left margin next to metal chloride compounds that have similar chemical behaviors.

Are the metal ions grouped together in the table also grouped together in vertical columns in the periodic table? ________

____

2. Nonmetal Ions: halides (group VIIA Ions), phosphates, and carbonates.

Compound	Behavior with $AgNO_3$	Behavior with HNO_3
NaCl		
NaBr		
NaI		
Na_3PO_4		
Na_2CO_3		

Do the compounds containing halide ions have similar chemical behaviors? Explain on the basis of your observations.

Compare the chemical behaviors of the compounds containing phosphate and carbonate ions with the compounds containing halide ions. Are the chemical reactions of the phosphate and carbonate ions similar to or different from the chemical reactions of the halide ions? Explain on the basis of your observations.

3. Salt containing a group IA, IIA, or IIIA metal ion (*cation*) and a halide, carbonate, or phosphate negative ion (*anion*).

Reaction with	Behavior observed
1 M NaOH	
1 mL 6 M NaOH	

The unknown salt contains a cation from Group ____________ in the periodic table.

Compound	Behavior with $AgNO_3$	Behavior with HNO_3
Unknown Salt		

The unknown salt contains an anion from Group ____________ in the periodic table.

D. Plots of Properties of the Elements

1. At what intervals do maxima in ionization energies occur (how many elements between maxima)?

2. What elements are represented by these maxima?

3. At what intervals do minima in ionization energies occur (how many elements between minima)?

4. What elements are represented by these minima?

5. At what intervals do the elements F, Cl, Br, and I occur (how many elements between)?

6. How do these elements compare in ionization energy with the elements having maxima?

7. What chemical characteristics would you expect of elements having high ionization energy?

8. What chemical properties would you expect of elements having low ionization energy?

9. At what intervals do maxima in atomic radii occur (how many elements between maxima)?

10. What elements are represented by these maxima?

11. How does the atomic radius of atoms of the elements change as the atomic number increases within a group?

12. How does the atomic radius of atoms of the elements change as the atomic number increases within a period?

EXPERIMENT 11

Reaction Stoichiometry and Theoretical Yield

Purpose: To carry out and observe a series of reactions by which copper metal is converted to a coordination compound, $[Cu(NH_3)_4](NO_3)_2$, followed by another series of reactions by which the coordination compound is converted back to copper metal; to measure observed yields and to calculate theoretical and percent yields for both $[Cu(NH_3)_4](NO_3)_2$ and the final copper product.

Materials: Copper gauze or wire (B & S #26 or smaller), 6 N HNO_3, concentrated aqueous ammonia (ammonium hydroxide), methanol, 6 N HCl, magnesium metal (Mg turnings) vacuum filtration apparatus consisting of Buchner funnel, suction flask, and thick-wall flexible tubing.

Safety Precautions: During the initial reaction of copper with nitric acid, nitrogen dioxide is produced. This reaction must be carried out in the hood. Care should be taken with nitric acid, hydrochloric acid, concentrated aqueous ammonia, and methanol.

Hazardous Waste Disposal: All waste materials may be disposed of in sinks or solid waste receptacles. The final copper product should be turned in with the report sheet for evaluation by the instructor.

INTRODUCTION

Some of the most interesting research in organic chemistry involves the preparation and study of a large class of substances known as coordination compounds. These compounds, sometimes called complexes, are typically salts that contain complex ions. A complex ion is an ion that contains (1) a central metal ion (usually a transition metal ion) and (2) small polar molecules or simple ions (called ligands) that are bonded to the central metal ion. The bonding between the central metal ion (electron-deficient) and the ligands (electronic-rich) is called coordinate covalent bonding. The small polar molecule or simple ion donates both electrons involved in bonding. Coordinate covalent bonds are relatively weak.

Examples of complex ions:

$[Fe(CN)_6]^{4-}$ $[Ag(NH_3)_2]^{1+}$ $[Cu(NH_3)_4]^{2+}$

Structural diagram of $[Cu(NH_3)_4]^{2+}$:

$$\left[\begin{array}{ccc} & NH_3 & \\ & | & \\ H_3N - & Cu & - NH_3 \\ & | & \\ & NH_3 & \end{array}\right]^{2+}$$

The coordination compound that will be prepared in this experiment is:

$$[Cu(NH_3)_4](NO_3)_2, \text{ tetramminecopper(II) nitrate.}$$

The complex ion in this compound is enclosed in brackets in the formula. The positive complex ion contains the Cu^{2+} (central metal ion) and four ammonia, NH_3, molecules (small polar molecules). The negative ions are nitrate ions.

The coordination compound will be prepared in several steps using various chemical reactions. In ***step 1,*** copper metal will be dissolved in nitric acid. The chemical equation that represents this process is:

$$Cu_{(s)} + 4\ HNO_{3(aq)} \rightarrow \underset{\text{(blue solution)}}{Cu(NO_3)_{2(aq)}} + 2\ H_2O_{(l)} + \underset{\text{(brown gas)}}{2\ NO_{2(g)}} \qquad \text{Equation 1}$$

After the copper has dissolved, the beaker will contain a blue-colored aqueous solution of copper (II) nitrate. Another product, nitrogen dioxide, will have escaped during the dissolution process as a brown gas.

Step 2 will involve two chemical changes. (a) When aqueous ammonia is first added to the copper (II) nitrate solution prepared in ***step 1,*** the solution becomes basic and copper (II) hydroxide is formed. This chemical change is represented by the following chemical equation.

$$\underset{\text{(blue solution)}}{Cu(NO_3)_{2(aq)}} + 2\ NH_{3(aq)} + 2\ HOH \rightarrow \underset{\text{(light-blue precipitate)}}{Cu(OH)_{2(s)}} + 2\ NH_4NO_{3(aq)} \qquad \text{Equation 2}$$

The addition of more aqueous ammonia results in a reaction that produces the soluble coordination compound, $[Cu(NH_3)_4(NO_3)_2$, and is represented by Equation 3.

$$\underset{\text{(light-blue precipitate)}}{Cu(OH)_{2(s)}} + 2\ NH_4NO_{3(aq)} + 2\ NH_{3(aq)} \rightarrow \underset{\text{(dark-blue solution)}}{[Cu(NH_3)_4](NO_3)_{2(aq)}} + H_2O \qquad \text{Equation 3}$$

In ***step 3,*** the water-soluble coordination compound will be rendered insoluble by the addition of methanol. We could represent the change by the following chemical equation.

$$\underset{\text{(dark-blue solution)}}{[Cu(NH_3)_4](NO_3)_{2(aq)}} \xrightarrow{\text{methanol}} \underset{\text{(dark-blue solid)}}{[Cu(NH_3)_4(NO_3)]_{2(s)}} \qquad \text{Equation 4}$$

Once recovered, the blue solid, tetramminecopper(II) nitrate, will be changed back to copper metal by a series of reactions involving a number of procedural steps. Following weighing and the removal of a small amount to hand in to the instructor, the dark-blue solid prepared in part A will be dissolved in water. ***Step 4,*** the dissolving of the compound in water, is represented by Equation 5.

$$[Cu(NH_3)_4](NO_3)_{2(s)} \xrightarrow{H_2O} [Cu(NH_3)_4](NO_3)_{2(aq)} \qquad \text{Equation 5}$$

In ***step 5,*** the NH_3 ligands will be detached from the Cu^{2+} ion by adding hydrochloric acid. As hydrochloric acid is added, two consecutive reactions take place and are represented by the following two chemical equations. First, copper(II) hydroxide is formed.

$$\underset{\text{(dark-blue solution)}}{[Cu(NH_3)_4](NO_3)_{2(aq)}} + 2\ HCl_{(aq)} \rightarrow \underset{\text{(light-blue precipitate)}}{Cu(OH)_{2(s)}} + 2\ NH_4NO_{3(aq)} + 2\ NH_4Cl_{(aq)} \qquad \text{Equation 6}$$

Then as excess HCl is added, the light-blue copper(II) hydroxide disappears and blue-green copper(II) chloride is formed.

$$\underset{\text{(light-blue precipitate)}}{Cu(OH)_{2(s)}} + 2\ HCl_{(aq)} \rightarrow \underset{\text{(blue-green solution)}}{CuCl_{2(aq)}} + 2\ H_2O \qquad \text{Equation 7}$$

In ***step 6,*** magnesium metal is used as a reducing agent to produce solid copper according to the following equation:

$$CuCl_{2(aq)} + Mg_{(s)} \rightarrow MgCl_{2(aq)} + Cu_{(s)} \qquad \text{Equation 8}$$

During this step, a secondary reaction occurs that is responsible for the evolution of hydrogen gas and some noticeable heat as excess HCl reacts with magnesium.

$$2\ HCl_{(aq)} + Mg_{(s)} \rightarrow MgCl_{2(aq)} + H_{2(g)} \qquad \text{Equation 9}$$

PROCEDURE

A. Preparation of $\left[Cu(NH_3)_4\right](NO_3)_{2(s)}$ from Cu(s)

Step 1. Reaction of copper with nitric acid. PERFORM THIS STEP IN THE HOOD. Weigh approximately 1 gram of copper wire on a milligram balance. The weight should not be exactly 1.00 g, just somewhere in that neighborhood. Record the exact weight of the copper wire. This weight will be used to calculate a theoretical yield.

Place the $Cu_{(s)}$ in a 150-mL beaker and 10 mL of 6 N nitric acid.

Set up a ring stand with iron ring and wire gauze in the hood. The wire gauze platform should be adjusted to a height so that the Bunsen burner may be placed beneath the wire gauze. Place the beaker containing the copper and nitric acid on the wire gauze platform. Add another ring as a safety ring around the beaker.

Heat the mixture in the beaker gently until the evolution of brown gas becomes fairly rapid. Do not heat the mixture to the point where it spatters or is evaporated to dryness. If the evolution of brown gas appears to stop before all the Cu has reacted (dissolved), add a few more milliliters (no more than 5 mL) of 6 N HNO_3.

What is the name of the brown gas produced in this reaction? What is the color of the resulting solution in this chemical reaction? What is the name of the chemical compound that produces this color?

When all the copper has dissolved, allow the beaker and contents to cool for a few minutes.

Step 2. Addition of aqueous ammonia. PERFORM THIS STEP IN THE HOOD. Obtain 10 mL of aqueous ammonia (labeled NH_4OH) in a small graduated cylinder. While stirring the contents of the beaker with a glass rod, slowly (1 mL portions) add the aqueous ammonia (*in a fume hood*) to the contents of the beaker. You should first observe the formation of a copper (II) hydroxide precipitate. Describe the color of this precipitate. As more aqueous ammonia is added, a second color change should be observed and the precipitate formed initially should dissolve. Describe the second color change. Continue adding aqueous ammonia (more than 10 mL may be required) until all of the precipitate has dissolved. At this point all of the $Cu_{(s)}$ will have been converted to the $Cu(NH_3)_4^{2+}$ complex ion. What is the principal negative ion in the solution?

Step 3. Recovery of tetramminecopper(II) nitrate, $[Cu(NH_3)_4](NO_3)_2$. Remove the solution from the hood. At the lab bench, slowly add 40 mL of methanol (methyl alcohol) while stirring. **Caution: Methanol is extremely poisonous and flammable. Avoid inhaling it. Ingestion can cause blindness. Keep it away from open flames.** Set up a vacuum filtration system, as shown in Figure 11.1, with filter paper placed in a clean Buchner funnel. Use a wash bottle to add a few drops of water to the filter paper so that the paper will lie flat in the bottom of the funnel. The suction flask should be connected by means of rubber tubing to a vacuum source such as the side arm (aspirator) on the water faucet. Turn on the water faucet to produce suction through the funnel.

Stir the mixture containing the blue crystalline product with a stirring rod and quickly transfer it to the funnel so that most of the solid product is removed from the beaker. After the liquid phase has passed through the filter paper, wash the solid by pouring 10 mL of methanol over it.

Use the methanol rinse to wash any solid from the beaker into the funnel. When the methanol has passed through the filter, repeat the washing process two more times with 10-mL portions of methanol. (If the solid is not washed well with methanol, it will be impossible to dry it within a reasonable period of time.) Dry the solid product by allowing the air to be pulled through the solid in the filtration system for 10 to 15 minutes. Scrape the dried solid from the filter paper into a preweighed 250-mL beaker. Weigh the beaker with the solid compound.

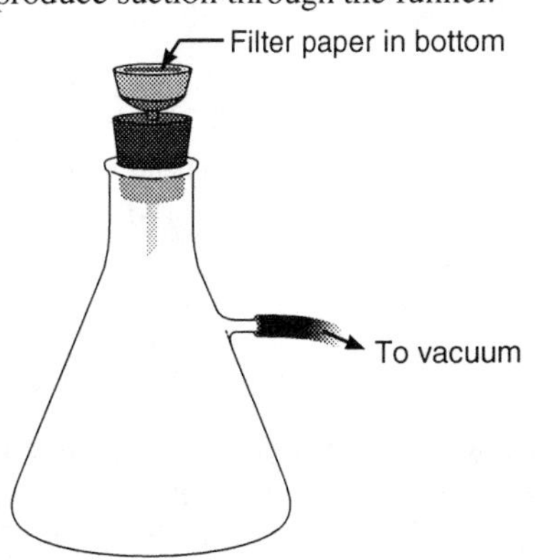

FIGURE 11.1
Vacuum filtration system.

B. Preparation of Cu from $\left[Cu(NH_3)_4\right](NO_3)_2$

Step 4. Dissolving the compound in water. Remove a small amount of $\left[Cu(NH_3)_4\right](NO_3)_{2(s)}$ from the beaker and place it in a labeled small test tube. Turn in the test tube to the instructor. (Reclaim the test tube during the next lab period.) Reweigh the beaker with the remaining solid compound. Add 20 mL of distilled water to the contents of the beaker and stir with a clean stirring rod until the solid is completely dissolved. If the solid does not dissolve in 20 mL of water, add additional 5-mL portions of water while stirring, until *all the solid is dissolved.* Describe the color of the resulting solution.

While stirring the contents of the beaker, add 6 M HCl dropwise until a light-blue precipitate appears. Identify this precipitate by both formula and name. Continue to add 6 M HCl slowly while stirring, until this precipitate dissolves.

Step 5. Reduction of copper ion to copper metal. Put about 30 pieces of magnesium turnings on a piece of paper. Add 5 to 10 pieces of Mg turnings to the contents of the beaker. Stir. When the Mg has reacted, add more magnesium. Repeat this procedure until little or no copper appears to be formed when Mg is added. The solution should be practically colorless when the process of forming copper metal is complete. This reaction process is slower than you may want it to be, but this is characteristic of many oxidation-reduction reactions of the single replacement type. If the reacting mixture in the beaker gets "cloudy" (precipitate forms) during this step, you have probably added too much Mg. To dissolve this precipitate, just add 6 M HCl until the solution is clear.

Step 6. Recovery of copper. Begin the recovery and purification of copper metal by carefully pouring off the liquid in the mixture to another container. Discard the liquid. Wash the mixture of $Cu_{(s)}$ and unreacted $Mg_{(s)}$ that remains in the beaker three times with distilled water. To do this, use the wash bottle to transfer distilled water to the beaker. Stir well. Pour off and discard the wash water. (Repeat two more times.)

After you have thoroughly washed the solid with distilled water, add 10 mL of distilled water and 10 mL of 6 M HCl. Stir. Any magnesium will react (dissolve) with the HCl at this point but the $Cu_{(s)}$ will remain unreacted. The reaction of $Mg_{(s)}$ with HCl produces hydrogen gas bubbles and is a relatively slow process. If the mixture has a lot of unreacted $Mg_{(s)}$, more HCl may be required to dissolve it all.

When all of the residual magnesium has reacted, filter the $Cu_{(s)}$ using the suction filtration system. Then wash the copper by pouring three 10-mL portions of methanol over the $Cu_{(s)}$. Allow the copper to dry 5 to 10 minutes as air is pulled through the filtering system. While the copper is drying, weigh a clean, dry watch glass on the milligram balance. When the copper has dried, transfer it to the preweighed watch glass. Weigh the watch glass plus copper. Put the sample of copper in a clean, small, labeled test tube. Turn in the sample to the lab instructor.

Calculations

A. Preparation and Recovery of $[Cu(NH_3)_4](NO_3)_2$

1. **Theoretical yield of $[Cu(NH_3)_4](NO_3)_2$.** This is the maximum amount of product that could be produced from the amount of solid copper used as starting material. This would be the yield if every atom reacted and if no product were lost in the filtration or transfer of solid. This calculation is based on the assumption that copper was the limiting reagent and all other reactions were added "in excess" during the preparation process. For our calculation purposes, we can summarize the various reaction steps of the preparation in the following way.

$$Cu_{(s)} \xrightarrow{\text{step 1}} \xrightarrow{\text{step 2}} \xrightarrow{\text{step 3}} \left[Cu(NH_3)_4(NO_3)_{2(s)}\right]$$

Thus, for every mole of $Cu_{(s)}$ used as starting material, theoretically 1 mol of $[Cu(NH_3)_4](NO_3)_2$ could be produced. (For easier labeling in the calculations, $[Cu(NH_3)_4](NO_3)_2$ will be referred to as compound.) The calculation of a theoretical yield based on the amount of copper you used as starting material can be outlined as follows.

$$(\text{g Cu used}) \times \frac{1 \text{ mo Cu}}{(\text{AW}) \text{ g Cu}} \times \frac{1 \text{ mol compound}}{1 \text{ mol Cu}} \times \frac{(\text{MW}) \text{ g compound}}{1 \text{ mol compound}} = (\text{ theoretical yield of compound})$$

where AW and MW are the atomic weight and molecular weight, respectively.

2. **Actual yield (experimental yield).** This is the amount of product actually recovered at the completion of the procedure. Calculate this value by subtracting the weight of the empty beaker from the weight of the beaker plus solid.
3. **Percent yield.** This is a ratio of the amount of product recovered to the amount of product that could be recovered theoretically.

$$\frac{\text{actual yield compound}}{\text{theoretical yield compound}} \times 100 = \%\text{ yield of compound}$$

B. Preparation and Recovery of Cu

1. **Theoretical yield of Cu.** This is the maximum amount of copper that could be produced from the amount of $[Cu(NH_3)_4](NO_3)_2$ used as starting material. This would be the yield if every atom reacted and if no product were lost in the preparation or recovery processes. This calculation is based on the assumption that the compound, $[Cu(NH_3)_4](NO_3)_2$, was the limiting reagent and that all other reactants were added "in excess" during the preparation process. For our calculation purposes, we can summarize the various reaction steps of the preparation in the following way.

$$[Cu(NH_3)_4](NO_3)_{2(s)} \xrightarrow{\text{step 4}} \xrightarrow{\text{step 5}} \xrightarrow{\text{step 6}} Cu_{(s)}$$

Thus, for every mole of $[Cu(NH_3)_4](NO_3)_2$ used as starting material, theoretically 1 mol of Cu could be produced. The calculation of a theoretical yield based on the amount of compound you used as starting material can be outlined as follows.

$$\text{g compound used} \times \frac{1 \text{ mol compound}}{(\text{MW}) \text{ g compound}} \times \frac{1 \text{ mol Cu}}{1 \text{ mol compound}} \times \frac{(\text{AW}) \text{ g Cu}}{1 \text{ mol Cu}} = \text{theoretical yield of Cu}$$

2. **Actual yield.** This is the amount of $Cu_{(s)}$ actually recovered at the completion of the procedure. Calculate this value by subtracting the weight of the empty watch glass from the weight of the watch glass plus $Cu_{(s)}$.
3. **Percent yield.** This is a ratio of the amount of $Cu_{(s)}$ recovered to the amount of $Cu_{(s)}$ that could be recovered theoretically.

$$\frac{\text{actual yield Cu}}{\text{theoretical yield Cu}} \times 100 = \%\text{ yield of Cu}$$

Experiment 11

Reaction Stoichiometry and Theoretical Yield

Name______________________________

Date______________________________

REVIEW QUESTIONS

1. You were directed to carry out the procedure of allowing copper metal to react with nitric acid in a fume hood. Why was this procedure not carried out in the open laboratory?

2. Identify, by name *and* formula, the copper compound produced when:

a. Copper metal reacts with hot nitric acid in aqueous solution.

b. The acidic solution (from 2a) is neutralized with a base, such as ammonia, and a light-blue precipitate forms.

c. Excess ammonia is added to a copper solution (from 2a) to produce a deep-blue color.

d. Aqueous copper(II) chloride reacts with magnesium metal.

3. A student prepares a compound by heating 0.503 g of copper metal with excess sulfur. The process is represented by the following balanced equation:

$$2\ Cu_{(s)} + S_{(s)} \rightarrow Cu_2S_{(s)}$$

a. Calculate the theoretical yield of $Cu_2S_{(s)}$.

b. If the student recovers 0.500 g of Cu_2S, what is the percent yield?

Experiment 11

Reaction Stoichiometry and Theoretical Yield

Name________________________________

Date________________________________

REPORT SHEET

Data and observations

A. Preparation of $[Cu(NH_3)_4]\ (NO_3)_{2(s)}$ from $Cu_{(s)}$

Step 1. Reaction of copper with nitric acid.

Weight of copper initially used ____________________

Name of brown gas produced as copper reacts with nitric acid ____________________

Color of solution after $Cu_{(s)}$ has reacted ____________________

Name of chemical producing this color ____________________

Step 2. Addition of aqueous ammonia.

Color of Cu(II) hydroxide precipitate ____________________

Final color when excess ammonia has been added ____________________

Negative ion in solution along with $[Cu(NH_3)_4]^{2+}$ ____________________
(Hint: Look carefully at the formula for the compound being produced.)

Step 3. Recovery of tetramminecopper(II) nitrate, $[Cu(NH_3)_4](NO_3)_2$.

Weight of empty 250-mL beaker ____________________

Weight of beaker + product ____________________

Weight of $[Cu(NH_3)_4](NO_3)_2$ product ____________________

B. Preparation of Cu from $[Cu(NH_3)_4](NO_3)_2$

Step 4. Dissolving tetramminecopper(II) nitrate in water.

Weight of empty 250-mL beaker ____________________

Weight of $[Cu(NH_3)_4]\ (NO_3)_2$ + beaker ____________________

Weight of $[Cu(NH_3)_4]\ (NO_3)_2$ used in this part of experiment ____________________

Color of $[Cu(NH_3)_4]\ (NO_3)_2$ in aqueous solution ____________________

Identity of light-blue precipitate when HCl is added Formula: ____________________

Name: ____________________

Step 5. Reduction of copper ion to copper metal.

Balanced equation for the formation of Cu when Mg is added to a $CuCl_2$ solution

Why does the blue copper color in solution become less intense as magnesium reacts?

____

____

____

Identify, by name, the gas evolved when magnesium reacts with the acidic solution containing $CuCl_2$.

Step 6. Recovery of copper.

Weight of empty watch glass ____________

Weight of watch glass + copper product ____________

Weight of copper ____________

Calculations

A. Preparation and Recovery of $[Cu(NH_3)_4](NO_3)_2$

1. Theoretical yield of $[Cu(NH_3)_4](NO_3)_2$.

Weight of copper initially used ____________

Molecular weight (MW) of $[Cu(NH_3)_4](NO_3)_2$ ____________

Calculation of theoretical yield of $[Cu(NH_3)_4](NO_3)_2$ (use dimensional analysis and show work)

2. Actual yield (experimental yield).

Actual yield (weight of $[Cu(NH_3)_4](NO_3)_2$ product) ____________

3. Percent yield.

Percent yield of $[Cu(NH_3)_4](NO_3)_2$ ____________

B. Preparation and Recovery of Cu

1. Theoretical yield of Cu.

Weight of $[Cu(NH_3)_4](NO_3)_2$ used in part B of experiment ____________________

Calculation of theoretical yield of copper (use dimensional analysis and show work)

2. Actual yield (experimental yield).

Weight of copper product ____________________

3. Percent yield.

Percent yield of Cu ____________________

EXPERIMENT 12

Synthesis of Alum

Purpose: To prepare an inorganic salt, $KAl(SO_4)_2 \cdot 12H_2O$ called *alum*. The experiment will also allow you to practice the techniques used in synthesis.

Materials: Piece of aluminum from beverage can (brought into lab by student), 1.4M KOH, 9M H_2SO_4, ethanol

Apparatus: Hot plate, stirring rods, 50-mL or 100-mL graduated cylinder, suction-filtration flask, Buchner funnel, filter paper, ice water bath, scissors

Safety Precautions: Safety glasses must be worn at all times in the laboratory. Take special precautions to protect your hands when handling the sharp aluminum pieces and scissors. In the first step, care must be taken to avoid exposing the hydrogen produced to any open flame. Finally, the KOH and H_2SO_4 are corrosive to the skin. If you spill any on yourself, immediately wash with water.

Hazardous Waste Disposal: All the solutions may be washed down the drain with plenty of water.

INTRODUCTION

Much has been said and written lately on the topic of recycling. One of the substances to receive early attention in this area was aluminum. Aluminum beverage cans are easily recycled into new materials. However, this process is principally a physical change—the aluminum is melted down (m.p. 660°C) and then recast into new containers.

A chemical *synthesis* requires a *chemical change*. A synthesis is the production of a compound by a chemical reaction or series of reactions where the main objective is to maximize the purity and quantity of the product. This is different from other lab experiments you might do where the emphasis is on making an accurate measurement of some property, or on deducing a natural law. In this experiment you will synthesize a *compound*, alum, from the *element*, aluminum, and some common laboratory reagents.

Aluminum (Al) is the most abundant metal, and the third most abundant of all elements in the earth's crust, but it is never found in the "native state," as the pure, uncombined element in nature. The element aluminum contains one *p* and two *s* electrons in the valence shell, which are easily lost to form the Al^{+3} ion, which is the form of aluminum that is always observed in nature. In fact, the loss of the three valence electrons to form the cation is so energy-favorable that elemental aluminum (as in the familiar soft drink can) was not seen by people until about 1886 when Charles Hall in the U.S. and Paul Heroult (France) independently developed a commercially feasible process for extracting pure aluminum from aluminum ore. Before this time, the tedious and costly method for producing elemental aluminum had the metal selling for about $545 per pound. It was considered a precious metal, like gold or silver, and was generally used only in jewelry. The Hall process uses *electrolysis*, running an electric current through molten aluminum ore. The electrons are forced into the empty *s* and *p* orbitals to complete the element's neutral electron configuration. It is easy to see why it took so long to isolate aluminum—chemists had to wait for physicists to harness electrical energy!

The current required to produce aluminum is enormous—about 15 kwh (kilowatt hour) for every one kilogram of aluminum, not counting the energy used to keep the furnace hot. This accounts for recent interest in recycling. It is much less energy expensive to simply melt and recast the metal than to expend the electrical energy necessary to make more metal from the ore.

In this experiment, you will take advantage of the reactivity of aluminum metal to produce an aluminum compound, alum, $KAl(SO_4)_2 \cdot 12H_2O$. (The element got its name from the well-known compound.) For further information, consult your textbook or the *CRC Handbook of Chemistry and Physics*.

PROCEDURE

A. Synthesis of Alum

1. *Before coming to lab:* Obtain a piece of aluminum roughly 7.5 cm x 5 cm. Cut this piece (or slightly larger) from an aluminum beverage can. Be careful of sharp edges!! Using the side of the scissors, scrape off as much of the paint as possible from the outside and as much plastic as possible from the inside. (If not for the thin layer of plastic, aluminum would dissolve into your soft drink.) Bring the sample to lab. You will need about one gram.
2. Cut the rectangle of aluminum into tiny pieces and drop into a small weighing dish. This is easiest done in a checkerboard fashion. The pieces should all be smaller than □ (but do not spend too much time on this step).
3. Weigh the aluminum pieces on the top-loader balance to the nearest 0.01 g. Record.
4. Place the clippings in a 250-mL beaker and take it to the hood. Add 50 mL of 1.4 M KOH (potassium hydroxide). Place on the medium-warm hot plate (share with others) and you should soon see bubbles of hydrogen gas forming. Hydrogen is explosive (remember the Hindenburg!) so keep all open flames away. The gas is safely vented up the hood. Do not heat so much that the solution *boils*. In fact, once it gets started the reaction should continue without much additional heat.
5. When the first reaction is finished, the bubbling will cease and the solution may take on a dark-gray color. This results from decomposed paint and plastic. While still warm, the solution must be filtered to remove this residue. You will use suction filtration to speed this step. The filter flask should be rinsed with distilled water. Then use a clean piece of filter paper in the Buchner funnel. Moisten the paper slightly with your water bottle and turn on the suction by turning on the water (full). Slowly pour your solution through. The filtrate should be completely clear. If it is not, repeat the process with a new piece of filter paper.
6. Transfer the filtrate to a *clean* 250-mL beaker. SLOWLY and CAREFULLY add 20 ml of 9M H_2SO_4 with constant stirring. Soon you will notice the precipitation of $Al(OH)_3$. You will also notice the heat evolved from the neutralization reaction. Keep going. By the time the last of the sulfuric acid is added, the solid should have redissolved. If not, keep stirring and/or try warming the solution on a hot plate. All solid must dissolve or the crystallization step will be unsatisfactory. If necessary, filter the solution as in step 5.

B. Isolation of Alum by Crystallization

1. Prepare an ice water bath by filling a 600-mL beaker half-full with crushed ice. Add tap water. Chill the solution in the 250-mL beaker in the ice bath undisturbed for at least 10 minutes. Watch the solution for the formation of crystals.
2. If no solid has formed, or it seems incomplete after 15 minutes, consult your instructor or TA. Alum has a strong tendency to form supersaturated solutions. There are various ways to encourage crystallization. For alum, the best way to encourage precipitation is to scratch the bottom or sides of the beaker with the roughened end of a stirring rod. Clean and reassemble the filtration apparatus. Filter the alum crystals from the solution. Rinse your beaker with a small amount of chilled ethanol. Allow air to flow through the crystals for 5 to 10 minutes.
3. Remove the filter paper and dry the crystals as well as you can on fresh filter paper. Record the weight of your product.

CALCULATIONS

The overall reaction can be written

$$2\ Al + 2\ KOH + 4\ H_2SO_4 + 22\ H_2O \rightarrow 2\ KAl(SO_4)_2 \cdot 12H_2O + 3\ H_2$$

Calculate the theoretical yield (to nearest 0.01 g) of alum possible, assuming all reagents in excess and the weight of aluminum as the limiting amount. Then calculate the percent yield (to the nearest 1%). (Be careful! Don't forget the twelve waters in the formula for alum.)

Experiment 12
Synthesis of Alum

Name____________________________________
Date_____________________________________

REVIEW QUESTIONS

1. What is the name of the principal ore of aluminum? Its formula?

2. How many grams of hydrogen were produced when your mass of aluminum reacted completely? Show your work.

3. Given the mass of hydrogen calculated in #2, what volume would this occupy at STP? At 25°C and 755 mm Hg?

4. Occasionally, a student will obtain a % yield greater than 100%. Aside from simple error in calculation, what is the most likely cause of this in this experiment?

5. Despite the fact that aluminum metal is easily and rapidly oxidized by air to aluminum oxide, we do not see aluminum cans and foils decomposing before our eyes (as a nail might rust). Explain.

6. How many moles of sulfuric acid H_2SO_4 would be required to make your mass of alum? How many moles of H_2SO_4 did you use? Is the H_2SO_4 in excess?

7. How much electricity (in kilowatt hours) does it take to produce one ton of aluminum, not counting the energy required to keep the furnace hot?

8. Give at least three reasons why aluminum metal is so popular in manufacturing and technological uses.

Experiment 12
Synthesis of Alum

Name____________________________________

Date_____________________________________

REPORT SHEET

Mass of aluminum ______________________ g

Mass of alum obtained ______________________ g

Theoretical yield of alum ______________________ g

% Yield ______________________ g

Calculations:

Description of product:

EXPERIMENT 13

Impurities in Natural Water

Purpose: To study natural water, some of its common impurities, and some common methods of purification.

Materials: Solid $CoCl_2$, 14 M aqueous ammonia (NH_4OH), phenolphthalein indicator, saturated $Ca(OH)_2$ solution, 6 N HNO_3, 6 N HCl, 0.05 N $AgNO_3$, 0.2 N $BaCl_2$, and 0.2 N $Al_2(SO_4)_3$, 0.0100 M EDTA solution, pH = 10 buffer solution, EBT indicator, ammonium molybdate reagent, $SnCl_2$ in glycerol solution, 5.0 mg/L phosphate standard, muddy water.

Apparatus: Model water softener containing cation exchange resins, demineralizer (cation and anion exchange), microscope light or other narrow beam source, 50-mL buret.

Safety Precautions: Have the instructor check the distillation apparatus before it is used. When evaporating water to dryness, remove the beaker (with tongs) just as the last traces of water disappear to avoid cracking the beaker.

Hazardous Waste Disposal: Silver nitrate, cobalt chloride, barium chloride solutions, and any water sample to which these reagents have been added should be disposed of in the heavy metal waste container.

INTRODUCTION

Water covers nearly three-quarters of the earth and is the most abundant substance on the planet's surface. Over 97% of the water is unfit for human consumption or municipal use because it is "saline water" found in oceans containing 3.6% (36,000 ppm) dissolved salts. Less than 3% of the earth's water is "fresh water," which is necessary for living things to survive, and over two-thirds of the fresh water is frozen in glaciers and polar regions. Consequently, less than 1% of the water on the earth is available as useful fresh water. Most of this is groundwater found in porous underground strata known as "aquifers."

Natural water always contains dissolved or suspended impurities. There is no such thing as "pure" spring water. If it has been in contact with the earth, it holds minerals and, frequently, gases in solution. Fine soil particles and bacteria will be in suspension. Often contaminants are present for which humans are responsible. Rainwater holds dissolved gases as well as suspended dust particles and bacteria.

Impurities in natural water consist of any substances other than the H_2O molecule. They may be classified into five groups.

1. **Suspended solids (detritus):** sand, clay, mud, silt, and organic particles (such as bits of leaves).
2. **Dissolved gases:** O_2, N_2, CO_2, NH_3 (ammonia), H_2S (hydrogen sulfide), and oxides of nitrogen.
3. **Dissolved ionic substances (salts):**
   cations including H^+, Li^+, Na^+, K^+, Mg^{2+}, Mn^{2+}, Fe^{2+}, and Ca^{2+}.
   anions including Cl^-, F^-, OH^-, NO_3^-, SO_4^{2-}, HCO_3^-.
4. **Dissolved organic substances:** vegetable and animal matter and their decay products.
5. **Microorganisms:** bacteria, viruses, protozoa, algae.

The intended use of the water determines its purification treatment. For household use, water is often "softened" with a household water softener. The water passes through an ion-exchange cartridge in which ions with a dipositive charge such as Ca^{2+} and Mg^{2+} are exchanged for monopositive sodium (Na^+) ions. Excessive amounts of these ions cause a problem in wash water because they form an insoluble product with soap that reduces the ability of soap to clean effectively.

For laboratory use water is sometimes distilled. The process of boiling the water and collecting the condensed vapors eliminates essentially all impurities that do not vaporize. Dissolved gases are largely driven out of the boiling water but may redissolve in the distillate. The components of glass dissolve very slightly, so for distilled water of very high purity, a tin or tin-lined condenser system is used.

Modern water purification systems often use deionization (demineralization) or reverse osmosis (RO) units because they are much less energy-intensive than distillation. Deionizers produce water of sufficient purity for many laboratory purposes. The water passes through a bed containing synthetic cation and anion exchange resins that work in pairs, exchanging virtually all positive and negative ions for hydrogen and hydroxide ions. They then combine to form water. In a reverse osmosis unit, water is passed, under high pressure, through multiple layers of a semipermeable membrane. The membrane allows H_2O molecules to pass through its pores, but suspended particles, microorganisms, and dissolved ions and molecules are not able to travel through the membrane. RO units typically remove around 90% of the dissolved salts in water, whereas deionizers approach 100% efficiency for the removal of ions. However, RO systems also remove nonionic materials that are not touched by deionizing systems.

Today, research laboratories use distillation or reverse osmosis in combination with deionization, carbon adsorption, and membrane filtration to produce ultrapure water that is both sterile and free from dissolved and suspended impurities.

Drinking water must be free of harmful bacteria, suspended matter, odor, color, and objectionable taste. Dissolved minerals and gases, unless present in large amounts, usually do not need to be removed. Small amounts of calcium, fluoride, phosphate, iron(II), and other ions are beneficial to a person's health, but excessive amounts may significantly reduce water quality. For example, calcium and magnesium ions (Ca^{2+} and Mg^{2+}) are essential for skeletal development, yet these and other dipositively charged ions cause "hard water" if their concentrations become too high. Iron is essential in blood, yet too much iron in drinking water causes brown stains of iron oxide in sinks and may produce a rusty turbidity in water. Phosphate is important for the development of bones and teeth, but it is also a plant nutrient that causes lakes and rivers to become green with algae.

If water is to be used for industrial purposes, appreciable concentrations of minerals, which are chiefly the salts of calcium or magnesium, must be removed. If they remain in the water, they form deposits in boiler tubes or cause foaming of boiling water. Several softening methods are in use, all of which either precipitate the harmful minerals before they reach the boiler or tie them up as harmless complex ions. In the laundry these minerals sometimes interfere with cleansing action and may deposit a dirty curd.

In this experiment we investigate several methods for purifying water, including removal of dissolved substances by distillation, removal of suspended matter by sedimentation and filtration, and removal of hardness and other minerals by ion exchange or by chemical precipitation. We also test for a variety of impurities found in natural drinking water, including calcium and magnesium ion hardness, chloride ion, phosphate ion, sulfate ion, and bicarbonates. These impurities are not harmful to a person's health at concentrations found in natural water.

Tests for chloride, sulfate, phosphate, and bicarbonate involve chemical reactions that produce visible colored products or precipitates. These tests, as used in this experiment, are qualitative in that they allow us to see whether or not a particular chemical species is present at detectable levels. However, in other situations, these reactions can be used for quantitative estimation of the concentration of these ions in solution.

Quantitative analysis of the hardness caused by dissolved calcium and magnesium ions can be made by titrating a known volume of water with a known concentration of EDTA (ethylenediaminetetraacetic acid) solution. One mole of EDTA links with one mole of Ca^{2+} or Mg^{2+} ions to form a complex ion. The concentration of the EDTA is such that, when a 100-mL water sample is used, 1.0 mL of the EDTA solution indicates 10 ppm (parts per million) of hardness as dissolved $CaCO_3$. The indicator Eriochrome Black T (EBT) turns the solution from wine-red to blue as soon as one drop of excess EDTA has been added, and marks the end point of the reaction.

PROCEDURE

A. Distillation

1. **Distillation of a solution containing a dissolved solid ($CoCl_2$).** Set up the apparatus shown in Figure 13.1. The boiler may be a small Florence or Erlenmeyer flask (250–500 mL) and the combined condenser and receiver is a test tube immersed in a beaker of cold water.

Add 100 mL of water and a few crystals of $CoCl_2$ to the boiler to give the water a slight pink color. Put in a few boiling chips to prevent uneven boiling and distill slowly until half an inch or so of water has accumulated in the receiver. Note the color of the distillate as compared with the mixture being distilled. Is the pink-colored impurity present in the distillate? Explain.

2. **Distillation of a solution containing a dissolved gas (NH_3).** Empty the still, rinse it, and put in a fresh sample of water. Add 5 mL of concentrated aqueous ammonia (NH_3) and distill slowly, using a fresh receiver. Note the odor of the distillate as compared with that of the solution being distilled.

   A more sensitive test for ammonia in the distillate can be made by adding a drop of phenolphthalein indicator solution. Ammonia in water produces a basic solution that will turn the indicator pink.

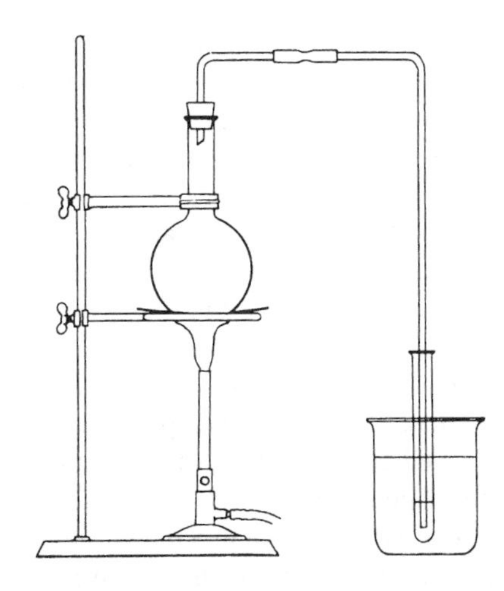

FIGURE 13.1
Distillation apparatus.

3. **Evaporation of water to concentrate dissolved material and for examination of the residue.** Before proceeding with Part B, measure about 125 mL of tap water into a beaker and start it boiling gently. Watch it from time to time as you proceed with Part B, and when the volume has been reduced to about half, withdraw a 20-mL sample. Boiling away some of the water concentrates ions that may be present in a very small amount of tap water. Set this sample aside for use in Parts C2 and 3, and boil the balance in the beaker to dryness. Remove the flame as soon as the water is gone and save the residue for use in Part C5.

B. Purification by Coagulation of Suspended Matter and Sedimentation or Filtration (Drinking Water Treatment)

1. Shake a bottle of muddy water and take out a 50-mL sample. When most of the dirt appears to have settled, decant the water into a test tube and note the remaining turbidity. Hold the sample in a strong but narrow beam of light and note if the beam is visible in the mixture (Tyndall effect). Light reflected from suspended particles, even those too small to be seen with a microscope or trapped by a filter. The molecules or ions present in a true solution cannot reflect light, so the Tyndall effect provides a way to distinguish between suspended and dissolved matter.
2. Divide the muddy water sample into two parts and save one for comparison. Add to one sample 15 mL of limewater, $Ca(OH)_2$, and 2 mL of $Al_2(SO_4)_3$ solution. Shake well and allow to stand. What is the precipitate? Write the equation for its formation. Set the samples aside for 15 minutes and then observe.
3. Treat an amount of tap water similar to one of the portions in Part B2, with $Ca(OH)_2$ and $Al_2(SO_4)_3$ solution. After 15 minutes compare the color of the residue with that of the first sample. Explain.
4. Filter all three samples, using separate filter papers, and compare the liquids under the Tyndall beam. Explain.

C. Minerals in Tap Water (or other potable water)

1. **Calcium and magnesium hardness.** Rinse a 50-mL buret with about 10 mL of EDTA solution, mount it vertically in a ring stand, and fill it above the zero mark with EDTA solution. Open the valve and carefully withdraw solution into a waste beaker until the meniscus stands at exactly 0.0 mL. No bubble should remain at the tip. With a graduated cylinder measure as accurately as possible 100.0 mL of the water to be tested and pour it into a 250-mL beaker. Add 1 mL of the buffer solution and 10 drops of EBT indicator. Allow EDTA solution to run into it from the buret with active stirring until the wine-red color of the indicator turns blue. You should not add more than 1 drop of excess. It may save time to establish the approximate value of EDTA solution needed by a rapid preliminary titration. Then make a second titration, letting the EDTA solution run rapidly until within about 1 mL of the required volume. From that point the solution should be added dropwise with active stirring until the end point is reached. The volume of the EDTA solution is read to the nearest 0.1 mL. For each 1.0 mL of solution required, the water contains 10 mg/L of hardness, reported as mg $CaCO_3$ per liter (mg $CaCO_3$/L). Leave the buret set up for further hardness tests in Part E.
2. **Chloride ion test.** Divide the 20-mL sample of water saved from Part A3 into two parts and to the first one add a few drops of HNO_3 and 1 mL of $AgNO_3$ solution. A precipitate or cloudiness is AgCl, which indicates that Cl^- ion was present

in the sample. The Tyndall effect can often be used to decide a doubtful test. Write the equation. Nitric acid, HNO_3, need not be included in the equation because it merely serves to prevent a false test. Try the test on water directly from the tap.

3. **Sulfate ion test.** To the other half of the sample from Part A3 add a few drops of HCl and about 1 mL of $BaCl_2$ solution. Heat the sample just to boiling and with the Tyndall test determine if there is a precipitate or cloudiness. The substance is $BaSO_4$ and indicates that SO_4^{2-} ion was present. Write the equation. The HCl need not be included in the equation. Try the test directly on tap water.
4. **Phosphate ion test.** Transfer 25 mL of the water sample to be tested to a clean Erlenmeyer flask or beaker. **Note:** This test is particularly interesting if done on surface water that becomes green with algae in the summer because of excessive phosphate in the water. Before delivering the water, the beaker or flask should be cleaned with warm water followed by a thorough rinse with laboratory (distilled or deionized) water. Cleaning with soap is not recommended because soaps and detergents often contain phosphate.

   Add 20 drops of ammonium molybdate solution to the 25-mL water sample, and swirl to mix. Add 2 drops of stannous chloride (in glycerol) solution and mix again by thorough swirling. If phosphate is present, a blue color will develop to a maximum in 5 minutes. For the sake of comparison, run the same phosphate test on a "standard" water sample containing 5.0 mg/L of phosphate. The intensity of the blue color is directly related to the concentration of phosphate in the water. How does your water sample compare in phosphate content to the 5.0 mg/L standard? Record on the report sheet.
5. **Indication of carbonates.** Inspect the solid residue from the evaporation of Part A3. It consists for the most part of $CaCO_3$ and $MgCO_3$ produced by the reactions:

$$Ca(HCO_3)_{2(aq)} \rightarrow CaCO_{3(s)} + H_2O_{(l)} + CO_{2(g)}$$

$$Mg(HCO_3)_{2(aq)} \rightarrow MgCO_{3(s)} + H_2O_{(l)} + CO_{2(g)}$$

   Put a few drops of HCl solution on the residue and tilt the beaker so as to let it flow slightly. A slight bubbling at the leading edge indicates the carbonate ion CO_3^{2-}. A typical equation is:

$$2\,HCl_{(aq)} + CaCO_{3(s)} \rightarrow H_2O_{(l)} + CO_{2(g)} + CaCl_{2(aq)}$$

D. Water Softening

1. **Household type (cation-exchange resin in which cations are replaced by Na^+).** Pour a beaker of tap water through a single resin water softener like the model shown in Figure 13.2. By ion exchange the resin captures all metal ions and releases Na^+ ions in their place. Hence the water now contains increased concentrations of sodium salts such as NaCl. A typical equation for this ion exchange process is:

$$Na_2R + CaCl_2 \rightarrow CaR + 2NaCl$$

   In this equation, the symbol "R" represents the negatively charged cation exchange resin. Make the hardness test on resin-softened water. Also, make the chloride, sulfate, and phosphate ion tests and record your results.

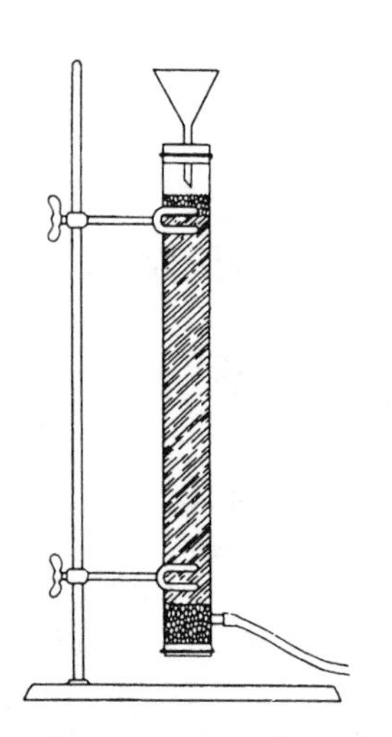

FIGURE 13.2
Working model of a cation-exchange water softener.

2. **Deionizing type (ion exchange using both cation- and anion-exchange resins).** Combinations of ion exchange resins are used in deionizing units. The cation-exchange resin removes all metal ions and replaces them with H^+ ions, and the anion-exchange resin removes negative ions and replaces them with OH^- ions. The neutralization reaction $H^+ + OH^- \rightarrow H_2O$ follows. Since the total positive charge must equal the total negative charge in solution, all cations and anions are effectively replaced by water molecules. Preparation of deionized water is less expensive than distillation of water due to lower energy requirements. Deionized water is often substituted for distilled water in scientific applications. Obtain a sample of deionized water and test for hardness, chloride ions, sulfate ions, and phosphate ions. Record your results.

Experiment 13
Impurities in Natural Water

Name________________________________
Date________________________________

REVIEW QUESTIONS

1. What reaction between soil, water, and limestone rock deposits accounts for the relatively large amounts of $Ca(HCO_3)_2$ and $Mg(HCO_3)_2$ found in groundwater? Where does the necessary CO_2 come from?

2. What natural geological structures are formed through the same reaction in question 1?

3. Reserve supplies of distilled water are often stored in covered, but not sealed, tanks lined with tin. Tin is used because it does not become an impurity in the water. What impurity is likely to be present after the water has been stored for some time?

4. How do boiling chips prevent uneven boiling (known as "bumping") in water?

5. What reaction between soap and minerals in water produces soap curd? It is not necessary to write a balanced equation. Simply describe the kind of mineral involved and the kind of reaction produced.

6. Why will some detergents lather in hard water when soap will not?

7. What advantages does a deionizing unit have over a household-type water softener?

8. What advantages might a "separate-bed" deionizing system have over a "mixed-bed" unit?

Experiment 13
Impurities in Natural Water

Name____________________________________

Date____________________________________

REPORT SHEET

A. Distillation

1. Distillation of a solution containing a dissolved solid ($CoCl_2$).

Is the $CoCl_2$ impurity present in the distillate? ____________________

Explain from your observations:

2. Distillation of a solution containing a dissolved gas (NH_3).

Is the NH_3 impurity present in the distillate? ____________________

Explain from your observations:

B. Purification by Coagulation of Suspended Matter and Sedimentation or Filtration

1. Is suspended matter present?

2.What is the precipitate? ____________________ Write the equation for its formation:

What is observed after 15 minutes?

3. How does the color of the precipitate compare with that of part B2? ____________________

Explain any color difference:

4. Do the filtered samples still show the Tyndall effect? ____________________

Can you see any difference between the untreated sample and the others? ____________________

Explain:

Does the lime-alum treatment aid in the removal of dissolved matter? ____________________

C. Minerals in Tap Water (or other potable water supply)

1. Calcium and magnesium hardness

Final buret reading =

Initial buret reading = ______________________

Volume EDTA used =

Hardness of tap water = ______________________ (include correct concentration units)

2. Chloride ion test

Evaporated water ______________________ Tap water ______________________

Equation for test: ______________________

3. Sulfate ion test

Evaporated water ______________________ Tap water ______________________

Equation for test: ______________________

4. Phosphate ion test

Does tap water produce a blue color? ______________________

How does it compare to the 1.0 mg/L standard? ______________________

5. Indication of carbonates

Is CO_3^{2-} ion present in the residue? ______________________

How do you know?

Write a balanced equation for the reaction of HCl with any $MgCO_3$ that may have been present:

D. Water Softening

1. Household type

Final buret reading =

Initial buret reading = ______________________

Volume EDTA used =

Hardness of softened water = ______________________

Is chloride ion present? ______________________ Is sulfate ion present? ______________________

Is phosphate ion present? ______________________

2. Deionizing type

Final buret reading =

Initial buret reading =

____

Volume EDTA used =

Hardness of deionized water = ________________________

Is chloride ion present?________________________

Is sulfate ion present?________________________

Is phosphate ion present?________________________

EXPERIMENT 14

Oxidation-Reduction Reactions and the Activity Series

Purpose: To carry out typical decomposition, combination, and single-replacement redox reactions.

Materials: Potassium chlorate, manganese dioxide, magnesium ribbon, powdered sulfur, charcoal, litmus paper, steel wool; granular Fe, Cu, Zn, Al, Mg, and Pb metals; strips of Cu, Pb, and Zn; 0.2 N solutions of $Pb(NO_3)_2$, $Zn(NO_3)_2$, $Cu(NO_3)_2$, and $AgNO_3$; and 6 N HCl.

Apparatus: Pneumatic trough, gas bottles, glass or plastic square plates for covering the mouths of the gas bottles, crucibles, and wire triangles.

Safety Precautions: Potassium chlorate is a powerful oxidizing agent. If molten potassium chlorate is allowed to contact paper or other organic matter, combustion is likely to start spontaneously. Therefore, handle $KClO_3$ carefully and rinse any leftover material down the sink with lots of water. Be aware that some metals react with surprising violence when put in 6 N HCl. Wear safety goggles and be careful.

Hazardous Waste Disposal: The solutions containing lead, zinc, copper, or silver nitrate should be disposed of in the heavy metal waste container. Metal strips should be placed in a solid-waste-metal container.

INTRODUCTION

Oxidation-reduction (redox) reactions are those in which valence electrons are transferred from one atom, ion, or molecule to another. In some reactions, such as single replacement reactions, the electrons are completely transferred from one atom or ion to another. In others, such as many decomposition reactions, the control of electrons is only partially transferred, and the change is reflected by a change in **oxidation number.** The loss of electrons (oxidation number increases in a positive direction) is **oxidation** and the gain of electrons (oxidation number becomes less positive or more negative) is **reduction.** The loser of electrons is **oxidized** by the gainer of electrons, which is **reduced.** The gainer of electrons is therefore called the **oxidizing agent,** and the loser of electrons is called the **reducing agent.**

Decomposition reactions, combination reactions, and single-replacement reactions are subgroups of oxidation-reduction reactions. The general forms of these types of reactions can be written as follows:

decomposition reactions	$AB \longrightarrow A + B$
combination reactions	$A + B \longrightarrow AB$
single-replacement reactions	$A + BC \longrightarrow AC + B$

We cannot see electrons, and therefore, we cannot observe the actual transfer of electrons from one substance to another in oxidation-reduction reactions. To determine if a reaction is a redox reaction, we compare the oxidation numbers (oxidation states) of a substance as a reactant and as a product. If a particular species has changed oxidation number during the reaction, we assume that a redox reaction has occurred. One substance in a system does not change oxidation state without a balancing, complementary change in another substance. Therefore, we rely on formulas and equations to interpret redox reactions.

We can write the following reaction for the decomposition of water: $2\ H_2O \longrightarrow 2\ H_2 + O_2$. As a reactant, hydrogen has an oxidation number of 1+ in H_2O; as a product, it has an oxidation number of 0 (zero). Therefore, in the decomposition of water, hydrogen is **reduced** (gains one electron per atom). The oxidation number of oxygen changes from 2– (reactant) to 0 (product); therefore, the oxygen is **oxidized** (loses two electrons per atom) when water decomposes in this fashion.

A combination (also called composition or synthesis)-type redox reaction can be represented with the following reaction equation: $2\ Na + Cl_2 \longrightarrow 2\ NaCl$. The oxidation number of sodium has changed from 0 to 1+ (loses one electron per atom); therefore, it is **oxidized** in the reaction. The oxidation number of chlorine has changed from 0 to 1– (gained one electron per atom); chlorine is **reduced** in the reaction.

When a zinc metal strip (elemental zinc) is placed in a solution of silver nitrate, some silver metal forms on the surface of the strip and flakes off. Simultaneously, some zinc dissolves as it reacts with the silver nitrate solution, but this change is less obvious. Both the zinc metal and the silver ion have changed characteristics. We can write the balanced reaction equation that represents this process in several ways.

1. Word reaction equation:
   zinc + silver nitrate $\longrightarrow$ zinc nitrate + silver
2. Formula reaction equation:
   $Zn + 2\ AgNO_3 \longrightarrow Zn(NO_3)_2 + 2\ Ag$
3. Total ionic reaction equation:
   $Zn + 2\ Ag^+ + 2\ NO_3^- \longrightarrow Zn^{2+} + 2\ NO_3^- + 2\ Ag$
4. Net ionic equation:
   $Zn + 2\ Ag^+ \longrightarrow Zn^{2+} + 2\ Ag$

As we study the reaction equations more carefully, we notice several things. Zinc atoms carry no charge as a reactant (zinc's oxidation number is 0); zinc exists in the 2+ state as a product. Silver exists as 1+ ions as a reactant; silver metal as a product carries no charge, which means its oxidation number is zero. (Throughout the process the nitrate ions remain unchanged.) The reaction involves zinc and silver. When the reaction takes place, we interpret this to mean that the zinc metal transfers two electrons to silver ions, producing zinc ions and silver metal. Or we might say, the zinc atoms become 2+ ions by the loss of two electrons per atom. This is the oxidation process. The silver ions accept one electron per ion to become silver atoms. This is the reduction process. Metals differ in their ability to donate electrons to another species. If you put a piece of silver metal into a zinc nitrate solution, no chemical change would occur because silver atoms do not give up electrons to zinc ions. Metals that can easily give up electrons undergo very rapid reactions.

The activity series of the metals is based upon the relative ease with which the metals give up electrons (are oxidized). Metals that are oxidized very easily are listed at the top. Succeeding metals are oxidized less and less easily as one reads down the table. Metals near the bottom have very little driving force to give up electrons. The oxidation of metal atoms is reversible—the ion can regain electrons under the proper conditions. These reduction (electron gain) reactions occur more and more strongly as one reads down the table.

A practical application of oxidation-reduction reactions is in electrochemical cells such as automotive "wet-cell" batteries and flashlight "dry-cell" batteries. In an electrochemical cell, the actual transfer of electrons from one species to another is controlled in such a way as to produce an electric current through a wire.

PROCEDURE

A. Decomposition Reactions

Potassium chlorate decomposes to produce oxygen gas and potassium chloride at temperatures above 400° C. The rate of decomposition is slow but can be accelerated by the addition of a catalyst (a substance that accelerates the rate of reaction but is not chemically changed at the end of the reaction). In this procedure MnO_2 is a good catalyst. Assemble the apparatus as illustrated in Figure 14.1. Use a large test tube (ignition tube) to hold the $KClO_3$ and MnO_2. The gas collection bottles, rubber tubing, and right-angle bends will be provided. Be sure that the rubber stopper fits securely into the test tube and the rubber tubing fits tightly onto the glass tubing. If the connections are not tight, the oxygen gas may leak out at these points and make the gas collection very slow.

Fill a pneumatic trough (the large metal container in the common equipment cabinets) about one-half full of water. Then fill three gas collection bottles with water. (One of these is an extra in case you "goof.") Transfer a bottle to the pneumatic trough

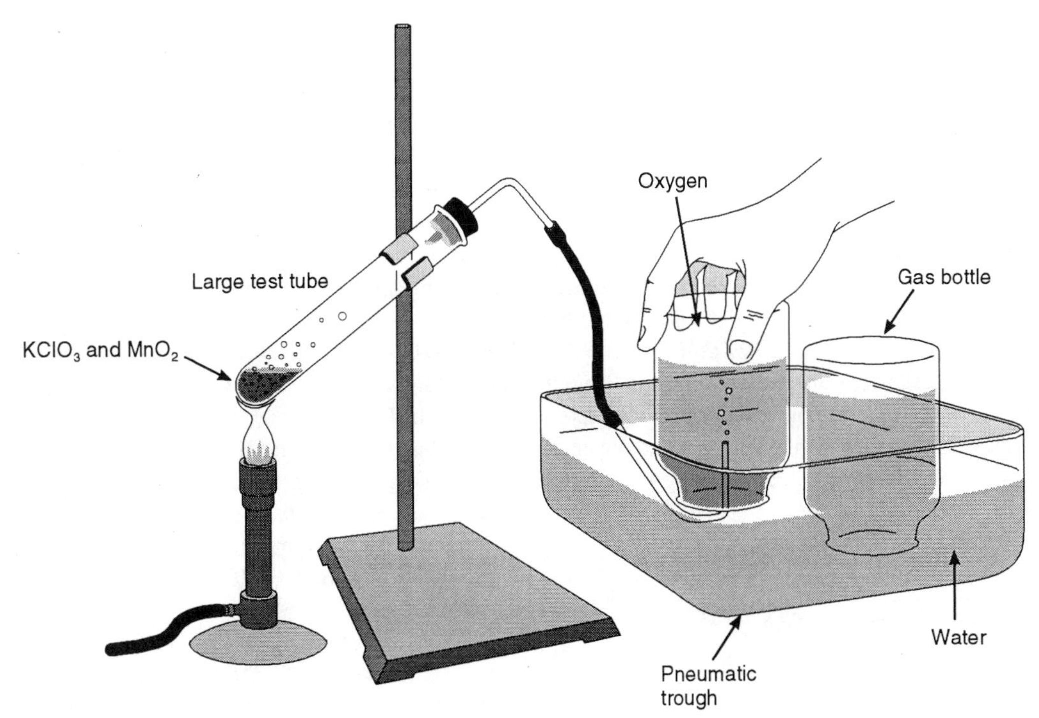

FIGURE 14.1
Apparatus used to collect oxygen gas from potassium chlorate decomposition.

by covering its mouth with a glass plate, inverting it, and lowering it into the water. Remove the glass plate below the water level in the trough. Start by putting two bottles in the trough. Leave one standing ready for transfer to the trough when needed.

Weigh approximately 12 g of $KClO_3$ (potassium chloride) and 2 g of MnO_2 (manganese dioxide) into a small, dry beaker. Mix thoroughly and transfer to the large, dry test tube. Spread the mixture out so that it covers about the lower one-third of the test tube and there is space above the $KClO_3$. Reconnect the stopper and tubing and start heating the mixture. One person should be tending the Bunsen burner at all times. Heat gently so that oxygen is evolved slowly but steadily. If the test tube is heated too strongly, the oxygen will be evolved very rapidly and you will not be able to remove bottles from the trough and replace them fast enough. The $KClO_3$ will be "used up" and you will have to begin again. If the test tube is heated too gently, you may be producing oxygen gas for a week before you collect three bottles of gas. Your lab instructor will help you in determining the correct rate of oxygen production. Regulate the rate of oxygen production by holding the burner closer or farther from the test tube.

When one bottle of gas is collected, immediately move the tube to another bottle. (Be careful not to pinch the rubber tubing with the edge of the bottle.) Then cover the mouth of the gas-filled bottle with a glass plate and remove it from the water. Collect three bottles of oxygen in this manner. Set the bottles on the lab bench with mouth upward and covered with a glass plate. (See Figure 14.2). The oxygen will not escape easily, since it is slightly more dense than air. (The density of air is about 1.3 g/L, that of oxygen about 1.4 g/L, and that of hydrogen about 0.1 g/L.) Save these bottles and use the oxygen gas in the next section.

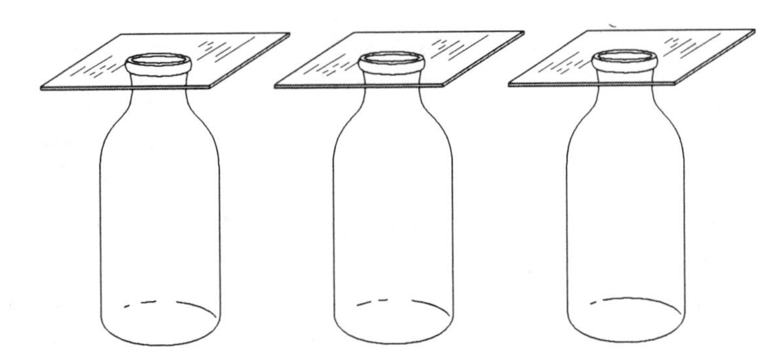

FIGURE 14.2
Storing collected oxygen for future use.

B. Combination (Synthesis) Reactions

1. **Magnesium oxide.** *CAUTION! DO NOT LOOK AT BURNING MAGNESIUM! EYE DAMAGE CAN RESULT!* Put a strip of magnesium about 2 to 3 cm long in a clean crucible. Place the crucible in a wire triangle suspended on a ring stand assembly. Heat with a Bunsen burner in the hottest part of the flame for about 15 minutes. Remove the heat and allow the crucible to cool. Gently scrape the white powder (magnesium oxide) into a small beaker. Leave any remaining magnesium in the crucible. Add 25 mL of distilled water to the beaker and stir to dissolve the magnesium oxide.

   Test for the acidic or basic nature of this solution by dropping small pieces of red litmus paper and blue litmus paper into the solution. Note any color changes in the litmus.
2. **Sulfur dioxide.** *CAUTION! PERFORM THIS STEP IN THE HOOD.* Clean a deflagrating spoon by burning out any residue it contains. Ignite the residue in the hood with a Bunsen burner and allow it to burn until the spoon is clean. Put a *small* amount of sulfur (about one fourth of the spoon) into the spoon and ignite the Bunsen burner *in the hood.* Thrust the burning sulfur into a glass bottle filled with oxygen gas and cover all but a small part of the bottle opening with a glass plate as the sulfur burns (Figure 14.3). After combustion slows, indicating that most of the oxygen has been used up, remove the deflagrating spoon and quickly cover the gas bottle completely with the glass plate. *DO NOT BREATHE THE GAS.* It is very irritating to the human respiratory system.

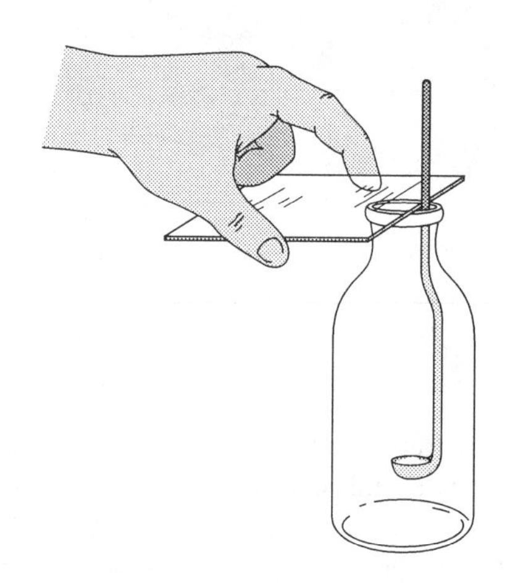

FIGURE 14.3
Burning sulfur or carbon in oxygen gas.

   After removing the spoon from the bottle, burn any excess sulfur out of the deflagrating spoon (*in the hood*), allow the spoon to cool, and retain it for the next part of the experiment.

   Add about 25 mL of distilled water to the bottle containing sulfur dioxide and *quickly* replace the cover. Shake gently to dissolve the gas in the water. Test for the acidic or basic nature of this solution by dropping small pieces of red and blue litmus paper in the solution. Replace the cover. Note any color changes that occur immediately. Some oxides dissolve very slowly, so keep shaking the solution if no immediate response is noted. Record any changes that occur over the course of about 15 minutes.
3. **Carbon dioxide.** Ignite a small piece of charcoal in a clean deflagrating spoon as shown in Figure 14.3. Thrust the deflagrating spoon with the glowing charcoal into a bottle of oxygen and *quickly* cover the mouth of the bottle as much as possible with a glass plate. Remove the spoon, add 25 mL of distilled water, cover the bottle, and shake to dissolve the gas in the water.

   Test for the acidic or basic nature of this solution by dropping small pieces of red and blue litmus paper into the solution. If no immediate color change of the litmus is noted, keep shaking the bottle. Record any color changes you observe over a period of about 15 minutes.

C. Single Replacement Reactions

1. **Reactions of metals with an acid.** The reaction between some metals and the hydrogen ions (H^+) of acids are examples of simple redox reactions. Physical evidence that suggests a chemical reaction is taking place in a reaction of this type is gas evolution (H_2) and disappearance of the solid metal. For example, chromium metal reacts with hydrochloric acid, and we can write the following reaction equation to represent the change. (You will not mix chromium with hydrochloric acid but with different metals.)

$$2\,Cr + 6\,HCl \longrightarrow 2\,CrCl_3 + 3\,H_2$$

   CAUTION! DO NOT HOLD TEST TUBES WITH REACTION MIXTURES IN YOUR HAND. In separate, clean, large labeled test tubes put *small* portions of granular iron, lead, zinc, aluminum, copper, and magnesium to a depth of approximately 1 mm. *Carefully* add about 5 mL of 6 M HCl solution to each test tube. Observe whether a reaction (evolution of H_2 gas) occurs in each case. If no reaction is apparent after several minutes, warm the reaction mixture in a warm water bath, but do not heat the mixture to boiling. Observe again after several minutes.

In all cases in which a reaction occurs, the metal atoms have donated electrons to the hydrogen ion and the H^+ is reduced to H_2 gas. Write a balanced formula equation for each reaction that occurred. Also arrange the metals tested and the element hydrogen in a preliminary "activity series" as follows. Group "above H_2" those metals that react with H^+ to evolve hydrogen gas, and group "below H_2" those metals that do not react with hydrogen ion to evolve hydrogen gas.

2. **Reactions of metals with metal ions.** Both hydrogen ions and metal ions can be reduced by metals that are higher on the activity series. When hydrogen ion is reduced by reaction with a metal, hydrogen gas is evolved at the surface of the metal. When a metal ion is reduced, a black deposit forms on the surface of the metal that is serving as the reducing agent. The black deposit consists of small crystals of a free metal, which forms as metal ions in solution gain electrons from the solid piece of metal.

   Prepare 12 clean test tubes. Into each of three test tubes put about 5 mL of $Zn(NO_3)_2$ solution. Into each of three test tubes put about 5 mL of $Pb(NO_3)_2$ solution. Put about 5 mL of $Cu(NO_3)_2$ solution into each of three additional test tubes. Into each of the remaining three test tubes put about 5 mL of $AgNO_3$ solution. Label each test tube so that you know which solution it contains. Each test tube now contains a solution of positive metal ions and negative nitrate ions.

   Obtain four strips each of Cu, Pb, and Zn metals. Clean each strip with steel wool. (Do this on a piece of paper so that the bench top is not scratched.) Put a different metal into each tube of $Zn(NO_3)_2$ solution. Do the same with each of the other three solutions. Label each test tube so that you know which metal it contains. Observe what is happening (or not happening) in each test tube. Record your observations on the report sheet. Note that the order of reacting mixtures on the report sheet may not correspond to the order of mixing described in the procedure.

   In all cases in which a reaction occurs, the metal atoms have donated electrons to the metal ion and the metal ion is reduced to a free metal. Your instructor will tell you if you should write formula reaction equations or net ionic reaction equations to represent the reactions that occurred.

   Using the information from part C1 and this section (part C2), you should be able to arrange four metals and the element hydrogen in a short activity series.

3. **Identification of an unknown metal ion.** Obtain a solution of an unknown salt, which may be $Pb(NO_3)_2$, $Zn(NO_3)_2$, $Cu(NO_3)_2$, or $AgNO_3$. Identify the cation by using the reactions observed in the previous section of the experiment.

Experiment 14

Oxidation-Reduction Reactions and the Activity Series

Name____________________

Date____________________

REVIEW QUESTIONS

1. If you had collected H_2 instead of O_2 in the bottles, how would you have stored the bottles for future use of H_2? Why?

2. Complete and balance the following equations for redox reactions (if no reaction occurs write "no reaction"):

a. $Mg + HBr \longrightarrow H_2 \quad + $

b. $Mg + Cu(NO_3)_2 \longrightarrow \quad + $

c. $Mg + Mg(NO_3)_2 \longrightarrow \quad + $

3. Does the following reaction equation represent a redox or a nonredox reaction?

$$3\ NaOH + AlCl_3 \longrightarrow Al(OH)_3 + 3\ NaCl$$

If this represents a redox reaction, which element is oxidized and which is reduced in the reaction?

Oxidized:____________________ Reduced: ____________________

4. Determine the element oxidized and the element reduced in the following reaction equation:

$$Ca + 2\ H_2O \longrightarrow Ca(OH)_2 + H_2$$

Oxidized:____________________ Reduced: ____________________

5. What would you expect to observe if a piece of solid silver metal were placed in a solution of 6 M hydrochloric acid?

6. What would you expect to observe if a piece of solid iron metal were placed in a solution of silver nitrate?

7. An unknown metal is believed to be silver, lead, zinc, or copper. A piece of the metal is placed in a 6 M HCl solution and no bubbles of hydrogen are evolved at the surface of the metal. Another sample of the metal is then placed in 0.05 M silver nitrate solution and a dark deposit slowly forms on the surface of the metal. Which of the four metals is it?

Experiment 14

Oxidation-Reduction Reactions and the Activity Series

Name____

Date____

REPORT SHEET

A. Decomposition Reaction

Write the balanced equation for the decomposition of potassium chlorate.

Give the oxidation number of each element in the reaction as a product and as a reactant.

Element	As a Reactant	As a Product
Potassium		
Chlorine		
Oxygen		

Which element is oxidized in the reaction?________________________

Which element is reduced in the reaction? ________________________

B. Combination (Synthesis) Reactions

1. Magnesium oxide

Write the balanced equation for the synthesis of magnesium oxide from magnesium and oxygen.

Give the oxidation number of each element in the reaction as a product and as a reactant.

Element	As a Reactant	As a Product
Magnesium		
Oxygen		

Which element is oxidized in the reaction?________________________

Which element is reduced in the reaction? ________________________

Litmus test with solution:

Blue litmus color ________________________

Red litmus color ________________________

2. Sulfur dioxide

Write the balanced equation for the synthesis of sulfur dioxide from sulfur and oxygen.

Give the oxidation number of each element in the reaction as a product and as a reactant.

Element	As a Reactant	As a Product
Sulfur		
Oxygen		

Element oxidized ____________________ Element reduced ____________________

Litmus test with solution:

Blue litmus color ____________________ Red litmus color ____________________

3. Carbon dioxide

Write the balanced equation for the synthesis of carbon dioxide from carbon and oxygen.

Give the oxidation number of each element in the reaction as a product and as a reactant.

Element	As a Reactant	As a Product
Carbon		
Oxygen		

Element oxidized ____________________ Element reduced ____________________

Litmus test with solution:

Blue litmus color ____________________ Red litmus color ____________________

What similarity in acidity exists between water solutions of nonmetal oxides?

Do metal oxides exhibit the same kind of acidic tendency when dissolved in water?

C. Single Replacement Reactions

1. Reactions of metals with an acid

Reactants	Observation	Balanced Equation
Fe + HCl		
Pb + HCl		
Cu + HCl		
Zn + HCl		
Al + HCl		
Mg + HCl		

Activity series

Metals above H_2: ____________________

H_2

Metals below H_2: ____________________

On the basis of your observations, why do you think copper is often used to transport water in plumbing systems? (Water is often acidic owing to dissolved CO_2.)

2. Reactions of metals with metal ions

Reactants	Observation	Balanced Equation
$Zn + Zn(NO_3)_2$		
$Zn + Pb(NO_3)_2$		
$Zn + Cu(NO_3)_2$		
$Zn + AgNO_3$		
$Pb + Zn(NO_3)_2$		
$Pb + Pb(NO_3)_2$		
$Pb + Cu(NO_3)_2$		
$Pb + AgNO_3$		
$Cu + Zn(NO_3)_2$		
$Cu + Pb(NO_3)_2$		
$Cu + Cu(NO_3)_2$		
$Cu + AgNO_3$		

On the basis of the pattern of replacement in the reactions observed, predict what you would observe if a piece of silver metal (Ag) were placed in a solution of $Cu(NO_3)_2$.

Activity series: Arrange copper, lead, silver, zinc, and hydrogen in order of decreasing ability to transfer electrons to a metal ion or to a hydrogen ion. Use observations from parts B1 and B2.

__________ > __________ > __________ > __________ > __________

(most active) (least active)

3. Identification of an unknown metal ion

Unknown number ________________ Identity of cation ________________

Equation of reaction used for identification:

EXPERIMENT 15

Physical Properties of Liquids and Solids

Purpose: To investigate some common properties of solids, liquids, and amorphous materials and to differentiate between these three forms of matter.

Materials: Thermometer, hexane, dry ice, solid magnesium sulfate, table salt, microscope slides, construction paper, cotton balls, cotton swabs, iodine, evaporating dish, sulfur, solder, soda-lime (soft glass) rod, thermometer with cork or stopper, boiling chips, ice, syrup, glycerine, NaCl-water saturated solution, methanol, bobby pins, magnifying lenses, and binocular microscope.

Safety Precautions: Remember that glass remains hot for a long time. Be sure to wait until glass that has been melted is thoroughly cool before touching it.

INTRODUCTION

Liquids and solids are the states of matter that are called condensed states. The latter term seems to have originated from the observation that liquids and solids can be obtained by condensing gases. The kinetic-molecular theory allows us to explain condensed states as those in which the basic units (atoms, molecules, ions) of the substance are about as close together as they can get and in which the basic units are not capable of free and independent movement. This explanation gives rise to the concept that there are interactions between the basic units. These interactions are called intermolecular forces. Both repulsive and attractive forces operate among basic units of a liquid or solid sample. A mathematical description of the liquid or solid state analogous to the gas state is very complicated, because terms must be included in a math relationship for all interactive forces. At this time, we will work with more qualitative descriptions of liquids and solids.

As you grew up and learned about the world, you became able to distinguish between those substances called liquids and those called solids on the basis of appearance and feel. But since the terms "liquid" and "solid" are both labels for a particular physical state (the condensed state) we need definitions to distinguish them from each other. Liquids are defined as having definite volume and indefinite shape. As a consequence of their indefinite shape, liquids are able to flow. Solids are defined as having definite volume and definite shape. As a consequence of their definite shape, solids are not able to flow. Each aspect of the definitions of liquids and solids can be explained in terms of the kinetic-molecular theory, but that will be done in lecture rather than in this lab activity.

During this lab activity you will examine some physical properties and changes associated with liquids and solids. You will explore the concept of vapor pressure of liquids and solids, the crystalline structure of solids, the melting points and boiling points of pure substances and mixtures, the viscosity of liquids, and the volume of solids. You should review the discussion of these concepts in your textbook, since the presentation there is very complete.

From your everyday experiences you should be familiar with several phase changes of matter: melting and freezing, evaporation, and condensation. You could probably give an example of each change if asked to do so. You also know that there are temperatures associated with each of the changes. The temperature associated with the solid-liquid phase change is called the melting point. And when you hear the term "boiling point" you know immediately that this refers to a temperature at which evaporation of a liquid or condensation of a gas takes place. The temperature at which a liquid boils is very dependent on the gas pressure above the liquid. Because liquids may boil at different temperatures depending on the gas pressure, it is necessary to have a standard reference temperature of boiling, which is called the normal boiling point or normal boiling temperature. This normal boiling point refers to the boiling temperature at standard pressure (1 atm, 760 torr, or 760 mm Hg). The normal boiling point of water is 100° C, but water can boil at temperatures above or below 100° C depending on the gas pressure above the water. The temperature of melting is not so dependent on pressure conditions, but we still use the concept of normal boiling point to specify standard pressure conditions.

Another phase change that should be familiar to you is **sublimation.** Some solids change directly to the gas phase under normal atmospheric conditions. The reverse process of sublimation is called deposition.

The term **volatile** should be familiar to you. Volatile liquids or solids are those whose basic units can readily move into the gas phase under normal conditions. Volatility is useful for relative comparisons. It is also possible to measure and express the pressure created by the gas in quantitative terms. This pressure is called vapor pressure. Liquids or solids that are volatile have relatively high vapor pressures, while nonvolatile liquids or solids have low vapor pressures. Relative degrees of volatility and vapor pressure are explained in terms of the attractive intermolecular forces among the molecules of the liquid or the solid. High vapor pressure of a liquid or solid is explained in terms of relatively weak intermolecular forces among the basic units of the liquid or solid (molecules or atoms can escape the liquid or solid rather easily and enter the gas phase). Low vapor pressures are explained in terms of relatively strong intermolecular forces among the basic units of the liquid or solid. You must understand that we use the terms "volatile" and "nonvolatile," "high" and "low," "strong" and "weak" in a relative manner. We do not have a "cutoff" value that separates volatile from nonvolatile, high vapor pressure from low vapor pressure, or strong intermolecular forces from weak intermolecular forces. These adjectives must be understood as being opposite ends of continuums for the ranges of volatility, vapor pressure, and intermolecular forces of substances.

Liquids can be compared in terms of a physical property known as **viscosity.** Viscosity is defined as internal resistance flow. Again, viscosity of liquids should be thought of as lying on a continuum and not just being one extreme or the other (viscous or nonviscous). In advertisements for syrup or catsup, the word "thick" is used instead of viscous. (Apparently, the general public does not like to hear the word "viscous" describing things that they eat.) This substitution of terms is not satisfactory in chemistry, because "thick" is an adjective that refers to distance, such as a thick wall.

True **solids,** such as ice, salt, or metals, have a definite temperature at which they melt (a definite melting point). Such solids have a crystal structure (a regular arrangement of atoms or molecules) and have a crystalline appearance (although a strong microscope is sometimes necessary to see the individual crystals). **Amorphous** materials are substances such as plastic, glass, rubber, and tar. They do not have a definite melting point. They simply become softer and more pliable as the temperature increases, but never reach a temperature at which they suddenly change from a solid to a liquid. They do not have a crystal structure and do not appear crystalline.

It has long been known that the physical properties of metals can be changed by various treatment processes. For example, the making of samurai swords has been an art for centuries. Such swords have a long, flexible shaft and a hard edge that can be made very sharp. Normally it is not possible to get a sharp cutting edge on a flexible piece of metal, and if a piece of metal is hard enough to get a sharp cutting edge, its flexibility is low. However, through a long process of bending, pounding, heating, and cooling it is possible to make from one metal sample a sword with both properties. We now attribute differences in properties of metals to differences in crystal size and structure that result from different treatment processes of the metal. When two samples of the same substance—one flexible and one hard—are compared, differences are found in their internal structures. In the samurai sword, the crystal structures are different in the shaft and at the edge. During this laboratory activity you will perform several treatment processes on metal samples and observe any changes in physical properties that may result.

This activity involves making many comparisons; it may be classified as descriptive or qualitative in nature. This does not mean, however, that you can be sloppy with lab techniques and measurement. The procedures require careful attention to such details as controlling variables if your comparisons are to be reliable.

PROCEDURE

A. Vapor Pressure (Demonstration)

1. **A liquid (hexane).** We can memorize and recite the definition for vapor pressure, and all of us know some common things that are volatile (for example, ammonia). But how do we know that there really are atoms or molecules in the gas phase having the same identity as those in the liquid or solid phase? This might be difficult to determine if the gas phase

is colorless and odorless. Your instructor will demonstrate that hexane vaporizes at room temperature by making use of a known chemical property for hexane to detect any gas that may be present. Observe the demonstration and record your observations.

2. **A solid (carbon dioxide).** The instructor will demonstrate that some solids, such as carbon dioxide (dry ice), produce vapor and considerable vapor pressure without ever melting. The instructor will make use of carbon dioxide's lack of flammability to prove that CO_2 vapor is produced by the solid material. Observe the demonstration and record your observations.

B. Structure

1. **Magnesium sulfate (epsom salts) crystals.** Prepare a piece of glass, such as a microscope slide, by spreading a drop of liquid hand soap solution all over the surface. Allow to dry for 5 minutes. Add about 15 grams of Epsom salts (magnesium sulfate) to about 25 mL of hot water. Stir. If some solid remains on the bottom, the solution is saturated (which is ok). Dip a cotton ball into the solution and spread the liquid onto the microscope slide. Allow to dry. Draw a picture of the $MgSO_4$ crystals formed on the glass.
2. **Sodium chloride (table salt) crystals.** Add NaCl to 15 to 20 mL of warm water. Stir until some solid remains undissolved. This should take nearly 10 grams of salt in 20 mL of water. With a cotton swab, use the salt solution to "paint" your name on a piece of colored construction paper. Allow to dry and then examine the residue with a magnifying lens or a microscope. Describe and/or draw a picture of the salt crystals which form.
3. **Iodine.** PERFORM THE FOLLOWING PROCEDURE IN THE HOOD. Set an evaporating dish in an iron ring attached to a ring stand. Choose a watch glass that fits on the evaporating dish like a lid. Put a very small amount of iodine into the evaporating dish. Cover the evaporating dish with the watch glass and put a small amount of ice in the watch glass. Gently heat the evaporating dish with the Bunsen burner. Control the amount of heat delivered to the dish by moving the Bunsen burner rather than by turning the gas on and off. Watch the contents of the evaporating dish as well as the substance that is deposited on the underside of the watch glass. After all the solid has collected on the watch glass, carefully take the watch glass off the evaporating dish, discard the water, turn the watch glass upside down, and examine the substance collected with a magnifying glass or a binocular microscope. If a magnifying glass is not available, move the watch glass and see how the light is reflected off the surface of the substance. Classify the substance as **crystalline** or **amorphous** (glassy) in nature.
4. **Sulfur.** Put solid sulfur into a disposable test tube to a depth of about 5 cm. Set a beaker of cold water near your working place in the hood. Hold the test tube with a test tube holder, and IN THE HOOD slowly heat the sulfur over a Bunsen burner flame. Heat the sulfur until it melts and is deep red in color. Then quickly pour the liquid into the cold water. After the substance has solidified in the cold water, take it out, examine it, and try to stretch it. Classify this sulfur polymer as a crystalline or amorphous (glassy) solid.

C. Melting and Boiling

1. **Melting characteristics of crystalline and amorphous materials.** Obtain a piece of solder several inches long. Although its composition can vary, solder is a mixture (alloy) of tin and lead (usually 50% Sn and 50% Pb). Hold the solder with a pair of forceps over an evaporating dish. With your other hand direct the hottest part of a Bunsen burner flame at the solder until it melts and a few drops of molten metal fall into the evaporating dish. Record the melting characteristics of this crystalline solid.

   Obtain a similar length of solid glass rod. Glass is a mixture of Na_2O, CaO, and SiO_2, but it is amorphous rather than crystalline. Upon looking at glass versus solder, it is not necessarily obvious that one is a crystalline solid and the other is amorphous. For one thing, the metal crystals are too small to be seen without a powerful microscope. Hold the glass rod with forceps and hold in the hottest part of a burner flame. Soda-lime glass (also called "soft glass") begins to soften at about 600 °C. Heat the glass as hot as possible and push the softened glass against the bottom of the evaporating dish to get a feeling for the viscosity of the glass compared to solder in the liquid state. Compare the melting of glass to that of solder based upon your observations.
2. **Boiling point of water.** Obtain a thermometer with a rubber stopper attached near the top of the thermometer. The rubber stopper is used to mount the thermometer in a clamp on a ring stand. Set up an apparatus to measure boiling points like that illustrated in Figure 15.1, with a 500-mL Florence flask. Put about 200 mL of distilled or deionized water into the flask and add several boiling chips. Suspend the thermometer in the water so the mercury bulb does not touch the glass of the flask. The mouth of the flask must be left open. When you have this setup together, heat the water gently

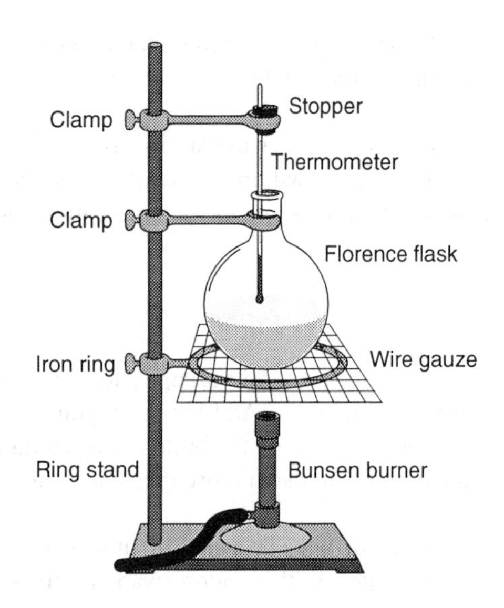

FIGURE 15.1
Boiling point apparatus.

until it boils steadily (but not rapidly) for a minute or two. If you heat the water too rapidly, you could run into a problem called superheating. If the temperature goes up much above 100°C, let the system cool and start over with the heating. Watch the thermometer, and when it has reached a constant temperature, record that temperature.

One definition of the boiling point of a liquid refers to the gas-liquid equilibrium. You just measured the temperature of a boiling liquid, but to validate the definition we should be able to compare the temperatures of the liquid and vapor when they are in equilibrium with each other. To do this, raise the thermometer so that the mercury bulb is two to three centimeters above the liquid in the flask. Heat gently to avoid superheating. When the liquid has boiled gently for a minute or two and drops of condensed vapor fall from the thermometer bulb back into the liquid, equilibrium should be reached. Record the constant temperature reached by both the vapor and the liquid.

Read and record the barometric pressure in millimeters of mercury. The normal boiling point of water is 100.0°C, but the corrected boiling temperature for the existing barometric pressure may be estimated using a boiling point depression of 1.0°C for each 30 mm of pressure below 760 mm. Use the corrected boiling point of water to determine your thermometer error at the boiling point of water.

Carefully remove the thermometer from the flask and add about 45 g of common table salt (NaCl) to the water. Wash any salt caught in the neck of the flask into the flask with a small amount of distilled water from your wash bottle. Swirl to dissolve the salt. Put the thermometer back into the flask and heat the solution gently to boiling, allowing it to boil gently for several minutes. Record the constant temperature reached by both the vapor and the liquid as was done earlier.

3. **Melting point of water.** Determine the melting point of water in the following way. Put about 200 mL of crushed ice into a 250-mL beaker. Add some tap water. Stir the ice-water mixture for a few minutes. Read and record the temperature of the ice-water mixture. This should be the melting point (freezing point) of water. If the thermometer varies from 0.0°C, calculate a correction factor for the thermometer at the freezing point of water. Add about 25 g of NaCl to the ice-water mixture. Stir for at least 1 minute to dissolve the salt. Read and record the temperature of this mixture.

D. Liquids

1. **Viscosity.** Obtain a funnel (glass is better than plastic) with a small-diameter delivery tube (the smaller the better). Put a finger at the bottom of the funnel tube and fill the funnel with water. Transfer this water to a 100-mL graduated cylinder to determine the volume of the funnel. Record this volume. This will be the standard sample size for each substance to be examined in this procedure. You will use the same funnel for all the substances, so it must be washed and dried after each sample is used. All substances tested are water-soluble. Besides tap water, several substances on the reagent shelf will be used for viscosity measurements. They are glycerin, saturated NaCl in water, and methanol. Record and identify these substances on the report sheet.

   Note: Do the viscosity and density measurements (part D2) simultaneously to save time and chemicals.

   Perform the following procedure with the four different substances. With a weighed graduated cylinder, obtain a 15.0-mL sample of the first liquid. Weigh the graduated cylinder again, this time with the liquid in it. Record the volume and the graduated cylinder weights in the appropriate spaces in parts D1 and D2 of the report sheet. Put your finger at the bottom of the funnel tube and pour the sample into the funnel. Watch the clock and at some convenient time take your finger away so the liquid can drain through the funnel. Determine and record the length of time it takes for the liquid to drain completely through the funnel. After the procedure has been completed for the four substances, list the substances in order of increasing viscosity. Use the weight and volume figures for density measurement.
2. **Density.** Measure the density of each of the liquids by first weighing an empty, small graduated cylinder. Then add about 15.0 mL of the liquid and weigh it again. This should already have been done in conjunction with the viscosity measurements. Calculate the densities of all four liquids. Is there any obvious relationship between the viscosity of the liquids and their densities?

E. Solids: Metal Temper

1. **Flexibility versus hardness of a metal.** Obtain three bobby pins made of uncoated metal. This metal is a high-carbon steel. Bend them out to a "V" shape. Hold one of the bobby pins with your crucible tongs and heat it in the hottest part of the Bunsen burner flame for several minutes to red-hot temperature. This process is called tempering. Plunge the red-hot bobby pin into cold water immediately after removing it from the Bunsen burner flame. This process "freezes" the metal in a particular crystal structure that exists when the metal is glowing red in the flame. The rapid cooling in water is called quenching. Repeat the heating process again, but allow the hot metal to cool slowly to room temperature. This allows the metal to alter its crystal structure as it cools slowly. The bobby pins as purchased had been tempered and quenched but the tempering temperature was not the same as that of the Bunsen burner flame. After the two heated pins have cooled, test both of them and an untreated one for flexibility by trying to bend them back to their original shape. Record your observations.

Experiment 15

Physical Properties of Liquids and Solids

Name________________________________

Date________________________________

REVIEW QUESTIONS

1. Why are there signs at gasoline pumps that say something like "Turn off engine before pumping gas"?

2. In this activity you worked with two solids that changed directly to the gas phase without going through the liquid state. Identify the two solids using formulas, and name the process involved.

3. In measuring viscosity, why were you directed to use the same funnel throughout the procedure?

4. Plastics, such as polyethylene, gradually become more brittle as the temperature is lowered and gradually become more pliable as the temperature is increased. This is why plastic containers are more likely to crack in the cold of the winter than in the warm summer. Would you classify plastics as solids or as amorphous materials? Explain.

Experiment 15
Physical Properties of Liquids and Solids

Name____
Date____

REPORT SHEET

A. Vapor Pressure (Demonstration)

1. A liquid (hexane)

Observed physical properties of hexane

Property used to detect hexane gas

2. A solid (carbon dioxide)

Observed physical properties of carbon dioxide

Property used to detect CO_2 gas

B. Structure

1. Magnesium sulfate (epsom salts) crystals

Describe what happens to the glass surface as the water evaporates from the magnesium sulfate solution.

Drawing of the magnesium sulfate crystal structure on the glass

2. Sodium chloride (table salt) crystals

After examining the sodium chloride crystals that form on the construction paper, describe them and draw a picture of the crystals.

3. Iodine

Observations when iodine is heated and allowed to cool

Is iodine crystalline or amorphous? Give evidence for your conclusion.

4. Sulfur

Observations when sulfur is heated and allowed to cool

Is sulfur crystalline or amorphous? Give evidence for this conclusion.

C. Melting and Boiling

1. Melting characteristics of crystalline and amorphous materials

Describe the melting of solder which is a crystalline solid.

Describe the melting of glass, which is an amorphous material.

2. Boiling point of water

Thermometer reading with bulb in water ____________________

Thermometer reading with bulb in vapor ____________________

Barometric pressure ____________________

Corrected boiling point of water ____________________

Thermometer error ____________________

Temperature of salt water (liquid) ____________________

Temperature of salt water (vapor) ____________________

Compare the liquid temperatures of water and salt water. What effect does a solute have on the boiling point of water?

Compare the vapor temperatures of water and salt water. What can you conclude about the vapor composition of both liquids?

3. Melting point of water

Temperature of ice-water mixture ____________________

Temperature of ice-salt-water mixture ____________________

What effect does a solute have on the melting point of water?

D. Liquids

1. Viscosity

Sample size ________________________ mL

Substance	Time (sec)

Order of increasing viscosity:

____________ < ____________ < ____________ < ____________

2. Density

Water:

weight of graduated cylinder plus contents ________________________

weight of graduated cylinder ________________________

volume of liquid ________________________

Calculation of density of water:

Glycerine:

weight of graduated cylinder plus contents ________________________

weight of graduated cylinder ________________________

volume of liquid ________________________

Calculation of density of glycerine:

Saturated NaCl solution:

weight of graduated cylinder plus contents ________________________

weight of graduated cylinder ________________________

volume of liquid ________________________

Calculation of density of acetone:

Methanol:

weight of graduated cylinder plus contents ________________________

weight of graduated cylinder ________________________

volume of liquid ________________________

Calculation of density of methanol:

Order of increasing density:

__________ <__________ <__________ <__________

Is there any obvious relationship between the viscosity of the liquids and their densities?

E. Solids: Metal Temper

1. Flexibility versus hardness of a metal

Observations for metal as produced by manufacturer

Observations for heated and slow-cooled metal

Observations for heated and rapid-cooled (quenched) metal

Can you infer that each treatment process altered the internal structure of the metal? Support your conclusion with observed physical evidence.

EXPERIMENT 16

Solutions

Purpose: To observe some of the properties of solutions and to note a few of their practical applications.

Materials: $Na_2S_2O_3 \cdot 5H_2O$, NaCl, sugar ($C_{12}H_{22}O_{11}$), I_2, $KClO_3$, NH_4Cl, NaOH, lard, ethanol, acetone, lubricating oil, methanol, hexane or heptane, styrofoam picnic cups.

Safety Precautions: Care should be taken when working with open flames and solvents such as heptane, hexane, acetone, or ethanol. These liquids are very flammable and should be kept away from open flames.

INTRODUCTION

Solutions resemble a mixture in which the components are of molecular size. They differ from ordinary mixtures in three main ways: (1.) There is, in most cases, a limit to the proportions that will mix; (2.) Some form of chemical change usually takes place; and (3.) They are homogeneous.

The term **solvent** can be applied to that substance present in the largest relative amount, and the term **solute** to the one or more others present. When a solution is prepared from a solid and a liquid, the solid is usually the solute and the liquid is usually the solvent, but this is not always true. A solution that contains less solute than equilibrium requires under existing conditions is **unsaturated**, because more solute will dissolve if added. A **saturated** solution contains just enough solute to be in equilibrium. A **supersaturated** solution contains more solute than can be held in stable equilibrium under existing conditions, and it is unstable or metastable.

Solubility has to do with whether or not a substance will dissolve in a given solvent. Many solutes are so slightly soluble that we cannot perceive that they have dissolved at all. In such cases we say that the substance is **insoluble** in the solvent. It is impossible to predict solubility precisely. However, the old saying that "like dissolves like" often proves a useful guide. In general, polar solvents such as water are more likely to dissolve ionic or highly polar materials, and nonpolar solvents such as benzene are better solvents for nonpolar substances.

Compounds having polar covalent bonds show opposite electrical charges in different parts of the molecule because of an unequal sharing of electron pairs in the covalent linkages. All degrees of polarity exist, from one extreme in which electrons are not shared at all but are transferred so as to produce separate positive and negative ions, to the other extreme in which electron sharing is nearly equal.

The temperature usually influences the amount of solute that will dissolve in a given amount of solvent. In most compounds temperature increase produces an increase in solubility, but with some substances it has little effect. In a few cases the solubility is even decreased by an increase in temperature.

The evolution or absorption of heat when a solution is formed indicates that some form of association between solvent and solute is occurring. In this experiment, a measurement is made of the heat evolved or absorbed by one mole of a solute when dissolved in 1,000 g H_2O.

PROCEDURE

A. Supersaturated Solution

To 5 mL of distilled water add 20 g of sodium thiosulfate, $Na_2S_2O_3 \cdot 5H_2O$. Heat without boiling until the crystals are completely dissolved. If even a trace of solid remains, the experiment will fail. Set the hot solution away until it has completely cooled, then shake it. If nothing happens, "seed" the solution by dropping in one tiny crystal of $Na_2S_2O_3 \cdot 5H_2O$. Describe the result and note if there is any temperature change.

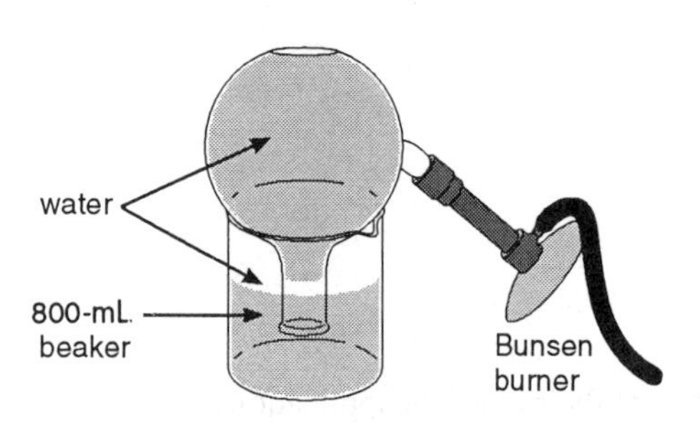

FIGURE 16.1
Heating water in inverted Florence flask.

B. Solubility of Gases in Water

Completely fill a 500-mL Florence flask with water, insert a one-hole stopper, hold a finger over the hole, and invert the flask in a 600-mL beaker containing some water. **No trace of air bubbles should be present when you remove the stopper from the flask.**

Heat the water *in the flask* (Figure 16.1) almost to boiling—but do not let it actually boil. The bubble that develops represents the air gases that had been dissolved in the water. Answer the questions on the report sheet.

C. Solubility in Different Solvents

1. **Solid solutes in liquid solvents.** The solubilities of four different substances of different degrees of polarity are examined in four solvents of different polarities. The solutes are:

   Sodium chloride, an ionic compound
   Sugar ($C_{12}H_{22}O_{11}$), a polar covalent compound
   Lard, an animal fat with both polar and nonpolar bonds
   Iodine, a nonpolar substance

   The solvents are heptane, acetone, ethanol, and water. **Note**: Acetone, ethanol, and particularly heptane are extremely flammable. Keep open flames away from them!

   In this part of the experiment, work with a partner in order to have enough test tubes available. Put a sample of each solid substance (match-head size) in four separate clean, *dry* test tubes. Add about 5 mL of heptane (or hexane) to each tube, stopper, and shake gently. Report whether the substance dissolves completely, partly, or not at all. Try the solubility of each of the four solids with each of the other three solvents (acetone, ethanol, and water) using only clean, dry test tubes. Report your observations.

   On the basis of your results, estimate the relative polarities of the four solvents.
2. **Liquid solutes in liquid solvents.** If a liquid completely dissolves in another liquid, the two liquids are said to be "miscible." Generally, liquids which have similar polarities are **miscible** and liquids which differ greatly in polarity are ***immiscible.***

   Put 5 mL of each of the solvents used in the previous section (C1) of the experiment into separate test tubes. Add about 20 drops of methanol to each test tube, stopper, and shake gently. Record whether the two liquids are miscible or immiscible. Is methanol polar or nonpolar?

   Put 5 mL of each of the solvents used in the previous section (C1) of the experiment into separate test tubes. Add about 20 drops of lubricating oil to each test tube, stopper, and shake gently. Record whether the two liquids are miscible or immiscible. Is oil polar or nonpolar?

D. Effect of Temperature on Solubility

1. **Sodium chloride.** Accurately measure 5 mL of distilled water into a small test tube. Weigh out on a piece of paper about 3 g sodium chloride to the nearest 0.01 gram. Add a spatula of sodium chloride to the water and shake. Does it all dissolve?

Continue adding sodium chloride a little at a time shaking until no more will dissolve. Take the temperature of the solution. Weigh the remaining sodium chloride, determine the amount dissolved, and calculate how many grams would dissolve in 100 mL of water at that temperature.

Raise the temperature of the solution to nearly boiling and determine if any undissolved salt remains. If it has all dissolved add a few more crystals, shake gently, and continue until no more dissolves. Measure the water temperature. Again weigh the remaining NaCl and calculate how many grams will dissolve in 100 mL of water at the higher temperature. Record the results on the report sheet.

2. **Potassium chlorate.** Repeat Part 1, starting with about 4.00 grams of potassium chlorate, $KClO_3$, and answering the same questions as those asked for sodium chloride.

E. Heat Effects in Dissolving

1. **Ammonium chloride solution.** Weigh out 2.68 g of NH_4Cl on the centigram balance. Measure into a styrofoam picnic cup 50.0 mL (50.0 g) of distilled water and take its temperature to the nearest 0.1°C. Add the NH_4Cl, stir with the thermometer until it has all dissolved, and again read the temperature. The amount of solute used in this procedure was chosen to produce a concentration of 1.0 mole of solute per 1,000 g of water. Calculate the number of moles of solute from the number of grams of solute used. Then calculate the number of calories that would be absorbed in dissolving one mole of NH_4Cl in 1,000 g of water.
2. **Sodium hydroxide solution.** Repeat the previous experiment and the calculations, using 2.00 g of sodium hydroxide, NaOH, and 50.0 mL of distilled water. Answer the questions on the report sheet.

Experiment 16

Solutions

Name________________________

Date________________________

REVIEW QUESTIONS

1. It is dangerous to mix H_2O and concentrated (98%) H_2SO_4 in any ordinary glass bottle. Why? How should the danger be eliminated?

2. A solution contains 225 g of Na_3PO_4 in 1,600 mL of solution. Calculate its molarity.

3. You need 675 mL of a 3 M K_2CO_3 solution. How would you make it?

4. A British Thermal Unit (Btu) is the heat required to raise one pound of H_2O by 1° F. Calculate the conversion factor between Btu and calorie.

5. How many calories are absorbed by 250 mL of water when its temperature rises from 25°C to 45°C?

6. What gas in water is constantly required by fish and other aquatic animals to remain alive and healthy? Is this gas likely to be in short supply in warm lake water in the summer or in cold lake water in the late fall?

Experiment 16
Solutions

Name________________________________
Date________________________________

REPORT SHEET

A. Supersaturated Solution

Is the solution free of crystals before shaking?________________________________

What changes occur on shaking?________________________________

If the solution requires "seeding" what then happens?________________________________

What temperature change, if any, results?________________________________

What kind of solution remains at the end?________________________________

Describe just how a supersaturated solution is formed.________________________________

____

____

____

____

B. Solubility of Gases in Water

What is seen as the temperature gradually rises?________________________________

What gases are present in the bubbles?________________________________

Are they present in the same proportions as in the air?________________________________

Explain:________________________________

____

Why do warm carbonated drinks froth more than cold ones when opened?________________________________

____

Why does even warm pop not froth so long as the bottle remains capped?________________________________

____

C. Solubility in Different Solvents

1. Solid solutes in liquid solvents

S = soluble PS = partly soluble I = insoluble

Increasing Polarity →

Substance	Iodine	Lard	Sugar	Salt(NaCl)
Heptane				
Acetone				
Ethanol				
Water				

The polarities of the solvents are as follows:

1. Most polar ____________ 2. ____________ 3. ____________ 4. Least ____________

2. Liquid solutes in liquid solvents

M = miscible IM = immiscible

Substance	Methanol	Motor Oil
Heptane		
Acetone		
Ethanol		
Water		

Is methanol polar or nonpolar?____

Is oil polar or nonpolar?____

D. Effect of Temperature on Solubility

Room temperature____________________ Highest temperature____________________

		NaCl	$KClO_3$
Original weight solute	=	____________	____________
Weight undissolved (room temperature)	=	____________	____________
Weight dissolved	=	____________	____________

Calculation of weight in 100 mL: ______________________________

____

$NaCl$ | $KClO_3$

Value from literature ______________ Value from literature ______________

		$NaCl$	$KClO_3$
Original weight solute	=	______	______
Weight undissolved (highest temperature)	=	______	______
Weight dissolved	=	______	______

Calculation of weight in 100 mL: ______________________________

____

$NaCl$ | $KClO_3$

Value from literature ______________ Value from literature ______________

E. Heat Effects in Dissolving

1. Ammonium chloride solution

Temperature of water ______________ Temperature of NH_4Cl solution ______________

Heat absorbed per mole NH_4Cl in 1,000 g H_2O ______________

Calculation: ______________________________

____

2. Sodium hydroxide solution

Temperature of water ______________ Temperature of NaOH solution ______________

Heat absorbed per mole NaOH in 1,000 g H_2O ______________

Calculation: ______________________________

____

EXPERIMENT 17

pH of Aqueous Solutions

Purpose: To investigate several pH measurement techniques; to introduce the concepts of strong versus weak acids and bases, hydrolysis of salts, and buffer solutions.

Materials: Cola soft drink, bleach solution, vinegar, laundry detergent solution, orange juice, milk, household ammonia, buttermilk, tap water, baking soda; 0.10 M HCl, 0.10 M NaOH, 0.10 M citric acid, methyl red/xylene cyanole mixed indicator (1.0 g methyl red, 1.4 g xylene cyanole, 1,000 mL H_2O); solid Na $C_2H_3O_2$, Na_3PO_4, $AlCl_3$, Na_2CO_3, NH_4Cl, NaCl; Hydrion pH test paper.

Apparatus: pH meters

INTRODUCTION

The chemistry exhibited by common acids is due mostly to the hydronium ion, H_3O^+, present in solution. Often we shorten the representation of the hydronium ion, H_3O^+, to that of the hydrogen ion, H^+. The concentration of the hydrogen ion in solution, abbreviated $[H^+]$, is a measure of the acidity of the solution and may be expressed in molarity (M) units. A more convenient expression of the hydrogen ion concentration is pH. The pH of a solution is related to the hydrogen ion concentration by the following expression.

$$\text{pH} = \log 1/[H^+] \text{ or } \text{pH} = -\log [H^+]$$

The use of pH has the advantage that one does not have to deal with the small fractions or negative exponents encountered in molarity expressions. For example, a solution with $[H^+] = 1 \times 10^{-6}$ M has a pH of 6.

$$\text{pH} = -\log [H^+]$$
$$\text{pH} = -\log (1 \times 10^{-6})$$
$$\text{pH} = -(0 + (-6)) = -(-6)$$
$$\text{pH} = 6$$

A solution with a measured pH of 9 has a $[H^+] = 1 \times 10^{-9}$ M (or 10^{-9} M), since $[H^+] = 10^{-pH}$.

Most aqueous solutions have a pH between 0 and 14. Because the pH scale is exponential, a difference in pH of one unit corresponds to a 10-fold difference in hydrogen-ion concentration. Although neutral water has a pH of 7, rainwater is acidic due to dissolved carbon dioxide, which produces a dilute carbonic acid solution. The pH of "normal" rain is about 5.6, but "acid" rain events, which are usually the result of human activities, are common. In parts of North America and Europe, the average rainfall pH is between 4 and 5. A pH of 4.6 would represent rainwater which is 10 times more acidic than normal. Rainwater with a pH of 3.6 (100 times more acidic than normal) has been observed. More examples of pH values and acidity may be found in your textbook.

neutral solution: $[H^+] = 10^{-7}$ M; pH = 7
acidic solution: $[H^+] > 10^{-7}$ M; pH < 7
basic solution: $[H^+] < 10^{-7}$ M; pH > 7

In this activity, the pH of various solutions is measured using pH test paper, acid-base indicators, and a pH meter. The pH test paper contains a universal indicator that results in color changes over a large range of $[H^+]$. A pH meter uses a "glass" electrode that is sensitive to changes in hydrogen ion concentration. The electrode (combined with a reference electrode such as the saturated calomel electrode) produces an electrochemical potential, measured in millivolts, that varies as the hydrogen-ion concentration changes. The pH meter is a specialized millivolt meter that reads the voltage produced by the combined glass-electrode and reference-electrode combination and converts the voltage to pH units.

According to the Arrhenius concept of acids and bases, an acid is a substance containing one or more hydrogen atoms that can become free hydrogen ions, H^+ (hydronium ions, H_3O^+) when dissolved in water. Examples include HCl, HNO_3, H_2SO_4, and $HC_2H_3O_2$. Bases contain one or more OH groups that can become free hydroxide ions when the substance is dissolved in water. Examples include NaOH, $Ca(OH)_2$, KOH, and $Mg(OH)_2$.

We can represent the dissolving and **dissociation** of an acid such as HCl in water with a reaction equation.

$$HCl_{(aq)} \longrightarrow H^+_{(aq)} + Cl^-_{(aq)}$$

A similar reaction equation can be written for the dissolving and dissociation of a base such as calcium hydroxide.

$$Ca(OH)_{2(s)} \longrightarrow Ca^{2+}_{(aq)} + 2\ OH^-_{(aq)}$$

We often write reaction equations that suggest that an acid dissociates completely (100% dissociation) in water. This is not always the case in actual solutions.

$$HCl_{(aq)} \longrightarrow H^+_{(aq)} + Cl^-_{(aq)} \qquad \text{100\% dissociated}$$

$$HC_2H_3O_{2(aq)} \longrightarrow H^+_{(aq)} + C_2H_3O_2^-{}_{(aq)} \qquad \text{1.8\% dissociated}$$

The *strength of an acid* depends on the degree of dissociation; we can say that the amount of H^+ in solution depends on the number of acid molecules that dissociate in water. The concentration of H^+ (strength of acid) is not always equal to the original concentration of an acid (concentration on the label of the bottle). Strong acids dissociate completely in water and the $[H^+]$ is equal to the acid concentration (M_{acid}) for acids having one H^+ per molecule. A 0.001 M solution of a strong monoprotic acid such as HNO_3 would be expected to have an $[H^+] = 0.001$ M and thus a pH = 3. Weak acids do not dissociate completely in water and the $[H^+]$ in solution is less than the acid concentration. A 0.001 M solution of an acid such as H_2S would be expected to have an $[H^+] < 0.001$ M and thus a pH > 3. Calculation of the exact pH of a weak acid solution is possible if the concentration of the acid is known.

The *strength of a base* is related to solubility. Strong bases dissociate completely when their solid form dissolves in water, so there is a stoichiometric amount of OH^- produced for the amount of solid that it dissolves. A 0.001 M solution of KOH would have an $[OH^-] = 0.001$ M and thus a pH = 11.

A **salt** is one of the products of a chemical reaction between an acid and a base, (usually called a neutralization reaction).

$$\underset{\text{base}}{KOH} + \underset{\text{acid}}{HCl} \longrightarrow \underset{\text{salt}}{KCl} + \underset{\text{water}}{HOH}$$

When a salt that has been formed from the reaction between a strong base and a weak acid is dried and dissolved in water, the pH of the salt solution is different from the pH of the original water. This is because the salt reacts with the water in a reaction known as **hydrolysis**. The formation of such a salt, sodium acetate, is represented by the neutralization of acetic acid (weak acid, mostly undissociated) by sodium hydroxide (strong base, completely dissociated).

$$Na^+_{(aq)} + OH^-_{(aq)} + HC_2H_3O_{2(aq)} \longrightarrow Na^+_{(aq)} + C_2H_3O_2^-{}_{(aq)} + HOH$$

Sodium acetate reacts with water (a hydrolysis reaction) to produce sodium hydroxide, which is completely dissociated, and acetic acid, which is slightly dissociated.

$$NaC_2H_3O_{2(s)} + HOH \longrightarrow \underbrace{Na^+_{(aq)} + OH^-_{(aq)}}_{\text{strong base}} + \underbrace{HC_2H_3O_{2(aq)}}_{\text{weak acid}}$$

This process leads to an excess of OH^- relative to H^+. The resulting salt solution has a higher pH than that of water. Another way of interpreting this is to say that the strong base dominates over the weak acid. Sodium acetate is classified as a *basic* salt.

When a salt that has been formed from a reaction between a weak base and a strong acid is dissolved in water, hydrolysis occurs. Ammonium chloride (NH_4Cl) is formed when ammonium hydroxide (a weak base) is neutralized by hydrochloric acid (a strong acid). When dissolved in water, hydrolysis occurs.

$$NH_4Cl_{(s)} + H_2O \longrightarrow \underset{\text{weak base}}{NH_4OH_{(aq)}} + \underset{\text{strong acid}}{H^+_{(aq)} + Cl^-_{(aq)}}$$

The product, NH_4OH, is more correctly thought of as aqueous ammonia ($NH_3 + H_2O$), which is a weak base. The other product is hydrochloric acid, a strong acid. In the final salt solution there are more H^+ ions relative to OH^- ions, the strong acid dominates over the weak base, and the solution has a lower pH than that of the water in which it was dissolved. Ammonium chloride is classified as an *acidic* salt.

Certain solutions have the capacity to resist changes in pH (i.e., maintain a relatively constant $[H^+]$). These solutions are called **buffers**, and may be defined as solutions that resist change of pH more effectively than solutions of either acids or bases having the same pH as the buffer. Buffer systems are very important in biochemical functions such as oxygen transport in blood and prevention of lactic acid buildup in muscles as a result of exercise.

A solution that resists a pH change when acid (H^+) or base (OH^-) is added to it is said to be buffered. However, the amount of H^+ or OH^- that can be added to a given buffer without a drastic pH change is limited. The amount of acid or base that a buffer system can absorb without drastic pH changes is called buffer capacity. Systems with a high buffer capacity can buffer (absorb) more H^+ (acid) and OH^- (base) than systems with a low buffer capacity. Buffer systems are generally produced when a weak acid and a salt containing the anion of the weak acid are mixed. Thus, acetic acid and sodium acetate would be a good buffer mixture. We examine the citric acid/citrate buffer system in this activity. The chemical reactions and mathematics associated with buffers are complex and are not needed to complete this activity. These topics will be left for study at a later time.

PROCEDURE

The acidity of a solution may be determined by measuring the pH of the solution. The pH test paper we use is called Hydrion paper. In the later sections of this experiment, pH meters and chemical indicators are used to provide a wider variety of lab experiences. Hydrion paper turns a different color for each whole number of pH value and should be precut into 2-cm strips. This paper is expensive, so conserve. Put a piece of pH paper on a clean, dry watch glass. With a stirring rod, place a small drop of solution to be tested on the strip of pH paper. Immediately, compare the resulting color with the standard color code for the pH paper.

A. pH, $[H^+]$, and Acid Strength

1. **Hydrochloric acid.** Use *Hydrion paper* to determine the pH of 0.1 M HCl solution and record it. Record the hydrogen ion concentration $[H^+]$ that corresponds to this pH value. Dilute precisely 5.0 mL of the 0.1 M HCl solution with deionized water in a graduated cylinder to 50 mL in the following way. Put about 30 mL of distilled water into the cylinder. Add the 5.0 mL of acid and then bring the volume to exactly 50 mL with distilled water. Mix well. This should give you a 0.01 M HCl solution. Determine and record the pH of this solution and the corresponding $[H^+]$. (The color difference between pH 1 and pH 2 on the pH color comparison chart is a subtle one, but indicates a 10-fold dilution of the acid solution.)

   Perform a second 10-fold dilution, this time starting with the 0.01 M HCl to prepare 0.001 M HCl. Determine and record the pH and corresponding $[H^+]$ for this solution. On the basis of the information you just recorded, make inferences about the strength of hydrochloric acid and answer the questions on the report sheet.
2. **Acetic acid.** Use *Hydrion paper* to determine and record the pH and corresponding $[H^+]$ of 0.1 M $HC_2H_3O_2$. Then dilute precisely 5.0 mL of the 0.1 M $HC_2H_3O_2$ with deionized water to a final volume of 50 mL as directed previously. Mix well. This dilution should produce a 0.01 M $HC_2H_3O_2$ solution. Determine the pH and corresponding $[H^+]$ of this solution.

   Perform another 10-fold dilution starting with the 0.01 M $HC_2H_3O_2$ prepared in the previous step. This dilution should produce a 0.001 M acetic acid solution. Determine and record the pH and corresponding $[H^+]$ of this solution.

   On the basis of the information you just recorded, make inferences about the strength of acetic acid and answer the following questions on the report sheet.
3. **pH of common materials.** Use a pH meter to determine and record the pH and corresponding $[H^+]$ of each of the following solutions. Carefully immerse the bulb of the combination (glass and reference) electrode into each sample and record the pH reading. Rinse the electrode well with distilled water after each test. *Be careful! The electrode is fragile.*

Note: If the material is a solid (such as baking soda or detergent) add a spatulafull to 5-mL of deionized water and stir to mix. Then rank the solutions according to increasing acidity.

cola soft drink	bleach solution
vinegar	laundry detergent solution
orange juice	milk
household ammonia	buttermilk
tap water	baking soda

B. pH and Buffers

1. **Preparation of citric acid/citrate mixture.** Mix 10 mL of 0.1 M citric acid with 19 mL of 0.10 M sodium hydroxide. This solution contains citric acid and salts of citric acid produced by neutralization of citric acid by sodium hydroxide. The pH of this solution is approximately the same as that of deionized or distilled water that has absorbed carbon dioxide through contact with room air. Use Hydrion paper to measure the pH of deionized water and of this citric acid/citrate mixture. Record these pH values on the report sheet and save the solution to test its buffer capacity.
2. **Measurement of buffer capacity of deionized water.** Measure 10 mL of deionized water into a 50-mL beaker and add 1 drop of phenolphthalein indicator. With a medicine dropper, add 0.100 M NaOH solution dropwise until there is a color change. Record the number of drops required. Rinse out the beaker, measure in another 10-mL deionized water sample, and add 1 drop of methyl orange/xylene cyanole indicator. Then add 0.10 M HCl dropwise until the color becomes red-violet. Record the number of drops required.
3. **Measurement of buffer capacity of citric acid/citrate mixture.** Make the same tests done on deionized water using the solution prepared from citric acid and sodium hydroxide. Use 10-mL samples of the buffer solution. Record the results. Phenolphthalein is colorless below a pH of about 9 and red above that; methyl orange/xylene cyanole is green above a pH of about 5, gray at a pH of 4, and red-violet below that. What evidence indicates that the citric acid/citrate system used in this experiment is a buffer solution?

C. pH and Hydrolysis of Salts

Prepare solutions of each of the following substances by adding 2 g of each solute to 10 mL of distilled water in separate labeled test tubes.

$Na\ C_2H_3O_2$	Na_3PO_4
$AlCl_3$	Na_2CO_3
NH_4Cl	$NaCl$

Mix well and divide each solution into two approximately equal parts. To one add 1 drop of phenolphthalein and to the other 1 drop of methyl orange/xylene cyanole. Decide which solutions remained approximately neutral (did not hydrolyze), which formed acidic solutions, and which hydrolyzed to form basic solutions. Choose one of the samples that hydrolyzed and write the hydrolysis reaction. Classify each salt as a neutral, a basic, or an acidic salt.

Experiment 17
pH of Aqueous Solutions

Name____
Date____

REVIEW QUESTIONS

1. pH gives us information about a specific chemical species in solution. What species is this?

2. What would be the pH of 0.001 M NaOH?

3. What is the hydrogen ion concentration of a solution with a pH of 4? Is this solution acidic, basic, or neutral?

4. Classify each of the following salts as acidic, basic, or neutral on the basis of their predicted behavior in aqueous solutions.

NH_4NO_3 ________________________ $CuSO_4$ ________________________

NaCl ________________________ $KC_2H_3O_2$ ________________________

5. What kinds of substances are needed for buffered solutions?

Experiment 17

pH of Aqueous Solutions

Name____________________

Date____________________

REPORT SHEET

A. pH, [H^+], and Acid Strength

1. Hydrochloric acid

HCl concentration (M)	pH	H^+ concentration (M)
0.100 M		
0.010 M		
0.001 M		

From the experimental data, how does the [H^+] compare to the concentration of HCl in each solution (same as or less than)?

On the basis of the answer to the previous question, does HCl dissociate completely in water? Explain.

Does the data indicate that HCl is a strong acid or a weak acid?

Write an equation for the dissociation of hydrochloric acid in water.

2. Acetic acid

$HC_2H_3O_2$ concentration (M)	pH	H^+ concentration (M)
0.100 M		
0.010 M		
0.001 M		

From the experimental data, how does the [H^+] compare to the concentration of $HC_2H_3O_2$ in each solution (same as or less than)?

On the basis of the answer to the previous question, does $HC_2H_3O_2$ dissociate completely in water? Explain.

Does the data indicate that $HC_2H_3O_2$ is a strong acid or a weak acid?

Write an equation for the dissociation of acetic acid in water.

3. pH of common materials

cola soft drink ______________________

vinegar ______________________

orange juice ______________________

buttermilk ______________________

tap water ______________________

bleach solution ______________________

laundry detergent solution ______________________

milk ______________________

household ammonia ______________________

baking soda ______________________

Characterize these solutions according to increasing acidity by arranging them according to increasing $[H^+]$ (decreasing pH).

______________________ Most basic

____

____

____

____

____

____

____

____

______________________ Most acidic

B. pH and Buffers

1. Preparation of citric acid/citrate mixture

Distilled water: pH ______________________

Citric acid/citrate mixture: pH ______________________

2. Measurement of buffer capacity of deionized water

Drops NaOH needed: ______________________ Drops HCl needed: ______________________

3. Measurement of buffer capacity of citric acid/citrate mixture

Drops NaOH needed: ______________________ Drops HCl needed: ______________________

What evidence indicates that the citric acid/citrate system used in this experiment is a buffer solution?

C. pH and Hydrolysis of Salts

Which salts did not hydrolyze? ____________________

Which salts hydrolyzed to form acidic solutions? ____________________

Which salts hydrolyzed to form basic solutions? ____________________

Explain why one salt did not hydrolyze.

Write an equation to represent the hydrolysis of one of the basic salts.

Write an equation to represent the hydrolysis of one of the acidic salts.

EXPERIMENT 18

Ion Combination Reactions

Purpose: To observe some ion combination reactions and then to use them in identifying an unknown.

Materials: Solution (0.1 M) of $Pb(NO_3)_2$, K_2CrO_4, $BaCl_2$, NaCl, $AgNO_3$, $FeSO_4$, $CuSO_4$, $FeCl_3$, KSCN, $K_3Fe(CN)_6$; 1 M H_2SO_4, 1 M Na_2CO_3, 6 N H_2SO_4, 6 M NaOH, 14 M NH_4OH. Solid unknowns (5 gram portions): NaCl, K_2SO_4, K_2CO_3, $FeSO_4$, $BaCl_2$, $Fe_2(SO_4)_3$.

Waste Disposal: Any mixture that contains lead, chromium, barium, or silver should be disposed of in the heavy metals waste container provided in the lab.

INTRODUCTION

In a large group of chemical reactions the atoms undergo no change in oxidation state (complete or partial transfer of electrons). These are sometimes called "ion combination" reactions because ions or molecules, often in solution, form stable new combinations. Ion combination reactions are also known as "metathesis" reactions. Metathesis (me-tath-e-sis) reactions are a group of reactions in which none of the substances change oxidation number as they interact to form new species. Precipitation and neutralization reactions are subgroups of metathesis reactions. Positive ions and negative ions (and occasionally neutral molecules) unite to form a substance that is (1.) virtually insoluble, or (2.) molecular (such as H_2O or CO_2), or (3.) other undissociated covalent species (such as complex ions). The insoluble material may be a precipitate (solid) or a gas. If a molecular substance forms, it may be insoluble or it may remain in solution, but there is usually evidence of reaction such as an energy change (absorption or evolution of heat) or a color change.

Ion combination reactions occur when ions from different substances combine to form new substances. You may also find the term "double replacement" or "double displacement" used to label this type of reaction. We sometimes write a general form of a reaction equation for this type of chemical change as $AB + CD \longrightarrow AD + CB$. The substances AB and CD are often strong electrolytes and form ions when dissolved in water. Examples of strong electrolytes are NaCl (sodium chloride), HCl (hydrochloric acid), and NaOH (sodium hydroxide). When water solutions of these compounds are mixed together, the ions can interact. Sometimes this interaction is only a momentary thing and no real change is noted by an observer. At other times a new compound may be formed that is not a strong electrolyte in water, and a change is observed. The specific combination of carbonate (CO_3^{2-}) compounds with strong acids (which release H^+ ions) produces H_2CO_3, which rapidly decomposes to H_2O and CO_2.

An example of a double replacement reaction is the combination of water solutions of lead(II) nitrate and aluminum chloride. Both of these substances are strong electrolytes, so a water solution of lead(II) nitrate contains lead(II) ions (Pb^{2+}) and nitrate ions (NO_3^-); a water solution of aluminum chloride contains aluminum ions (Al^{3+}) and chloride ions (Cl^-). When water solutions of lead(II) nitrate and aluminum chloride are mixed, a white solid (precipitate) forms. Analysis of this white solid shows that it is lead(II) chloride ($PbCl_2$). The following reaction equations can be written to represent this process.

1. **Word reaction equation:**

lead(II) nitrate + aluminum chloride ⟶ lead(II) chloride + aluminum nitrate

2. **Formula reaction equation:**

$$3\ Pb(NO_3)_2 + 2\ AlCl_3 \longrightarrow 3\ PbCl_2 + 2\ Al(NO_3)_3$$

3. **Total ionic reaction equation:**

$$3\ Pb^{2+}_{(aq)} + 6\ NO_3^-{}_{(aq)} + 2\ Al^{3+}_{(aq)} + 6\ Cl^-_{(aq)} \longrightarrow 3\ PbCl_{2(s)} + 2\ Al^{3+}_{(aq)} + 6NO_3^-{}_{(aq)}$$

A second product, aluminum nitrate, is soluble in water. The Al^{3+} and NO_3^- ions interact momentarily but remain as separate dissolved species. An observer would not notice this interaction. Ionic notation implies that the substance is dissolved in water.

4. **Net ionic reaction equation:**

$$3\ Pb^{2+}_{(aq)} + 6\ Cl^-_{(aq)} \longrightarrow 3\ PbCl_{2(s)}$$

reducing to the simplest ratio yields:

$$Pb^{2+}_{(aq)} + 2\ Cl^-_{(aq)} \longrightarrow PbCl_{2(s)}$$

PROCEDURE

A. Reactions That Produce a Virtually Insoluble Solid (Precipitate)

For this experiment it may prove helpful to use the list of common ions provided in Appendix B to help predict products and write reaction equations. **Note**: The instructor may require that certain precipitates filtered in section A of the experiment be given an "OK" before going on to the next section.

1. **Lead chromate.** In a clean test tube, mix 5 mL each of $Pb(NO_3)_2$ solution and K_2CrO_4 solution. (Volume measurements in this experiment need not be exact; estimate 5 mL as being a thumb's width of liquid in a standard test tube.) Stir the mixture and filter a portion of the precipitate. The substance is lead chromate, $PbCrO_4$, which is used as a pigment for outdoor paints such as those used for the center line of roads. Before going further write the word reaction equation, the formula reaction equation, the total ionic equation, and the net ionic equation.
2. **Barium sulfate.** Mix 5 mL each of $BaCl_2$ and $FeSO_4$ solutions in a clean test tube. Stir well, and filter part of the mixture. The precipitate is barium sulfate. This reaction is used to test for the presence of sulfate ion in water in **Experiment 13, Impurities in Natural Water**. Before going further write the word reaction equation, the formula reaction equation, the total ionic equation, and the net ionic equation.
3. **Iron(II) ferricyanide.** Mix 5 mL each of $FeSO_4$ and $K_3Fe(CN)_6$ solutions in a clean test tube. The $K_3Fe(CN)_6$ ionizes to produce 3 K^+ ions and 1 $Fe(CN)_6^{3-}$ ion. The precipitate is iron(II) ferricyanide, $Fe_3[Fe(CN)_6]_2$. Stir well and filter the mixture. Before going further write the word reaction equation, the formula reaction equation, the total ionic equation, and the net ionic equation.
4. **Silver chloride.** In a clean test tube mix 5 mL each of NaCl and $AgNO_3$ solutions. Avoid contact of $AgNO_3$ with skin or clothing. Stir well but do not filter the mixture. The precipitate is silver chloride. This reaction is used to test for the presence of chloride ion in water in **Experiment 13, Impurities in Natural Water.** Before proceeding write the ionic and net equation. Before going further write the word reaction equation, the formula reaction equation, the total ionic equation, and the net ionic equation.
5. **Copper(II) hydroxide.** Mix 5 mL of $CuSO_4$ and 5 drops of 6 M NaOH solutions in a clean test tube. Stir well; the precipitate is copper(II) hydroxide. Do not filter. Save this solution for part **B1**. Before going further write the word reaction equation, the formula reaction equation, the total ionic equation, and the net ionic equation.

B. Reactions That Form Complex Ions or Molecular Substances

1. **Tetramminecopper(II) ion.** Add 5 mL of 14 M aqueous ammonia ($NH_{3(aq)}$) to the mixture produced in Part A5. Stir well. What evidence of reaction is observed? The ammonia has reacted with copper(II) hydroxide to produce the covalently-bonded complex ion, $Cu(NH_3)_4^{2+}$, and aqueous hydroxide ion, OH^-. Before going further write the word reaction equation, the formula reaction equation, the total ionic equation, and the net ionic equation.

2. **Acid-base neutralization.** Into a small test tube put 5 mL of 6 M H_2SO_4 and into a large test tube put 5 mL of 6 M NaOH solution. Take the temperature of each solution, rinsing the thermometer well between tests. Pour the contents of the small test tube into the larger one and observe. Take the temperature of the mixture. Water, H_2O, a largely molecular substance, was formed. This kind of reaction, which occurs between acids and bases (substances producing H^+ ions in solution and substances producing OH^- ions in solution), is called neutralization. Before going further write the word reaction equation, the formula reaction equation, the total ionic equation, and the net ionic equation.
3. **Iron(III) thiocyanate.** Mix about 5 mL each of $FeCl_3$ and KSCN solutions in the same manner as you did in Part B2, taking the temperature before and after the reaction. Is there a temperature change? Is there a precipitate that can be filtered? What evidence of reaction is seen? Iron(III) thiocyanate, $Fe(SCN)_3$, which is largely molecular, was formed. Before going further write the word reaction equation, the formula reaction equation, the total ionic equation, and the net ionic equation.
4. **Carbon dioxide.** In a clean test tube mix 5 mL each of 1 M H_2SO_4 and Na_2CO_3 solutions. The gas that escapes is carbon dioxide, CO_2. Actually, the reaction is believed to occur in two stages: first a largely molecular substance, carbonic acid, H_2CO_3, forms; then this decomposes to H_2O and CO_2. Is there a temperature change? Before going further write the word reaction equation, the formula reaction equation, the total ionic equation, and the net ionic equation. Because H_2CO_3 is only an intermediate product, do not include it in your equations (use H_2O and CO_2 instead).

C. Identification of an Unknown

Obtain an unknown from the instructor, record its number, and dissolve all of it in 75 mL of water. Filter if solid remains. Try each of the following reagents in separate 5-mL portions of your sample: $AgNO_3$, $BaCl_2$, H_2SO_4, KSCN, $K_3Fe(CN)_6$. Use the results to identify at least one or both ions of your unknown. Write a balanced net ionic equation for the reaction you believe has taken place whenever a reaction is observed. Record the identity of the compound or at least one ion believed to be in the unknown.

Experiment 18

Ion Combination Reactions

Name_______________________________

Date_______________________________

REVIEW QUESTIONS

Recall that an ion combination reaction requires the formation of a virtually insoluble substance or a largely molecular one. Use the Table of Common Ions (Appendix C), General Solubility Rules (Appendix D), and the list of Common Strong and Weak Acids and Bases (Appendix E) to predict whether each of the following mixtures of ionic substances will react. Recall also that water, weak acids, and weak bases are largely molecular. In the following reactions predict the insoluble or molecular substances that form. Write the formula equation.

1. $AgNO_{3(aq)} + MgCl_{2(aq)} \longrightarrow$

2. $NH_4OH_{(aq)} + H_2SO_{4(aq)} \longrightarrow$

3. $ZnSO_{4(aq)} + H_2S_{(aq)} \longrightarrow$

4. $Al_2(SO_4)_{3(aq)} + NH_4OH_{(aq)} \longrightarrow$

5. $Ba(OH)_{2(aq)} + H_3PO_{4(aq)} \longrightarrow$

6. $KNO_{3(aq)} + NaCl_{(aq)} \longrightarrow$

7. $BaCO_{3(aq)} + H_2SO_{4(aq)} \longrightarrow$

8. $NH_4Cl_{(aq)} + NaOH_{(aq)} \longrightarrow$

9. $NaC_2H_3O_{2(aq)} + HCl_{(aq)} \longrightarrow$

10. $Na_2SO_{4(aq)} + NH_4OH_{(aq)} \longrightarrow$

Experiment 18
Ion Combination Reactions

Name______________________________________
Date_______________________________________

REPORT SHEET

A. Reactions That Produce a Virtually Insoluble Solid (Precipitate)

1. Lead chromate Instructor's "OK" (if necessary) __________

Observations:

Word reaction equation:

Formula reaction equation:

Total ionic equation:

Net ionic equation:

2. Barium sulfate Instructor's "OK" (if necessary) __________

Observations:

Word reaction equation:

Formula reaction equation:

Total ionic equation:

Net ionic equation:

3. Iron(II) ferricyanide Instructor's "OK" (if necessary) __________

Observations:

Word reaction equation:

Formula reaction equation:

Total ionic equation:

Net ionic equation:

4. Silver chloride Instructor's "OK" (if necessary) __________

Observations:

Word reaction equation:

Formula reaction equation:

Total ionic equation:

Net ionic equation:

5. Copper(II) hydroxide Instructor's "OK" (if necessary) __________

Observations:

Word reaction equation:

Formula reaction equation:

Total ionic equation:

Net ionic equation:

B. Reactions That Form Complex Ions or Molecular Substances

1. Tetramminecopper(II) ion Instructor's "OK" (if necessary) __________

Observations:

Word reaction equation:

Formula reaction equation:

Total ionic equation:

Net ionic equation:

2. Acid-base neutralization Instructor's "OK" (if necessary) __________

Observations:

Word reaction equation:

Formula reaction equation:

Total ionic equation:

Net ionic equation:

3. Iron(III) thiocyanate Instructor's "OK" (if necessary) __________

Observations:

Word reaction equation:

Formula reaction equation:

Total ionic equation:

Net ionic equation:

4. Carbon dioxide Instructor's "OK" (if necessary) __________

Observations:

Word reaction equation:

Formula reaction equation:

Total ionic equation:

Net ionic equation:

C. Identification of an Unknown

Unknown number ________________________

Test reagent	Observation	Net Ionic Equation
$AgNO_3$		
$BaCl_2$		
H_2SO_4		
KSCN		
$K_3Fe(CN)_6$		

Identity of ion or ions in unknown: ________________________

EXPERIMENT 19

Classification of Solid Substances

Purpose: To investigate the properties of polar covalent, nonpolar covalent, metallic, ionic, and network covalent solids.

Materials: Two or three representative examples of each category. Some suggestions:

Ionic: KCl, NaCl, $NaNO_3$, KNO_3
Polar Covalent: sucrose, glucose, urea
Nonpolar covalent: benzophenone, naphthalene, benzoic acid
Metallic: copper, tin, aluminum, iron
Network covalent: carbon, silicon dioxide (sand)
Polar solvent: water or ethanol
Nonpolar solvent: toluene or hexane

Apparatus: Test tubes, spatula, Bunsen burner, solid-state conductivity tester (any circuit tester from a hardware store is sufficient), conductivity tester for aqueous solutions (the type that is merely a lamp with a bulb is fine).

Safety Precautions: Avoid touching any of the materials. If you get any on your skin, wash thoroughly. Use the electrical equipment under the supervision of the assistant or instructor. Touching a live wire can result in electrocution. Be certain you understand all instructions. Wear safety glasses at all times.

Hazardous Waste Disposal: The water-soluble solutions may be discarded down the sink. The solids that dissolved in the nonpolar solvents should be collected in a waste container for organic wastes. The solids that do not dissolve may be placed in the garbage can.

INTRODUCTION

Up until the middle of this century, chemists were quite unable to explain why there were so many different kinds of substances. It was clear that there were many different kinds of materials, and like substances were easily grouped together, but the *explanation* for their properties was delayed until there was better understanding of atomic bonding.

The greatest achievements in describing atomic bonding must be attributed to Linus Pauling (*The Nature of the Chemical Bond*, 1939), winner of two Nobel Prizes and the author of the basic covalent bonding theory. Before beginning this experiment, you should review the sections on bonding types in your text.

As you know, electrons are transferred in ionic bonding, and in covalent bonding electrons are shared. An infinite number of intermediate possibilities exist in-between where electrons are shared, but unequally.

A : B	A : B	A : B	A: B
covalent (nonpolar)	polar covalent	(more) polar covalent or slightly ionic	ionic

It is important to remember that while most substances can be neatly placed in "ionic" or "covalent" categories, there are really many gray areas in-between. For covalent molecules, we also must consider whether they are polar or nonpolar, and how one molecule interacts with a like, adjacent molecule. To summarize:

I. *Ionic* Electrons transferred, ions of opposite charge are attracted to each other to form solids in crystalline form, soluble in polar or ionic solvents, often dissolves in water to yield free ions that will conduct electricity in solution, medium to high mp.

II. *Nonpolar Covalent* Electrons are generally shared within molecules; there is very little interaction between molecules, low mp and low bp, insoluble in polar solvents like water, soluble in nonpolar solvents like hexane.

III. *Polar Covalent* Electrons are unequally shared within molecules; there is interaction between molecules, the stronger the interaction, the more polar it tends to be, soluble in polar solvents like water but insoluble in nonpolar solvents like hexane.

IV. *Network Covalent* Electrons are shared, usually fairly equally. There is no "between molecules" as every atom is bonded to all adjacent atoms by strong covalent bonds. The material is one giant molecule sometimes called a *macromolecule.* Since covalent bonds are difficult to break, these substances are very unreactive. They are almost always solids and won't dissolve in anything.

V. *Metallic* Does not fit the covalent/ionic scheme. Describes bonding in elemental metals and alloys, a network of nuclei surrounded by a sea of electrons; typical metallic properties: lustrous, flexible, malleable, conducts electricity in solid state.

Dispersion There is no real bonding between or among units. Only a very weak random chance interaction holds the material together. These materials are almost always gases. Also called van der Waals forces, instantaneous dipole forces, or induced dipoles. None of these will be encountered in this lab.

The purpose of this lab will be to classify several known and unknown substances into one of the five categories (I, II, III, IV, V) based on physical properties and a few simple tests.

PROCEDURE

A. Solubility Tests

1. The items may be taken in any order, but you should test the knowns first. Only perform the test(s) that are required for identification. If you are certain a material is a metal, there is no point in trying to dissolve it in water.
2. For the solubility tests, place a half-pea-size amount in a small test tube and add 1-2 cm of solvent on top. Stir gently with a stirring rod for at least 20 seconds. If it doesn't dissolve, pour the solvent off and add the other one. Stir gently. The two solvents are (distilled) water for the polar, and toluene or hexane for the nonpolar. If one works, the other should not.

B. Melting Point

It is not necessary to measure this with a thermometer. The result need only be "high," "medium," or "low." Put a small, pea-size amount of solid in the bottom of a test tube. Using a cool flame heat gently by teasing the bottom of the test tube back and forth through the flame. If the substance melts under this effort, it is "low" melting. Next, hold the material in the flame for longer periods. If the substance melts under moderate heating, it is "medium" melting. If the material doesn't melt no matter how hard you try, it is "high" melting. Be careful not to overheat the test tube and don't point it at anyone while you are heating it.

C. Solid State Conductivity

If you believe the substance to be metallic, you should test it for conductance in the solid state. Use the circuit tester provided. If it lights up, the circuit is completed and the positive test confirms a metallic material.

D. Solution Conductivity

If a substance dissolves in water, it is probably either ionic or polar covalent, but only an ionic compound will produce free ions in solution that can conduct electricity. First dissolve a small amount of the substance in a *clean* 250-mL beaker in *distilled* water. Then bring it to the conductance tester. Follow all instructions; ask questions. UNDER NO CIRCUMSTANCES SHOULD YOU TOUCH THE BARE ELECTRODES!! YOU COULD BECOME PART OF THE CIRCUIT AND BE ELECTROCUTED!! Have the TA help you. If the light bulb lights up, this is positive for an ionic substance. Try distilled water alone to assure yourself that it is not a property of the water itself.

E. The Unknowns

The experiment is a process of eliminations and confirming tests. Remember, however, that there are all kinds of exceptions. For example, a great many ionic compounds do not dissolve in water.

The summary below may help you:

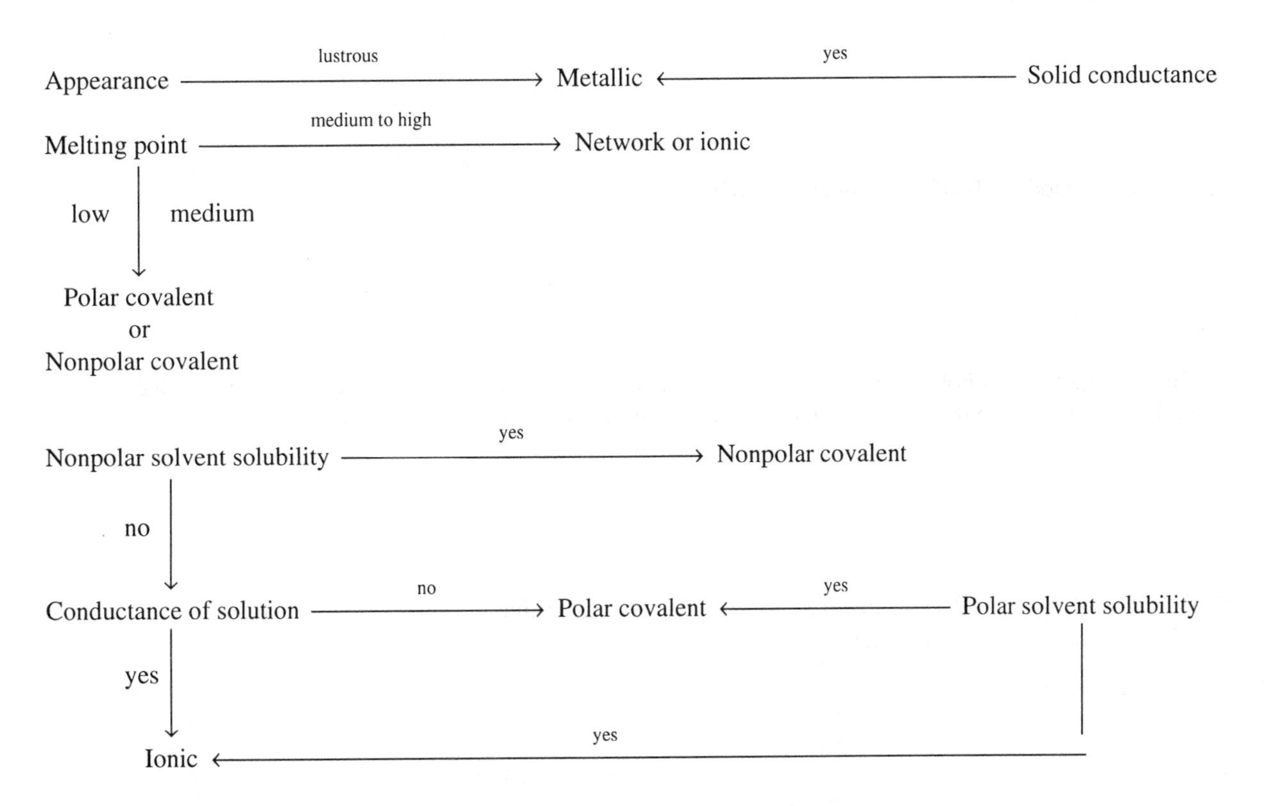

Experiment 19

Classification of Solid Substances

Name____________________________________

Date_____________________________________

REVIEW QUESTIONS

1. Describe the bonding in diamond. How would this be classified?

2. Use the theory of metallic bonding to explain how metals can conduct electricity in the solid state. Why can't ionic compounds?

3. Would solid iodine (I_2) dissolve in water? Why or why not?

4. Consider methyl alcohol, CH_3OH. What are the internal molecular forces? What kind of intermolecular (between molecules) forces are there? Which must be broken for methyl alcohol to evaporate?

5. Lead crystallizes in the face-centered cubic arrangement. If the radius of lead is 175.0 pm, calculate the density (g/cm^3) of lead. Compare this to a literature value.

6. A certain solid substance is high-melting, dissolves readily in water, and conducts electricity in aqueous solution. A student identified this substance as sucrose (table sugar). Explain why the student must be in error.

7. Glass is an amorphous mixture of silicon dioxide (and some contaminants). Based on its classification, explain the behavior of glass as it is heated. Does it melt? Explain.

Experiment 19
Classification of Solid Substances

Name________________________________

Date________________________________

REPORT SHEET

The knowns are labeled I (ionic), II (nonpolar covalent), III (polar covalent), IV (network covalent), V (metallic).

Knowns	Tests/Observations	Classification
I		ionic
II		nonpolar covalent
III		polar covalent
IV		network covalent
V		metallic

Unknowns	Tests/Observations	Classification
A		
B		
C		
D		
E		

EXPERIMENT 20

Reaction Rates

Purpose: To observe some of the factors which determine the rate of a chemical reaction.

Materials: 0.20 M KI (potassium iodide) solution, 0.1 M $(NH_4)_2S_2O_8$ (ammonium peroxydisulfate) solution, 0.002 M $Na_2S_2O_3$ (sodium thiosulfate) solution, starch indicator solution; 6 N solutions of NH_4OH, HCl, $HC_2H_3O_2$, H_3PO_4; 3% H_2O_2, 0.2 N $MnCl_2$, 0.1 M $CuSO_4$; powdered Fe, granular Fe and Zn (20 mesh), Fe nail, Al strips 1" × 1/4".

Waste Disposal: All solutions may be rinsed down the drain with water, but metals (granules, strips, nails, etc.) should be transfered to a waste basket.

INTRODUCTION

Some reactions occur virtually instantaneously, and little can be done to influence their rate. The explosion of dynamite or the neutralization reaction between a strong acid and a strong base are examples. Other reactions are slow, such as the rusting of iron, or very slow, such as the geologic changes that occur in a natural mineral. The rates of many reactions can, however, be accelerated or retarded. An industrial reaction, such as the synthesis of ammonia or the synthesis of common plastics, can often be made to proceed more rapidly, and so can be made economically practical; or the rate of deterioration of a product, such as the oxidation of steel or the loss of freshness in foods, can be slowed so as to maximize its shelf life or useful life.

The rate (or speed) of a reaction is usually defined in terms of the amount of reactants converted to products in a certain amount of time. The amount of reactants consumed and products formed in solution is generally expressed in concentration terms of moles per liter (molarity). The unit of time may be a second, a minute, an hour, or a longer period if the reaction is very slow. A typical unit for rate of reaction is moles per liter per second (mol/L · sec). Mathematical rate expressions can be written for chemical reactions, but this activity focuses on the variables that affect rates of reaction. You will collect and interpret information in terms of relative rates of reaction rather than absolute rates.

Many factors contribute to the occurrence or nonoccurrence of chemical reactions. The primary condition is that the particles of reacting substances (atoms, molecules, or ions) come into contact. This does not mean, however, that a chemical reaction occurs anytime two different substances come into contact with each other. The energy of reactants is another important variable in the rate of reaction.

Collision theory may be used to explain rates of chemical reactions. Fast rates are explained in terms of many collisions per unit time; slow rates as fewer collisions per unit of time. Certain methods can be used to increase or decrease the frequency of collisions between reacting substances; they result in an increase or decrease in the rate of reaction.

Some substances form natural protective coatings, usually oxides of the metal, that may hinder or delay reaction between the metal and another substance. Protective oxide coatings stick to the metal surface and act as a barrier between the metal and other substances. Such protective barriers decrease the number of collisions between metal and oxygen gas, thereby decreasing the rate of oxidation of the metal. Not all metal oxides are protective, however. The oxide of iron constantly flakes off the iron metal, reexposing "fresh" metal to the atmosphere, and more oxidation of the metal occurs.

When a solid is involved in a reaction with another substance, the reaction takes place only where the surface of the solid comes into contact with the other substance. Logically, one would reason that if the surface area of a solid could be increased, the probability of collisions would increase, resulting in an increased rate of reaction. A common method of increasing the surface area of a solid is to grind the solid into very fine particles.

Gases and dissolved salts exist in the smallest particle size, so reactions between such substances may be rapid if they occur. Some substances, such as weak acids and weak bases, do not dissociate completely into ions in an aqeous solution. If the reactive component in the solution is an ion such as H^+ or OH^-, then if all else is equal, substances that are dissociated to a small extent will exhibit slower rates of reaction than substances dissociated to a greater extent.

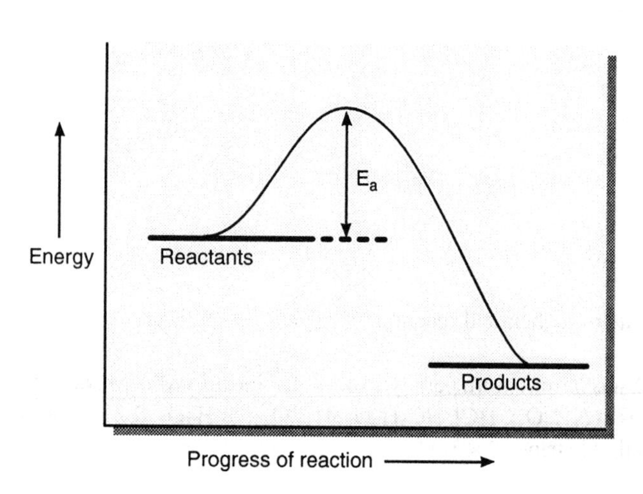

FIGURE 20.1
Energy plot for a single-step chemical reaction.

All collisions do not necessarily result in a chemical change, because energy requirements may not be met. The energy component is sometimes illustrated by energy profiles such as that in Figure 20.1. A collision will not result in formation of new products if the basic units of the reactant do not have sufficient energy to overcome the energy barrier to reaction. The amount of energy associated with this barrier is called activation energy (E_a). These energies vary widely among chemical reactions, but E_a is constant for a specified reaction. Sometimes the reactants have enough internal energy to overcome the energy barrier and form products. A common method of increasing the energy of reactants is to raise their temperature. Conversely, energy can be removed by lowering their temperature.

Catalysts can be used to alter the rate of reaction without being consumed in the reaction. There is some disagreement about the definition of a catalyst. An older definition describes a catalyst as an agent that can slow down or speed up the rate of a chemical reaction without being involved in the reaction. A newer system uses two words to describe these different functions. A catalyst is defined as a substance that speeds up the rate of reaction, and an inhibitor as a substance that slows it down. The newer system of defining catalysts and inhibitors is a recent development of increased work in the area of catalysis and is the system used in this activity. In energy terms, a catalyst acts by lowering the energy barrier to product formation. It does not affect the initial energy of reactants. Chemical kineticists are researchers who investigate the *mechanism* of how a catalyst interacts with the substances in a reaction to lower the energy barrier. Enzymes are examples of catalysts of biological systems. They are known to be necessary for almost all biochemical functions, but the exact mechanism by which they function is an area of intense research. Food preservatives, such as the antioxidants BHA and BHT, are examples of inhibitors used to slow down natural reactions that lead to the spoilage of food.

For a chemical reaction to occur, the reacting atoms, molecules, or ions must come into contact with each other. One might think that if the concentration of any of the reacting substances was increased, the rate of reaction would increase owing to increased probability of collisions. However, in a multistep chemical process this logic is *not* always correct. Only those reactants involved in the **rate-determining step** can affect the overall rate of reaction. Increasing the concentration of the reactant *not* involved in the rate-determining step will *not* change the overall rate of reaction. You can reason in the reverse order with these concepts. If increasing or decreasing the concentration of a reactant does not increase or decrease the observed rate of reaction, that substance is not in the rate-determining step of a multistep reaction.

The effects of temperature, concentration of reactants, and catalysts on rate of reaction will be examined using the oxidation-reduction reaction:

$$2\ I^-_{(aq)} + S_2O_8{}^{2-}{}_{(aq)} \longrightarrow I_{2(aq)} + 2\ SO_4{}^{2-}{}_{(aq)} \qquad \text{Equation 1}$$

The iodide ions will be obtained from potassium iodide (KI) solution and the peroxydisulfate ions ($S_2O_8{}^{2-}$) from ammonium peroxydisulfate [$(NH_4)_2S_2O_8$] solution. The method used for timing the rate of reaction involves what is frequently called a "clock" reaction. In addition to the reaction being studied, a second reaction represented by the following reaction equation will occur in the solution.

$$I_{2(aq)} + 2\ S_2O_3{}^{2-}{}_{(aq)} \longrightarrow 2\ I^-_{(aq)} + S_4O_6{}^{2-}{}_{(aq)} \qquad \text{Equation 2}$$

As iodine (I_2) is produced in equation 1, it reacts very quickly and completely with the thiosulfate ion ($S_2O_3{}^{2-}$) in equation 2 until all the thiosulfate ions have reacted. When all of the thiosulfate ions have reacted, the iodine produced in equation 1 remains in solution because no other substance is present in the reaction mixture. The presence of iodine at this point can be confirmed because iodine produces a yellow color in solution, but if starch indicator is also present in the reaction mixture, the formation of iodine is more easily observed as the deep-blue starch-iodine complex.

In this experiment the effects of temperature changes, changes in concentration, the use of a catalyst, the state of division in a heterogeneous reaction, formation of a passive coating, and the degree of ionization of a reagent are considered.

PROCEDURE

A. Effect of Temperature

Set up two clean burets. Fill one with 0.20 M KI (potassium iodide) solution, the other with 0.1 M $(NH_4)_2S_2O_8$ (ammonium peroxydisulfate) solution. Be sure that there are no air bubbles in the tips of the burets. (You will use these burets and solutions through the remaining procedures in this activity.) **Note**: Be sure that all test tubes have been cleaned and rinsed with deionized water before transferring reagents into them.

1. **Room temperature (20°C).** Put 10.0 mL of 0.002 M $Na_2S_2O_3$ (sodium thiosulfate) solution, measured carefully in a 25-mL graduated cylinder, into a clean, large test tube. Add 6 drops of starch indicator solution to the $Na_2S_2O_3$ solution. Fit the large test tube with a solid rubber stopper.

   Put 10.0 mL of 0.20 M KI, measured carefully with the buret, into a clean, dry, small test tube. Into another clean, dry, small test tube put 10.0 mL of 0.10 M $(NH_4)_2S_2O_8$, measured carefully with the buret. Stand all three test tubes in a water bath that has been adjusted to 20°C, which is just about or slightly below room temperature.

   Pour the 10.0 mL of KI solution into the large test tube containing sodium thiosulfate solution and starch. Then pour the 10.0 mL of ammonium peroxydisulfate solution into the large test tube. Begin timing. Stopper the large test tube and mix by inverting the test tube several times. Time the reaction from the point when the last solution was added to the large test tube until the dark starch-iodine color first appears.
2. **Low temperature.** Measure and prepare solutions of $Na_2S_2O_3$, KI, and $(NH_4)_2S_2O_8$ as directed for room temperature. *Before mixing* the solutions, allow the test tubes and their contents to stand in a beaker of cool water adjusted to about 10°C. When the solutions have reached 10°C, record the temperature, mix the solutions in the specified order for room temperature, and start timing as directed before. After mixing, put the large test tube into the cool water bath for timing.
3. **Elevated temperatures.** Measure and prepare solutions of $Na_2S_2O_3$, KI, and $(NH_4)_2S_2O_8$ as directed before. *Before mixing* the solutions, allow the test tubes and their contents to stand in a beaker containing water at 30°C. After the solutions have reached approximately 30°C, record the temperature, mix the solutions in the proper order, and start timing. After mixing, put the large test tube in the 30°C water bath for timing.

   Repeat the elevated temperature experiment using a water bath that has been adjusted to 40°C.

B. Effect of Concentration

The chemical reaction used to study temperature effects will also be used to study the effects of concentration of reacting sustances on the rates of a reaction. As before, the reagents should be placed in three test tubes and then mixed in the large test tube. All of the reactions will be run at room temperature. Prepare five different reaction mixtures, as outlined in Table 20.1. Record reaction times for each reaction mixture, as directed previously.

Does changing each concentration change the reaction rate equally? Suggest a possible explanation for the results you observed.

TABLE 20.1 Five reaction mixtures to investigate effect of concentration on rate.

Run Number	Large test tube 1 *0.002 M $Na_2S_2O_3$ (mL)*	Test tube 2 *0.20 M KI (mL)*	*Deionized H_2O (mL)*	Test tube 3 *0.10 M $(NH_4)_2S_2O_8$ (mL)*	*Deionized H_2O (mL)*
1	10.0 + 6 drops starch	10.0	5.0	5.0	10.0
2	10.0 + 6 drops starch	10.0	5.0	15.0	0.0
3	10.0 + 6 drops starch	10.0	5.0	10.0	5.0
4	10.0 + 6 drops starch	5.0	10.0	10.0	5.0
5	10.0 + 6 drops starch	15.0	0.0	10.0	5.0

C. Surface Area (Heterogeneous Reactions)

Put 5 mL of 6 M HCl into each of three clean test tubes. To the first test tube add a single iron nail that has been cleaned with steel wool. To the second add a full spatula of granular iron. To the third add a full spatula of iron powder. Observe any differences in rates of reaction that may occur. Order the three forms of iron in terms of increasing rate of reaction with HCl.

D. Action of a Catalyst

1. **Iodide/peroxydisulfate clock reaction.** Repeat the reaction of parts A and B of this experiment using reaction mixture 3 in Table 20.1. Before mixing the solutions, add 1 drop of 0.1 M $CuSO_4$ to the test tube containing 0.10 M $(NH_4)_2S_2O_8$. Mix thoroughly. Then mix all three solutions in the order described previously. Record the reaction time at room temperature.
2. **Decomposition of hydrogen peroxide.** Hydrogen peroxide slowly decomposes into water and oxygen by the following reaction:

$$2\,H_2O_2 \longrightarrow 2\,H_2O + O_2$$

Put about 5 mL of 3% H_2O_2 into a test tube and add 2–3 drops of 6 N NH_4OH. Shake the mixture and then add 1 drop of $MnCl_2$ solution. Observe and explain the result.

E. Effect of a Passive Coating

Obtain two strips of Al about 1/4 inch wide and 1 inch long. Burnish only one of them with steel wool. Drop each into a separate test tube containing 5 mL of 6 M HCl. Observe and report results. Explain.

F. Degree of Dissociation of a Substance Containing a Reactant

Put a spatula of granular zinc into each of three clean test tubes. To one zinc sample add 5 mL of 6 M HCl, to another add 5 mL of 6 M $HC_2H_3O_2$, and to the third zinc sample add 5 mL of 6 M H_3PO_4. Any chemical reaction that occurs is represented by the reaction equation: $Zn + 2\,H^+ \longrightarrow Zn^{2+} + H_2$. Wait a few minutes before making observations about relative reaction rates because there seems to be a slight induction period for one of the acids.

EXPERIMENT 20
Reaction Rates

Name____________________________________
Date____________________________________

REVIEW QUESTIONS

1. Iron dust may react with HCl more slowly than expected from the surface exposed. Suggest an explanation.

2. Look up some examples of catalysts in industrial reactions. See if you can find an industrial situation in which a catalyst is used to slow down a reaction.

3. What is meant by autocatalysis?

4. List some examples of catalysis that occur in life processes and discuss the role of enzymes in biological reactions.

5. Suggest a reaction by which aluminum becomes passive and write the equation. List some other metals with similar behavior.

6. The Haber process for manufacturing ammonia uses magnetic iron oxide, Fe_3O_4, as a catalyst. Under the same conditions of temperature and pressure does the use of a catalyst increase the yield of ammonia at equilibrium? Explain.

7. Many reactions require the application of activation energy in order to start them. Discuss the effect of a catalyst on activation energy.

8. Iron powder sometimes reacts with HCl more slowly (at first) than might be expected on the basis of the amount of surface area of the powder. Suggest an explanation.

9. Some reactions are autocatalytic. They start very slowly, but their rate increases dramatically as the reaction proceeds. Explain how this might occur.

10. Stainless steel is an alloy (a mixture) of iron, chromium, and nickel. Use one or more of the concepts studied in this experiment to suggest how chromium or nickel might help prevent the corrosion of stainless steel.

Experiment 20
Reaction Rates

Name__________________________________
Date___________________________________

REPORT SHEET

A. Effect of Temperature

	Low Temperature (10°C)	Room temperature (20°C)	Elevated temperature (30°C)	(40°C)
Time (sec)				

Write a short statement that expresses the relationship between temperature and the rate of reaction.

Approximately what temperature change is required to double the rate of this chemical reaction?

Explain (in terms of collision theory) why changing the temperature changes the rate of the reaction.

B. Effect of Concentration

Run Number	0.002 M $Na_2S_2O_3$ (mL)	0.20 M KI (mL)	0.10 M $(NH_4)_2S_2O_8$ (mL)	Time (sec)
1				
2				
3				
4				
5				

The following equation describes the reaction:

$$2\ I^-_{(aq)} + S_2O_8^{2-}{}_{(aq)} \longrightarrow I_{2(aq)} + 2\ SO_4^{2-}{}_{(aq)}$$

Write a short statement that relates iodide (I^- from KI) concentration to the rate of the reaction.

Write a short statement that relates peroxydisulfate ($S_2O_8^{2-}$ from $(NH_4)_2S_2O_8$) concentration to the rate of the reaction.

Why was the amount of $Na_2S_2O_3$ kept constant for all five experiments?

What function does the deionized water serve in each of the five reaction mixtures?

C. Surface Area (Heterogeneous Reactions)

Order the three forms of iron in terms of increasing rate of reaction with HCl.

Use collision theory to explain this order.

D. Action of a Catalyst

1. Iodide/peroxydisulfate clock reaction

Time required for color change when $CuSO_4$ is added to the reaction mixture: ____________ sec

Does $CuSO_4$ accelerate or inhibit the rate of the reaction?

What effect does $CuSO_4$ have on the activation energy of this reaction system?

2. Decomposition of hydrogen peroxide

Observation when $MnCl_2$ is added to H_2O_2:

Write the balanced equation for the reaction that takes place.

What effect does $MnCl_2$ have on the activation energy of this reaction system?

E. Effect of a Passive Coating

Observations of Al with HCl:

Burnished Al

Unburnished Al

Which strip started reacting first?

What is the probable formula of the passive coating?

F. Degree of Dissociation of a Substance Containing a Reactant

Arrange the three experiments in order of **increasing** reaction rate:

_______________________ _______________________ _______________________

least reactive most reactive

Explain the order of reactivity in terms of the degree of dissociation of the reacting acids; *i.e,. indicate which is least dissociated, which is most dissociated, etc.*

EXPERIMENT 21

The Iodine Clock Reaction: A Mathematical Approach

Purpose: To investigate the kinetics of the iodine clock reaction using formal mathematical and graphical methods.

Materials: 0.22M KI, 0.18M $K_2S_2O_8$, 0.22M KNO_3, 0.18M K_2SO_4, 0.010M $Na_2S_2O_3$ (The $Na_2S_2O_3$ solution must be made up using 100 mL of freshly prepared soluble starch solution for every 1 L of solution.)

Apparatus: 100-mL beakers, 5-, 10-, and 20-mL pipets, pipet bulbs, stopwatch or wristwatch that measures in seconds, hot (not boiling) water bath, ice water bath, thermometer.

Safety Precautions: Wear your safety glasses at all times in this lab. Handle the thermometers carefully.

Hazardous Waste Disposal: All the solutions may be washed down the drain with a large amount of water.

INTRODUCTION

Certain chemical reactions lend themselves readily to the study of their kinetics. The reaction between iodide ion and peroxydisulfate is a classic example. Its kinetics are so reliable that this reaction (and certain variations on it) has been called the Iodine "Clock" Reaction.

Actually, two reactions are taking place, only one of which we will study.

$$3I^- + S_2O_8^{-2} \rightarrow I_3^- + 2\ SO_4^{-2} \qquad \text{Equation 1}$$

$$I_3^- + 2S_2O_3^{-2} \rightarrow S_4O_6^{-2} + 3I^- \qquad \text{Equation 2}$$

Before we continue the discussion, a review of the nomenclature should be helpful.

I^-	iodide ion	SO_4^{-2}	sulfate ion
I_3^-	triiodide ion	$S_2O_3^{-2}$	thiosulfate ion
I_2	iodine	$S_4O_6^{-2}$	dithionite ion
$S_2O_8^{-2}$	peroxydisulfate ion, also commonly called "persulfate" ion		

Reaction (2) is virtually instantaneous. Reaction (1) takes place in a matter of seconds or minutes, and this is the reaction we will be investigating.

Solutions will be set up containing measured amounts of iodide, persulfate, and thiosulfate. The thiosulfate will be in the least amount; i.e., it is the limiting reagent. The fast reaction (2) will continue to cycle triiodide back into iodide as long as thiosulfate remains. When thiosulfate runs out, the first detectable amounts of triiodide are formed. Triiodide (also sometimes called just "iodine") reacts with starch to give a dark-blue to black color. This will be your end point. The reactions may be diagramed this way:

$$3I^- + S_2O_8^{-2} \rightarrow I_3^- + 2\ SO_4^{-2}$$

$$+\ 2\ S_2O_3^{-2} \qquad \left(\text{when } S_2O_3^{-2} \text{ used up}\right)$$

$$\rightarrow \text{Starch } I_3^- \ (\text{blue})$$

In kinetics, as opposed to most other chemical reactions, it is not possible to determine the exponents from the coefficients. The general rate expression for reaction (1) can be written

$$rate = k\left[I^-\right]^m\left[S_2O_8^{-2}\right]^n$$

where m and n are integers and k is a real number, which will all be determined in this experiment.

The rate constant k is valid only at the temperature of the solutions, which is approximately room temperature. If additional runs are made at other temperatures, it will be possible to calculate the energy of activation using the Arrhenius equation.

$$\ln k = A - \frac{E_a}{RT}$$

You will find m,n,k, and E_a for reaction (1).

PROCEDURE

A. Reaction Rates at Room Temperature

1. Work in pairs.

   This lab requires extensive pipetting. Please share the pipets and pipet bulbs. The solution concentrations are 0.22 M KI, 0.010 M $Na_2S_2O_3$, 0.18 M $K_2S_2O_8$, 0.22 M KNO_3, and 0.18 M K_2SO_4.

   You will need two clean, dry 100-mL beakers. Label one "A" and the other "B." Beaker A will always contain the KI, $Na_2S_2O_3$, and the KNO_3. Beaker B will only contain the $K_2S_2O_8$ and K_2SO_4. The starch indicator is already present in the thiosulfate, so you do not need to worry about adding it. The KNO_3 and K_2SO_4 are added to keep the ionic strength of the solutions constant.

   The volumes of each reagent are summarized on your report sheet. Be very careful that you do not mix up the reagents that have similar formulas. Pipet the appropriate amounts into the two beakers.
2. At some convenient time, start your watch and pour the contents from one beaker into the other and then back and forth several times to mix them. Do not spill. Set the beaker down on white paper and wait for the onset of the blue color. At room temperature, the moment of onset is unambiguous. One person should watch the beaker and be prepared to call "Now!" The other should operate the timepiece. If neither you nor your partner has a watch with a second timer, a laboratory stopwatch will be provided. Measure to the nearest second.

   Flush the solutions down the drain. Rinse and *dry* the beakers and repeat the run.
3. Continue for runs 2–7.

B. Reaction Rates at High and Low Temperatures—The Arrhenius Equation

1. For runs 8, 9, and 10, you will vary the temperature but not the concentration. Run #8 will be room temperature. Measure the temperature of any of the solutions and assume this temperature is uniform throughout. Record this for run #8.

   Now realizing that you are using the concentrations from run #3, you need not repeat the room temperature run. Since runs #8 and #3 are identical, simply copy down the two times from #3 into the spaces provided in #8.
2. Run #9 will be the "hot" solution. Pipet the reagents into beakers A and B, as before, and then set the beakers into the tub of hot water provided. The beakers will tend to float so don't let them turn over. After 2–5 minutes the solutions should be as hot as the bath. Check the temperature of both solutions and record the average of the two.
3. Pour one solution into the other and mix as before. Return the beaker to the hot bath to wait for the color change. It will be rapid.
4. Repeat run #9 for the hot temperature.

5. Now perform a cold run for run #10. Place your beakers into an ice bath. It will take somewhat longer for the solutions to reach cold temperatures. Swirl them a little to speed this. When the solutions are about 5–10° C mix and time them. Return the solution to the cold bath. You should swirl it occasionally to ensure a homogeneous solution. Record the time for the first appearance of any blue. If the reaction takes much longer than 4 minutes, abandon it and try again.
6. Repeat run #10. If different from the first trial, average the temperatures.
7. Rinse and dry any equipment. Return any beakers that belong to your partner. Return any thermometers and stopwatches.

CALCULATIONS

Part A—For Runs 1-7

1. Average the two times. Convert all times to seconds.
2. Now you must determine the concentrations of the two major reactants, I^- and $S_2O_8^{-2}$, in the combined beaker. The solutions went in as one concentration, but the addition of the other reagents diluted them. You must apply a *dilution factor* to the original molarity. Notice that in every case the final volume was 50 mL. The calculation will take the form

$$M_n = M_o \times \frac{V_o}{V_f}$$

M_n = new molarity　　M_o = original molarity

V_o = original volume　　V_f = final volume (here = 50 mL)

So for run #1, 20 mL of 0.22 M KI was diluted to 50 mL

$$M_n = 0.22\text{ M} \times \frac{20\text{ mL}}{50\text{ mL}} = 0.088\text{ M} = \left[I^-\right]$$

and 20 mL of 0.18 M $K_2S_2O_8$ was diluted to 50 mL

$$M_n = 0.18\text{ M} \times \frac{20\text{ mL}}{50\text{ mL}} = 0.072\text{ M} = \left[S_2O_8^{-2}\right]$$

In this manner, calculate $[I^-]$ and $[S_2O_8^{-2}]$ for runs 1–7.

3. Notice that runs 1, 2, 3, and 4 all have the same iodide ion concentration and that runs 1, 5, 6, and 7 all have the same persulfate concentration. Remember that the equation we are trying to solve is

$$rate = k\left[I^-\right]^m\left[S_2O_8^{-2}\right]^n$$

If we hold $[I^-]$ constant, the equation becomes

$$rate = k'\left[S_2O_8^{-2}\right]^n$$

If we then take the log of both sides

$$\log rate = \log k' + n\log\left[S_2O_8^{-2}\right]$$

$$y = b + mx$$

which is a straight line of the form $y = mx + b$ with a slope equal to the integer n. This is the exponent for $\left[S_2O_8^{-2}\right]$.

Therefore, you will need the logarithms of each of the concentrations. This is *common log* (log on the calculator, NOT ln). Find the logs of each concentration by entering the concentration in your calculator and pressing "log." Enter these values onto your report sheet.

4. But what about "rate" and "log rate"?

Rate can be defined as

$$rate = \frac{\Delta I_3^-}{\Delta time}$$

and I_3^- is dependent on $S_2O_3^{-2}$ in a 1:2 relationship (see eq. 2). Thiosulfate starts out as

$$0.01\text{ M} \times \frac{10\text{ mL}}{50\text{ mL}} = 0.002\text{ M}$$

and goes to zero (when it is used up, the blue appears). Therefore, I_3^- changes from zero and ends up as 0.001 M. ΔI_3^- *is* 0.001 − 0 = 0.001, so

$$rate = \frac{0.001}{time}$$

Calculate rate for runs 1–7 as $\frac{0.001}{time}$ using the average time in seconds. Then, as before, simply evaluate "log rate" with your calculator.

5. Now you are ready to find the exponents m and n. Plot only the four runs that apply:
   Runs 1,2,3,4 : Plot log rate *vs.* log $[S_2O_8^{-2}]$ to get the exponent n
   Runs 1,5,6,7 : Plot log rate *vs.* log $[I^-]$ to get the exponent m

   In each case, draw the best-fitting straight line and then calculate the slope of the line. Round off the value to the nearest whole number. Use this integer in all subsequent calculations.

   Refer to the remarks in the appendix on preparing graphs if you have any questions about the plots. You may put both plots on one sheet of graph paper. Turn in the graphs showing the calculations for slope with your report sheet.
6. Now you are ready to find the rate constant, k. This is just algebra. We have defined rate as

$$rate = k\left[I^-\right]^m\left[S_2O_8^{-2}\right]^n$$

and now you know "rate," $[I^-]$, m, $[S_2O_8^{-2}]$, and n for each run. Rearrange to solve for k:

$$k = \frac{rate}{\left[I^-\right]^m\left[S_2O_8^{-2}\right]^n}$$

Calculate k for runs 1–7. Be sure to use the rounded-off integers for m and n. Write in the appropriate units. Finally, calculate an average k. (This will apply only to room temperature.)

Part B—Energy of Activation—Runs 8, 9, 10

7. Calculate an average time for each run.
8. Change temperature in °C to K.
9. Calculate the rate as before: $rate = \frac{0.001}{time}$
10. Calculate a k for each run as you did in part 6. In each case, the concentrations are the same and are identical to the ones in run #3.
11. Evaluate ln k using your calculator. This is *"natural log"* and is "ln" or INV e^x on your calculator.
12. Evaluate $1/T$ using temperatures in K as T. (The numbers will be quite small.)
13. Using the Arrhenius equation

$$\ln k = A - \frac{E_a}{RT}$$

it is obvious that a plot of ln k *vs.* $1/T$ is a straight line with slope equal to $-E_a/R$.

Therefore, for runs 8, 9, 10:

$$\text{Plot } \ln k \text{ } vs. \frac{1}{T}$$

Find the slope of the line.

$$slope = -\frac{E_a}{R}$$

or

$$E_a = -(slope)(R)$$

where $R = 8.31$ J/mol · K
so E_a will have units of J/mol. Lastly convert this to kJ/mol. Remember that energy of activation is always a positive number. Make sure the signs work out.

Turn in your graph showing the calculation for slope.

Experiment 21

The Iodine Clock Reaction: A Mathematical Approach

Name____

Date____

REVIEW QUESTIONS

1. Given what you know about the structure of "peroxide" and the structure of "sulfate" ion, draw the structure of "peroxydisulfate."

2. What is the order of the reaction with respect to iodide ion? What is the order of the reaction with respect to peroxydisulfate ion? What is the overall order of the reaction?

3. E_a for a certain reaction is 9.32×10^4 J/mol. At 27° C, $k = 1.25 \times 10^{-2}$ L/mol · s. Calculate k at 200° C.

4. Draw an activation energy diagram for the iodide-persulfate reaction given that $\Delta H = -427$ kJ.

5. What is a catalyst? Redraw the energy diagram from #4 and add a curve for a catalyzed reaction.

6. Some chemical species slow down the rate of a reaction. These chemicals are very important to the field of biochemistry. What are they called? Give one example.

7. Use kinetic theory to explain why the rate slows as temperature decreases.

Experiment 21

The Iodine Clock Reaction: A Mathematical Approach

Name____________________

Date____________________

REPORT SHEET

Part A

Run #	Beaker A	Beaker B	Time(s)	Time(s)	Average Time(s)
1	20 mL KI 10 mL $Na_2S_2O_3$	20 mL $K_2S_2O_8$			
2	20 mL KI 10 mL $Na_2S_2O_3$	15 mL $K_2S_2O_8$ 5 mL K_2SO_4			
3	20 mL KI 10 mL $Na_2S_2O_3$	10 mL $K_2S_2O_8$ 10 mL K_2SO_4			
4	20 mL KI 10 mL $Na_2S_2O_3$	5 mL $K_2S_2O_8$ 15 mL K_2SO_4			
5	15 mL KI 5 mL KNO_3 10 mL $Na_2S_2O_3$	20 mL $K_2S_2O_8$			
6	10 mL KI 10 mL KNO_3 10 mL $Na_2S_2O_3$	20 mL $K_2S_2O_8$			
7	5 mL KI 15 mL KNO_3 10 mL $Na_2S_2O_3$	20 mL $K_2S_2O_8$			

Part B-Energy of Activation

Run #	Temp (°C)	Time(s)	Time(s)	Average Time(s)
8 (same as 3)				
9				
10				

Calculations-Part A

Run #	$[I^-]$	$[S_2O_8^{-2}]$	log $[I^-]$	log $[S_2O_8^{-2}]$
1				
2				
3				
4				
5				
6				
7				

Run #	1	2	3	4	5	6	7
Rate							
Log Rate							

$n =$ ____________________ $m =$ ____________________

Run #	k
1	
2	
3	
4	
5	
6	
7	

Average k at ________ °C is: ________________________

Part B–Energy of Activation

Run #	Temp (K)	*Rate*	k	ln k	$1/T$
8					
9					
10					

E_a = ________________________ J/mol = ________________________ kJ/mol

Write the rate expression below, filling in the values of m and n, the average of k at room temperature. Include units on k. Remember that concentrations are variables.

EXPERIMENT 22

First-Order Kinetics and Activation Energy

Purpose: To gather first-order kinetic data by monitoring the absorbance produced by a colored complex ion as it decomposes in aqueous solution over a 1-hour time period; to use the absorbance/time/temperature data to determine first-order rate constants and the activation energy of the reaction.

Materials: 1, 10-phenanthroline solution, 5.56×10^{-3} M (Dissolve 1.0 g of 1, 10-phenanthroline in about 800 mL of distilled water at about 80°C. Cool and dilute to 1,000 mL); iron (II) solution, 1.8×10^{-2} M (Dissolve 3.512 g of ferrous ammonium sulfate hexahydrate in about 100 milliliters of distilled water. Add 2 drops of concentrated hydrochloric acid, and dilute to 500 mL); 3 M sulfuric acid.

Waste Disposal: All solutions may be rinsed down the drain with water.

Note: This experiment requires 5 to 6 hours for students (working in pairs) to complete. All necessary data can be gathered in one laboratory period, but the plots and calculations require an additional 2 or 3 hours.

INTRODUCTION

Iron (II) ion, Fe^{++} combines with the chelating ligand, 1,10-phenanthroline, to form the complex ion tris(1, 10-phenanthroline)iron(II). The equation for the reaction forming this complex follows. Note that 1, 10-phenanthroline is abbreviated as "phen."

$$Fe^{++} + 3\ phen \longrightarrow Fe(phen)_3^{++} \qquad \text{Equation 1}$$

However, this complex ion is unstable in the presence of acids (pH < 3) and decomposes according to the reverse of equation 1 by means of a mechanism which yields a first-order rate law.

$$Fe(phen)_3^{++} \longrightarrow Fe^{++} + 3\ phen \qquad \text{Equation 2}$$

This experiment is designed to demonstrate a spectrophotometric method for obtaining kinetic data. Three $Fe(phen)_3^{++}$ decomposition experiments are run simultaneously, at different temperatures, and the absorbance of each reaction system is measured as a function of time. Since concentration is proportional to absorbance, the concentration of the iron complex can be determined from the absorbance data. The experiment also demonstrates the effect of temperature on the rate of a chemical change. It allows the student to plot first-order data and to calculate the half-life of a reaction.

Equation 3 is the rate law for the first-order disappearance of $Fe(phen)_3^{++}$ in which the rate constant is represented by the variable "k."

$$rate = -k[Fe(phen)_3^{++}] \qquad \text{Equation 3}$$

The mathematical expression for this rate law is commonly written in differential equation form as shown in equation 4.

$$d[Fe(phen)_3^{++}]/dt = -k[Fe(phen)_3^{++}] \qquad \text{Equation 4}$$

The integrated form of the rate law can be written as:

$$\ln [Fe(phen)_3^{++}] = -kt + \ln [Fe(phen)_3^{++}]_0 \qquad \text{Equation 5}$$

and the half-life is related to the rate constant according to equation 6

$$t_{1/2} = .693/k \qquad \text{Equation 6}$$

For first-order kinetics, the units of the rate constant are the reciprocal of time. In other words, if time had been measured in minutes, the units of k would be min^{-1} and the half-life calculated from k would, therefore, have units of minutes.

The activation energy can be determined graphically using the Arrhenius equation, in which R is the gas constant and T is the absolute temperature.

$$k = A\,e^{-Ea/RT}$$

Taking the natural logarithm of both sides of equation 6 gives equation 7

$$\ln k = \frac{-E_a}{RT} + \ln A \qquad \text{Equation 7}$$

A plot of ln k on the y axis versus 1/T on the x axis produces a straight line from which the activation energy, Ea, can be determined because the slope of the line is $-E_a/R$.

PROCEDURE

A. Data Collection

1. **Preparation of 9.0×10^{-5} M $Fe(phen)_3^{++}$ solution.** With a pipet, measure 10 mL of 1.80×10^{-2} M iron (II) solution into a 1,000 mL volumetric flask and dilute to 1.0 liter with distilled water. The concentration of iron in the solution is 1.8×10^{-4} M. With a graduated cylinder, measure 200 mL of the 1.80×10^{-4} M iron (II) solution, and transfer to a 600 mL beaker. Also, with the graduated cylinder, measure 200 mL of 5.56×10^{-3} M 1, 10-phenanthroline solution, and add this to the iron (II) solution in the beaker. This solution should contain a large excess of 1,10-phenanthroline so that it will react with and convert virtually all of the aqueous iron(II) ion into $Fe(phen)_3^{++}$. Label this solution "9.0×10^{-5} M $Fe(phen)_3^{++}$." On the report sheet show mathematically that the concentration of $Fe(phen)_3^{++}$ in this solution is 9.0×10^{-5} M. This solution will serve as the starting solution for rate experiments, and it will be used as a standard from which other more dilute standards will also be prepared.
2. **Kinetic runs.** Have a Spectronic 20 (or 21) set up and ready to make absorbance measurements at a wavelength of 510 nm. Use two cuvettes. One should be kept full of distilled water for zeroing the instrument, and the other should be used to measure absorbances. Before starting, check to see that the cuvettes are matched. They should give identical absorbance readings with distilled water.

   Obtain three 250-mL Erlenmeyer flasks. Place one in a warm water bath at 45°C; place the second in a water bath at 35°C; and keep the third flask at just above room temperature (25°C). Put 100 mL of the 9.0×10^{-5} M $Fe(phen)_3^{++}$ solution into each of the three flasks. Then add 20 mL of 3 M sulfuric acid solution (which should be equilibrated at the same temperature as the solution in the 250-mL flask). The decomposition reaction now begins! Immediately measure the absorbance of a small portion (use a pipet to remove about 4 mL from the flask) of 45°C mixture, and then discard the small amount used. Do the same thing with the other two solutions at the other two temperatures. **It is a good idea to stagger the starts (addition of sulfuric acid) of each of the three runs so that individual measurements will be taken 1 or 2 minutes apart from one another.** Record the three absorbances as the 0 time readings on the report sheet. Every 5 minutes for 1 hour, absorbance readings for each of the three mixtures should be measured. Discard the liquid used for absorbance readings because its temperature is likely to change while taking the reading, and this might change the temperature in the reaction flask.
3. **Standards.** While the reaction kinetics are being measured, one of the two people working together on this experiment should prepare three more standards by diluting 15.0 mL, 10.0 mL, and 5.0 mL of "9.0×10^{-5} M $Fe(phen)_3^{++}$" to a total of 20.0 mL in each case in a 25-mL graduated cylinder. Calculate the iron concentrations in these three standards and record.

B. Data Analysis

1. **Preparation of a standard curve and determination of $Fe(phen)_3^{++}$ versus time.** When all data has been recorded, use the standards and their absorbances to prepare a standard curve (which should be a straight-line plot of absorbance as a function of concentration) on a piece of graph paper. Plot the concentration of each of the standards (from Part A3) on the x-axis versus the absorbance of each standard on the y-axis. Then use the standard curve to determine the iron concentration for each absorbance measured during the decomposition runs.
2. **Plot of $Fe(phen)_3^{++}$ concentration as a function of time.** On a second piece of graph paper plot $Fe(phen)_3^{++}$ concentration (y-axis) for each decomposition experiment versus time (x-axis). Draw a smooth curve through each set of data points.
3. **Plot of natural log of $Fe(phen)_3^{++}$ concentration as a function of time.** On a third piece of graph paper, plot for each decomposition experiment, the natural log of $Fe(phen)_3^{++}$ concentration (y-axis) versus time. Use a ruler to draw a straight line through the points for each experiment. From equation 5 it can be seen that the slope of each of the log plots is equal to –k for that reaction. Thus, calculate and record the slope of each of the lines and record on the report sheet. Using the slopes of the three lines on plot 3, calculate the rate constant and half-life of each of the three reactions (use proper units). Remember to use significant figure rules discussed earlier.
4. **Arrhenius plot and activation energy determination.** Plot the natural log of k on the y-axis against 1/T (remember to convert temperature to °K) on the x-axis. Determine the slope of this line and calculate the activation energy using $R = 8.31 \text{ J °K}^{-1} \text{ mol}^{-1}$ or $R = 1.99 \text{ cal °K}^{-1} \text{ mol}^{-1}$.

Experiment 22

First-Order Kinetics and Activation Energy

Name________________________________

Date________________________________

REPORT SHEET

A. Data Collection

1. **Preparation of 9.0×10^{-5} M $Fe(phen)_3^{++}$ solution**
   Mathematical setup used to calculate the $Fe(phen)_3^{++}$ concentration

2. **Kinetic runs**

Time	Run 1 (45°) Absorbance	Run 2 (35°) Absorbance	Run 3 (25°) Absorbance
0	________	________	________
5	________	________	________
10	________	________	________
15	________	________	________
20	________	________	________
25	________	________	________
30	________	________	________
35	________	________	________
40	________	________	________
45	________	________	________
50	________	________	________
55	________	________	________
60	________	________	________

3. Standards

Concentration Fe	Absorbance

B. Data Analysis

1. **Preparation of a standard curve and determination of $Fe(phen)_3^{++}$ versus time.** Attach the standard curve to this report sheet. Use the standard curve and the absorbance values in section A2 of this report sheet to determine the $Fe(phen)_3^{++}$ concentration for the following table. Use a calculator to determine the natural logarithms of $[Fe(phen)_3^{++}]$.

	Run 1 (45°)		**Run 2 (35°)**		**Run 3 (25°)**	
Time (min)	$[Fe(phen)_3^{++}]$	Ln $[Fe(phen)_3^{++}]$	$[Fe(phen)_3^{++}]$	Ln $[Fe(phen)_3^{++}]$	$[Fe(phen)_3^{++}]$	Ln$[Fe(phen)_3^{++}]$
0						
5						
10						
15						
20						
25						
30						
35						
40						
45						
50						
55						
60						

2. **Plots of $Fe(phen)_3^{++}$ concentration as a function of time.** Attach graph paper (with three plots) to report sheet.
3. **Plots of natural log of $Fe(phen)_3^{++}$ concentration as a function of time.** Attach the graph paper and calculate the value of the rate constant (K) at each temperature.

	(45°)	(35°)	(25°)
K =	____________	____________	____________

4. **Arrhenius plot and activation energy determination.** Attach the Arrhenius plot and caluclate the value of the activation energy.

Ea = ____________

EXPERIMENT 23

Chemical Equilibrium

Purpose: To study several systems of chemical equilibrium and to observe how they are affected by changes in temperature, concentration, and competing equilibria.

Materials: 0.1 M $Ni(NO_3)_2$; 0.1 N $AgNO_3$; 0.033 M $FeCl_3$; 0.2 M $(NH_4)_2C_2O_4$; 0.2 M KSCN; 1 M thioacetamide (7.5 g CH_3CSNH_2 in 100 mL); 6 M NaOH; 6 M HCl; con. HCl; con. NH_4OH; solid NaCl; solid $CuSO_4 \cdot 5H_2O$; dimethylglyoxime solution (10 g of DMG per liter of ethanol); Hydrion paper, ice.

Waste Disposal: All test tube contents containing nickel or copper compounds should be disposed of in the heavy metal waste container near the reagent shelf.

INTRODUCTION

There are at least two very important reasons for our interest in chemical equilibrium. First, when substances that react spontaneously are brought into contact, the reaction that occurs brings the system nearer to equilibrium. The reactants are used up, and products appear until the system reaches equilibrium, at which point no further changes occur unless the system is disturbed by changing conditions. Thus, studying equilibrium is a key to understanding the nature of *spontaneous* change. Secondly, since the amounts of reactants and products present remain constant after equilibrium is reached, the maximum amount of product possible in a commercial process is often governed not by the "theoretical" or stoichiometric yield (the amount of product calculated from the balanced chemical equation) but by the *equilibrium* concentration of the product. In such a situation, conditions that shift the equilibrium toward the product or products result in a greater yield and hence a greater profit.

Imagine a system formed by combining two reactants, A and B. We begin with one mole of A, one of B. If the reaction is

$$A + B = C$$

can we prepare exactly one mole of C? Not necessarily—for the reaction may reach equilibrium while an appreciable quantity of A and B remain unreacted. Initially, when A and B are mixed, there is no C present. Only the *forward* reaction ($A + B \rightarrow C$) is possible. As quantities of C are formed, however, the *reverse* reaction ($C \rightarrow A + B$) becomes a possibility. Chemists believe that all reactions are reversible, at least to the extent of a few molecules or atoms undergoing the reverse reaction. In many reactions this tendency is so small as to be unmeasurable and such combinations go essentially to completion. In some cases the reverse reaction occurs to a greater extent. Let us imagine that as more C is formed the reverse rate increases. As A and B are used up, the forward rate slows. Eventually the two reactions proceed equally and we observe no further changes in the amount of A, B, and C in the system. Equilibrium has been reached, and further changes can be produced only by disturbing the balanced opposed rates of the forward and reverse reactions. A temperature change can accomplish this, since the two rates may not be equally affected. Changing the concentrations of A, B, or C can also disturb the equilibrium because the reaction rates depend on these concentrations.

Once an equilibrium state has been reached by a chemical system, it is possible to "stress" the system and establish a new state of equilibrium. On the basis of a large number of observations, Le Chatelier summarized the effect of stressing a chemical system at equilibrium. Le Chatelier's principle states: When some stress is applied to a system at equilibrium, the system will shift in such a direction to relieve the stress and establish a new state of equilibrium.

Stresses may be imposed on the system by adding or removing any substance (reactant or product) in the system. Adding a reactant causes the equilibrium to shift toward the products to use up some of the added reactant, and removing a reactant causes the equilibrium to shift toward the reactants to replace some of the reactant that has been removed. Similarly, adding a product shifts the equilibrium toward the reactants, while removing a product shifts the equilibrium away from the reactants and toward the products. Thermal energy may be looked upon as a reactant or a product. In exothermic reactions, heat is a product,

$$A + B \rightarrow C + \text{heat}$$

and in endothermic reactions, heat may be considered to be a reactant.

$$\text{heat} + A + B \rightarrow C$$

Consequently, raising the temperature of an equilibrium system moves the equilibrium in the direction that absorbs heat. For an exothermic reaction, the shift would be toward the reactants, and for an endothermic reaction, the shift would be toward the products.

The reactions we will observe in this experiment reach equilibrium rapidly—almost as soon as the reagents are mixed. We will disturb (place a "stress" upon) the resulting equilibrium mixture by changing the temperature or the concentration of one of the substances involved and observe the result. Le Chatelier's principle should be kept in mind as this experiment is undertaken. Two types of equilibria will be observed: heterogeneous, where a surface or phase boundary separates components; and homogeneous, where reactants and products are all in the same phase; for instance, all dissolved in a single solution.

PROCEDURE

A. Heterogeneous Equilibrium

The example we will observe is a saturated solution. Crystals of a solid, $Ni(OH)_2$, are in equilibrium, with dissolved ions.

1. Prepare $Ni(OH)_2$ by adding 10 drops of 6 M NaOH to 3 mL of 0.1 M $Ni(NO_3)_2$ solution in a test tube. Stir. Separate the gelatinous green $Ni(OH)_2$ from the solution by filtration. Pour about 5 mL of distilled water over the nickel hydroxide on the filter and let the water run through to wash off any excess sodium hydroxide.
2. Use a stirring rod to divide the green solid on the filter into three portions and place in three clean test tubes. Add 2 mL of distilled water to each tube, using the water to wash $Ni(OH)_2$ to the bottom. Stir each of the suspensions with a glass stirring rod for about 30 seconds. Each of these test tubes now contains a system at equilibrium: Solid $Ni(OH)_2$ and a saturated solution of Ni^{2+} and OH^- ions.
3. Pour one of the three samples through a filter paper, collecting the liquid in a clean test tube. Be careful not to get any of the solid $Ni(OH)_2$ in the solution. Test the liquid for the presence of OH^- ion by putting a drop on a piece of wide-range pH paper. Put a drop of distilled water on the paper for comparison. Does the test show that OH^- ions are present in the solution?

   Test the same solution for the presence of Ni^{2+} ions by adding 1 drop of 6 M NaOH and 10 drops of dimethylglyoxime (DMG) solution. The presence of nickel is shown by formation of a red precipitate. Since $Ni(OH)_2$ is only very slightly soluble, you will need to observe carefully to see the few specks of red solid formed. The reaction may take a few minutes. Is Ni^{2+} present in the solution?

   Write an equation for the equilibrium between solid $Ni(OH)_2$ and its ions in a saturated solution.
4. We will attempt to disturb this equilibrium concentration of the OH^- ion in the solution. Adding an acid uses up the OH^- ions by the reaction $H^+ + OH^- = H_2O$. In which direction do you predict the equilibrium will move when an acid is added? Add 8 drops of 6 M HCl to the second sample of $Ni(OH)_2$ in water, prepared in Part A2. Stir. Record the results.
5. The equilibrium can also be disturbed by changing the concentration of the Ni^{2+} ion. This can be done by using up the Ni^{2+} ions in solution to form $Ni(NH_3)_6^{2+}$, which is a soluble pale violet-blue complex ion. Predict whether the equilibrium will move to the right or left if Ni^{2+} ions are removed from solution. Test your prediction by adding 2 mL of concentrated NH_4OH (a saturated solution of NH_3 in water) to the third sample from A2. Stir. Record your observations.

B. The Common Ion Effect

1. Prepare a saturated solution of NaCl by adding 5 mL of water to 2.5 g of NaCl in a test tube. Stir until no more solid appears to dissolve. Allow the solid to settle to the bottom of the tube, then carefully pour the liquid into a clean test tube. Write

an equation representing the equilibrium between solid NaCl and its ions in the saturated solution. Predict what would happen if the concentration of Cl^- ion were greatly increased in this saturated solution. We can test the prediction by adding some concentrated HCl to the saturated solution of NaCl. Record your observations. What is the solid formed?

C. Homogeneous Equilibria

1. Measure 3 mL of 0.2 N KSCN into a beaker; add 100 mL of distilled H_2O. Note that neither K^+ nor SCN^- ion is colored in solution. Measure 6 mL of 0.1 N $FeCl_3$ solution into a second beaker; add about 100 mL of distilled H_2O. The yellow color is characteristic of iron (III) ions in solution. Combine the contents of the two beakers. The red color produced is evidence of the presence of a soluble complex ion, $(FeSCN)^{2+}$. Write an equation for the equilibrium that exists between Fe^{3+}, SCN^- (as reactants), and $Fe(SCN)^{2+}$ (as the product). The chloride ion and the potassium ion take no part in the equilibrium, although they are, of course, still present in the solution. The solution should be light-red and a pencil mark on paper should be legible through it. If necessary, add more water until it meets this description.

   Into each of eight clean, dry test tubes pour 5 mL samples of the solution. Set one sample aside as a standard for color comparisons.
2. Cool one sample in an ice bath for 5 minutes. Heat a second sample *nearly* to boiling. Compare the color intensity (representing the amount of $(FeSCN)^{2+}$) of these solutions with the standard. In which direction does the equilibrium move when the temperature rises? Which reaction, forward or reverse, uses up heat? Rewrite the equation for the equilibrium and add "heat" to the appropriate side.
3. To increase the concentration of Fe^{3+} ion, add 10 drops of 0.1 N $FeCl_3$ to one of the samples. Compare the color with the standard. Is more, or less, of the product $(FeSCN)^{2+}$ produced?
4. Add 10 drops of 0.2 N KSCN to another sample. Compare the color with the standard. Is more, or less $(FeSCN)^{2+}$ produced?

   Complete the following statement: Increasing the concentration of either of the reactants forces the equilibrium to move to the (*right, left*) and results in the production of (*more, less*) product.
5. To another sample of the equilibrium mixture add 5 mL of 6 M NaOH and stir. You will observe the formation of solid reddish $Fe(OH)_3$. What happens to the red color of $(FeSCN)^{2+}$? In which direction has the equilibrium moved? Why? Write an equation representing the reaction that has occurred.
6. To another sample of the equilibrium mixture add 10 drops of 0.1 N $AgNO_3$. The white precipitate that forms is AgSCN, a nearly insoluble solid. What happens to the red color of $(FeSCN)^{2+}$? In which direction has the equilibrium moved? Why? Write an equation representing the reaction that has occurred.
7. To the other sample of the equilibrium mixture add 2 mL of 0.2 M ammonium oxalate, $(NH_4)_2C_2O_4$, solution. Iron forms a soluble complex ion with oxalate ions, $Fe(C_2O_4)_3^{3-}$. Formation of the ion uses up the Fe^{3+} ions present and disturbs the equilibrium.

$$\begin{array}{l} Fe^{3+} + SCN^- \rightarrow (FeSCN)^{2+} \\ + \\ 3\,C_2O_4^{2-} \\ \Downarrow\Uparrow \\ Fe(C_2O_4)_3^{3-} \end{array}$$

The two equilibria, then, *compete* for the Fe^{3+} ions. From the results of your experiment, which of these two equilibria results in the smaller concentration of Fe^{3+} ions in solution?

D. Competing Equilibria

The following compounds of copper will be observed:

$CuSO_{4(s)}$	white solid	$Cu^{2+}_{(aq)}$	pale-blue
$Cu(Cl)_6^{4-}$	green	$Cu(NH_3)_4^{2+}$	bright-blue
$Cu(OH)_{2(s)}$	pale-blue solid	$CuS_{(s)}$	black solid

Dissolve a spatula full of $CuSO_{4(s)}$ in 20 mL of distilled H_2O in a 50-mL beaker. Record the color observed. Add concentrated HCl solution dropwise until the color changes. Record your observation. Add 20 mL of 6 M NaOH. Record your observations. Add 15 mL of concentrated NH_4OH. Stir. Record your observations. Add several drops of thioacetamide solution (a source of sulfide ion). Record your observations. Write an equation representing each reaction you have observed. Arrange the equations in order of decreasing concentration of Cu^{2+} ions in solution so that the reaction furnishing the *most* Cu^{2+} ions is at the top.

Experiment 23
Chemical Equilibrium

Name________________________________

Date________________________________

REVIEW QUESTIONS

1. Indicate the color of the solution when each of the following substances or chemical species is present in aqueous solution.

   a. SCN^- (thiocyanate) ion ____________________

   b. Fe^{3+} ion ____________________

   c. $FeSCN^{2+}$ (iron(III) thiocyanate) ion ____________________

   d. $Fe(C_2O_4)_3^{3-}$ complex ion ____________________

   e. $Cu(NH_3)_4^{2+}$ complex ion ____________________

2. Complete the following statement: Increasing the concentration of a reactant in an equilibrium system forces the reaction to move to the ____________________ (right/left) which results in a(n) ____________________ (increased/decreased) amount of products.

3. Given a homogeneous equilibrium reaction system represented by

$$A + B \rightleftharpoons C + D,$$

   indicate whether the following stresses would *increase* or *decrease* the concentration of product D.

   a. Substance A is added to the system. ____________________

   b. Substance C is added to the system. ____________________

   c. Substance B is removed from the system. ____________________

4. If a saturated solution of $NaNO_3$ in water were prepared, what acid would provide ions that would reduce the solubility of $NaNO_3$ thereby causing some of the salt to precipitate from solution?

5. On the basis of what you have learned in this experiment, explain why iron(III) hydroxide, $Fe(OH)_3$, is very soluble in acidic solutions but very insoluble in alkaline solutions.

Experiment 23
Chemical Equilibrium

Name____
Date____

REPORT SHEET

A. Heterogeneous Equilibrium

3. pH of solution____________________ pH of deionized water ____________________

   Results of test for Ni^{2+} ____

   Ions present in solution ____

   Equation for equilibrium between solid nickel(II) hydroxide and its saturated solution:

4. Predicted results of adding H^+ ____

   Observed results ____

5. Predicted results of adding NH_3 ____

   Observed results ____

B. The Common Ion Effect

Equation for equilibrium between solid NaCl and its saturated solution ______________________________

____

Predicted result of increasing Cl^- concentration______________________________________

Observed results____

What is the solid formed?____

C. Homogeneous Equilibria

1. Equation for equilibrium____

2. Compare the color intensity of the cool solution to that of the warm solution____________________

   ____

   Which solution contains more $FeSCN^{2+}$?____

   In which direction does the equilibrium move when the temperature rises?________________________

Which reaction uses up heat? ____________________

Write an equation for the equilibrium reaction including heat as either a reactant or a product:

3. Compare the color with the standard when additional Fe^{3+} is added ____________________

   When Fe^{3+} concentration is increased ____________________ $(FeSCN)^{2+}$ is produced.

4. Compare the color with the standard when additional SCN^{-} is added ____________________

   When SCN^{-} concentration is increased ____________________ $(FeSCN)^{2+}$ is produced.

   Increasing the concentration of either of the reactants forces the equilibrium to move to the ____________________

   and results in production of ____________________ products.

5. Results of adding NaOH ____________________

   Equilibrium moved to the ____________________ Explanation ____________________

   ____

   Equation:

6. Results of adding $AgNO_3$ ____________________

   Equilibrium moved to the ____________________ Explanation ____________________

   ____

   Equation:

7. Observation when oxalate ion is added to the equilibrium mixture ____________________

   ____

   ____

   equilibrium results in the smaller concentration of Fe^{3+} ions in solution.

D. Competing Equilibria

Copper Species Present	Reagent Added	Observation	Equation for Reaction
$CuSO_{4(s)}$	H_2O		
$Cu^{2+}_{(aq)}$	HCl		
$Cu(Cl)_6^{4-}$	NaOH		
$Cu(OH)_{2(s)}$	NH_4OH		
$Cu(NH_3)_4^{2+}$	Thioacetamide (S^{2-})		

Arrange the equations in order of decreasing concentration of Cu^{2+} ions in solution so that the reaction furnishing the *most* Cu^{2+} ions is at the top.

EXPERIMENT 24

K_{sp} for Calcium Hydroxide

Purpose: To measure the value of the solubility product constant of calcium hydroxide by titration of the base dissolved.

Materials: Solid $Ca(OH)_2$, 1M HCl, *standardized* 0.1M NaOH, Bromocresol Purple indicator.

Apparatus: 500-mL Erlenmeyer flask with rubber stopper, 50-mL buret, 25-mL pipet, pipet bulb, ring stand, funnel, filter paper, disposable plastic pipet or eyedropper, 50- or 100-mL graduated cylinder.

Safety Precautions: Avoid skin contact with the NaOH, the HCl, and the $Ca(OH)_2$. Wear safety glasses throughout this experiment.

Hazardous Waste Disposal: All the solutions may be flushed down the drain with a large amount of water. Place the leftover solid $Ca(OH)_2$ in the trash or in the waste container designated.

INTRODUCTION

You have studied a number of equilibrium constants, particularly those for weak acids and bases. In this experiment you will combine what you have learned about acid-base chemistry with a determination of K_{sp}.

The solubility product constant, K_{sp}, is one way of describing how much of a relatively insoluble compound will actually dissolve in solution. The constant is superior to just stating the solubility; e.g., 0.014 g/L, because it can take into account contributions to cation or anion concentrations from other sources. The K_{sp} is defined for a compound $M_n Y_x$ for which the equilibrium is

$$M_nY_{x(s)} \rightleftharpoons nM^{+x}_{(aq)} + xY^{-n}_{(aq)}$$

$$K_{sp} = \left[M^{+x}\right]^n \left[Y^{-n}\right]^x$$

Note that in keeping with the rules of equilibrium expressions, the solid compound does not appear.

Often a compound will be so insoluble that the value of K_{sp} will be 1×10^{-16} or less. This is true of many sulfides. Chemists use this fact in qualitative analysis to test for certain metal ions, and particularly, in the qualitative analysis scheme for cations.

Hydroxides are often moderately insoluble. You know that sodium hydroxide (NaOH) and potassium hydroxide (KOH) are very soluble and are useful as standard bases. This high solubility is a function of the Na^+ and K^+ cations. Going from the Group IA to Group IIA cations decreases the solubility of the hydroxides considerably. The K_{sp} for $Mg(OH)_2$ is 6×10^{-12}. A *suspension* (dissolved compound plus lots of undissolved solid floating in solution) of $Mg(OH)_2$ is familiar as the medication "Milk of Magnesia." The suspended particles constitute the milky appearance.

Calcium hydroxide is somewhat more soluble than magnesium hydroxide. The amount of hydroxide that will go into solution is just sufficient to be analyzed by acid-base titration. The results of the titration will allow you to calculate the K_{sp}. The equilibrium and solubility product constant expressions are

$$Ca(OH)_{2(s)} \rightleftharpoons Ca^{+2} + 2\ OH^-_{(aq)}$$

$$K_{sp} = [Ca^{+2}][OH^-]^2$$

PROCEDURE

A. Preparation of the Saturated $Ca(OH)_2$ Solution

1. Take 1 to 2 grams of solid calcium hydroxide and transfer to a large Erlenmeyer flask. Fill the flask about half-full with distilled water.
2. Close with a cork or stopper and shake the solution for about 5 minutes. You will tire easily doing this, but persist. It is essential to achieve a saturated solution, that is, one where equilibrium is established.
3. Set the flask aside. Keep it covered. Allow the solution to settle. You will titrate the solution later.

B. Preparation and Standardization of HCl Solution

1. With a graduated cylinder, measure out 20 mL of the stock HCl solution (about 1 M), transfer to a beaker, and dilute to approximately 200 mL (the exact volume is not important).
2. Rinse and fill a buret with your HCl solution.
3. Pipet 25 mL of the standardized NaOH solution provided into a small Erlenmeyer flask. Record the NaOH concentration on your report sheet.
4. Add 30–40 mL of distilled water, 2–3 drops of Bromocresol Purple indicator, and titrate with the HCl. The color change is purple (base) to yellow (acid). Record the initial and final buret readings to the nearest 0.01 mL.

   Rinse the solution down the drain and repeat the HCl standardization.

C. Titration of $Ca(OH)_2$ Solution

1. The $Ca(OH)_2$ solution should have settled out by now. Take care not to stir up any of the solid from the bottom. Set up a funnel with *two* pieces of filter paper and a small flask to catch the filtrate.

   Use a disposable plastic pipet or eyedropper to draw solution off the top of the saturated solution and transfer to the funnel for filtration. Do not allow the level of the solution to rise in the funnel above the edge of the paper. Collect 50 to 100 mL of filtrate and cover it.

   If the solution is still quite cloudy, repeat the filtration.
2. Now pipet 25 mL of the filtrate into a clean, small Erlenmeyer flask, add 30–40 mL of distilled water, 2–3 drops of Bromocresol Purple, and titrate with the standardized HCl. Record the initial and final buret readings to the nearest 0.01 mL.

   Repeat the titration with a second 25-mL aliquot. If time permits and there is poor agreement on the duplicate trials, perform a third trial of the standardization or the filtrate titration.

   All solutions may be discarded down the sink. Put the undissolved $Ca(OH)_2$ in the trash or designated waste container.

CALCULATIONS

1. Subtract the initial volume from the final volume to get the total volume HCl used.
2. Calculate the concentration of HCl using the equation $M_a V_a = M_b V_b$ where M and V refer to molarity and volume of acid (a) and base (b).
3. The calculations for the OH^- concentration are performed in exactly the same way.

4. K_{sp} – You know the value of $[OH^-]$ as just calculated. But what is $[Ca^{+2}]$? From the solution stoichiometry it can be seen that *one* mole of calcium hydroxide produces *one* mole of calcium ion and *two* moles of hydroxide ion. Therefore, $[Ca^{+2}]$ is exactly one-half $[OH^-]$.

$$[Ca^{+2}] = \frac{1}{2}[OH^-]$$

and K_{sp} then

$$K_{sp} = [Ca^{+2}][OH^-]^2$$

Compare your answer to a literature value. Cite your source.

Experiment 24

K_{sp} for Calcium Hydroxide

Name____________________________________

Date____________________________________

REVIEW QUESTIONS

1. Convert your K_{sp} answer to solubility of calcium hydroxide in grams per liter.

2. What is the pH at the equivalence point for each titration? Look up the end point for Bromocresol Purple indicator and explain why this works for this titration.

3. Calculate K_{sp} for $MnCO_3$ if 1.07 mg/L are required to make a saturated solution.

4. Why is it essential that the filtrate be as clear as possible?

5. Seawater contains 0.056 M Mg^{+2} and 0.010 M Ca^{+2}. If OH^- is slowly added to seawater, what will precipitate first? At what concentration of OH^-?

6. Why would it probably be impractical to calculate the K_{sp} of magnesium hydroxide using this method?

7. If the NaOH solution provided to you was not standardized, how would you go about standardizing it?

8. If the solution was poorly filtered and appeared cloudy when titrated, would the K_{sp} calculated from the data be higher or lower than the true value?

Experiment 24
K_{sp} for Calcium Hydroxide

Name______________________________

Date______________________________

REPORT SHEET

B. Preparation and Standardization of HCl Solution

Concentration of standard NaOH ______________________ M

Aliquot of standard NaOH used ______________________ mL

	Trial 1	Trial 2	Trial 3
Final Volume:	mL	mL	mL
Initial Volume:	mL	mL	mL
Volume HCl	mL	mL	mL
Molarity of HCl	______ M	______ M	______ M

Average concentration of HCl ______________________ M

C. Titration of $Ca(OH)_2$ Solution

Solution Titration – aliquot size ______________________ mL

Final Volume:	mL	mL	mL
Initial Volume:	mL	mL	mL
Volume HCl	mL	mL	mL

Average concentration of OH^- ______________________ M

K_{sp} calculation (show) : (compare answer to literature value)

EXPERIMENT 25

The Gas Laws

Purpose: To study the effect of pressure changes and temperature changes on the volume of a gas; to estimate the position of absolute zero on the Celsius temperature scale.

Materials: Apparatus shown in Figures 25.1, 25.2, and 25.3.

Waste Disposal: None

INTRODUCTION

Matter is usually defined as anything that has mass and occupies space. Different schemes may be used to classify matter into groups of substances with similar characteristics. One useful classification scheme is based on the physical state of matter. Using this scheme, we divide "matter" into subgroups of gases, liquids, and solids and then study the properties of each subgroup. Descriptions of and discussions about the states of matter can be found in your textbook.

Examine your textbook carefully and compare information about gases with that for liquids and solids. You should notice that the information about gases is very quantitative (e.g., gas laws), while information about liquids and solids is more descriptive and qualitative. The nature of the gas phase (constant, random motion of atoms or molecules) makes it easier to study and characterize gases than liquids or solids. At this time we know more about the behavior of gases than we do about materials in the condensed phases (liquids and solids). The information we have about the physical behavior of gases pertains to samples of pure substances in the gas phase as well as to mixtures of nonreacting gases.

A given sample of gas can be described by four parameters: the temperature, the pressure, the volume, and the mass of gas in the sample. If any three of these variables are held constant, the fourth can have only one value. This activity has been designed to allow investigation of the relationship among pressure, temperature, and volume of a gas. The amount of gas is controlled (held constant) by the design of the apparatus used. This experiment allows the investigation of the relationship among the pressure and volume of a gas at a constant temperature, and between the temperature and the volume of gas at a constant pressure.

A sample of gas (air in this case) was trapped at room temperature and pressure in the apparatus when the apparatus was constructed. The pressure may be changed by raising or lowering the mercury well (Unit A); the new volume is observed and determined from the measurements of the cylinder in which it is enclosed. The volume of a cylindrical container such as a glass tube can be calculated by the formula $V = \pi r^2 L$, where r is the radius of the tube and L is the length. Data obtained in this manner will be used to investigate Boyle's law.

Boyle's law deals with the effect of pressure changes on the volume of a given mass of gas at constant temperature. Since Boyle's law states that the volume of a confined gas sample is inversely proportional to the pressure exerted on the gas, it can be mathematically stated as:

$$\frac{V_1}{V_2} = \frac{P_2}{P_1} \text{ or } V = C(1/P) \text{ where C is a constant} \qquad \text{Equation 1}$$

Equations of the general form y = C(1/x) produce a hyperbolic curve when plotted. Volume versus pressure data obtained in this experiment will be used to plot such a curve.

The pressure on the confined gas sample may be held constant at atmospheric pressure by aligning the mercury level in the reservoir (Unit A) with the mercury level in the confined gas tube (Unit B). The temperature of the gas may then be changed over the 0°C to 100°C range by heating or cooling with water of the appropriate temperature. If the volume of the gas sample is determined as described, volume versus temperature data, at constant pressure, may be used to investigate Charles' law.

Charles' law deals with the influence of temperature on the volume of a given amount of confined gas at constant pressure. Since Charles' law states that the volume of a confined gas sample is directly proportional to the absolute temperature of the gas, it can be mathematically stated as

$$\frac{V_1}{V_2} = \frac{T_1}{T_2} \text{ or } V = C(T) \text{ where C is a constant} \qquad \text{Equation 2}$$

If the Celsius rather than the absolute temperature is used, Charles' law becomes

$$V = C(°C - A) \qquad \text{Equation 3}$$

where °C is the Celsius temperature and A represents the position of absolute zero. A plot of volume as a function of temperature in degrees Celsius will yield a straight line, which, when extrapolated to zero volume, will allow the estimation of A since the equation simplifies to A = °C when V = 0.

PROCEDURE

The apparatus should be sitting in some sort of pan to catch the mercury in case any connections come apart. IF AN ACCIDENT OCCURS AND MERCURY SPILLS, TELL YOUR INSTRUCTOR **IMMEDIATELY.** Become familiar with this equipment by first examining how it is constructed. Then examine the left arm of the apparatus, which is the mercury reservoir. This arm has an opening to the atmosphere and can be moved by loosening the thumbscrew. Move the reservoir up or down and refasten it. A movable marker is used to mark the mercury level in the reservoir. Move the marker so that the edge with the extending point is even with the top of the mercury.

Examine the right arm of the apparatus, which is also movable. A small sealed glass tube is inside the larger tube. This small tube is connected by flexible tubing to the mercury reservoir. The gas above the mercury in the small tube is the confined gas. You will change the pressure and the temperature (independent variables) of that confined gas and observe what happens to its volume (dependent variable). This arm has two movable markers. Set the top marker so that the edge with the extending point is even with the top of the confined gas column. (This marker will not be moved again throughout the experiment.) Position the bottom marker so that the edge with the extending point is even with the top of the mercury level in the small inner tube.

The apparatus allows you to manipulate the pressure of the confined gas. The pressure on the gas is equal to the pressure of the gas. Imagine a movable piston arrangement. If you push down on the piston, you increase the pressure on the gas. Simultaneously, as the volume of the gas decreases, its pressure increases. Before you make measurements with this apparatus it is important that you understand which movements increase the pressure on the gas and which decrease the pressure. Position both arms at the top of the apparatus, then adjust slightly so that the mercury levels in both arms are "even with each other" (same distance from the table top). With this arrangement (Figure 25.1) you can infer that the pressure of the confined gas is equal to the atmospheric pressure; that is, equal "push" is applied to both mercury surfaces. Lower the mercury reservoir (arm A) with respect to the confined gas column (arm B) and observe what happens to the length of the column of confined gas. This movement decreases the pressure on the confined gas (Figure 25.2). Place the mercury reservoir back up at the top of the apparatus. Now lower the confined gas column (arm B) relative to the mercury reservoir (arm A) and observe what happens to the column of confined gas. This operation increases the pressure on the confined gas (Figure 25.3).

Position both arms at the bottom of the apparatus so that the mercury levels in the arms are even with each other. Raise the mercury reservoir (arm A) with respect to the confined gas column (arm B) and observe what happens to the length of the column of the confined gas. Place the mercury reservoir (arm A) back at the bottom of the apparatus and raise the column of confined gas (arm B) with respect to the mercury reservoir, and observe what happens to the length of the confined gas column.

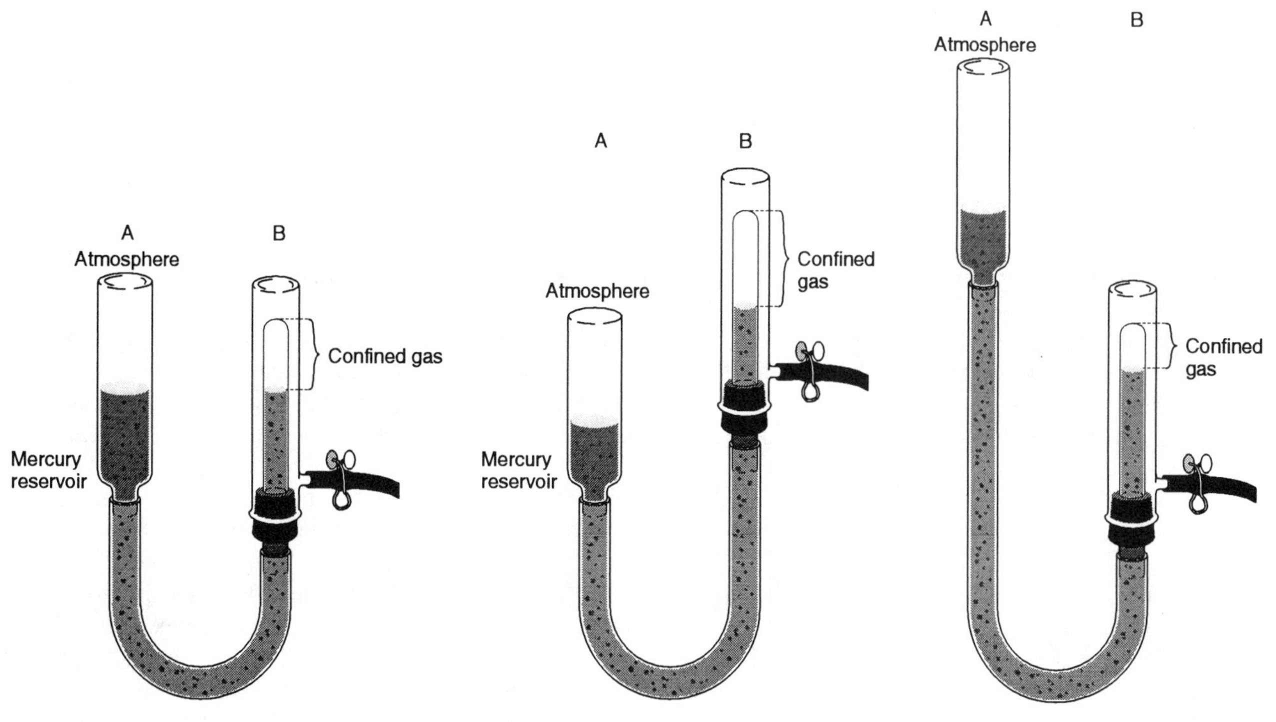

FIGURE 25.1
Pressure equal to atmospheric pressure.

FIGURE 25.2
Pressure less than atmospheric pressure.

FIGURE 25.3
Pressure greater than atmospheric pressure.

A. Effect of Pressure on the Volume of a Confined Gas at a Constant Temperature

1. Record the temperature of the room and the atmospheric pressure for the day on the report sheet. Set both arms of the gas law apparatus as low as possible. *Adjust the arms slightly so that the mercury level in the reservoir (Unit A) is level with the mercury in the small tube (Unit B),* making the distance between the mercury levels zero. Use the markers to locate the mercury levels. At this point, the pressure of the confined gas is equal to atmospheric pressure (Fig. 25.1). Record this pressure to accompany the zero distance between mercury levels. Measure the length (in centimeters) of the confined gas column in arm B and record it on the report sheet to go with the zero distance between mercury levels. You now have information to calculate one pressure-volume data point for the confined gas.

   Raise the mercury reservoir about 8 to 9 cm. The distance is not critical, since the objective is to raise this arm relative to arm B. Move the markers so that the edges with the extending points are even with the mercury levels in each arm. To obtain information for a new data point, carefully measure and record the distance (in centimeters) between the two mercury levels by measuring the distance between the points of the two mercury-level markers. (You should be able to put your ruler next to these markers to get accurate measurement, rather than trying to "eyeball" the mercury levels with the ruler.) To determine the pressure on the confined gas (also equal to the pressure of the confined gas), *add* the distance between the two mercury levels to the atmospheric pressure. Then measure and record the length of the confined gas column.

   To obtain information for more data points, repeat this procedure several times by *raising the mercury reservoir relative to the confined gas column.* The last measurement is made when arm A is raised to the upper limit with respect to arm B.
2. Set both arms as high as possible on the apparatus. *Adjust the arms slightly so that the mercury level in arm A is even with the mercury level in arm B* (the distance between mercury levels is zero). Record the pressure of the confined gas. Measure and record the length of the confined gas column.

   Lower the mercury reservoir with respect to the confined gas column about 8 to 9 cm. Again, the actual distance the arm is lowered is not critical, since the objective is to change the relative positions of arm A and arm B. Move the markers so that the edges with the extending points are even with the mercury levels in each arm. To obtain information for a data point, carefully measure and record the distance between the two mercury levels by measuring the distance between the points of the two mercury-level markers. To determine the pressure of the confined gas, *subtract* the distance between the mercury levels from the atmospheric pressure. Then measure and record the length of the confined gas column.

To obtain information for more data points, repeat this procedure several times by *lowering the mercury reservoir with respect to the confined gas column.* The last measurement can be made when arm A is at the lower limit of the apparatus with respect to arm B.

At this point in the experiment, you may wish to calculate the volume of the confined gas corresponding to each pressure measurement. See the calculations section for directions.

B. Effect of Temperature on the Volume of a Confined Gas at a Constant Pressure

1. Fill the water jacket (Unit B) with tap water at room temperature. *Adjust Unit A until the mercury level in both units is the same,* read and record both the confined-gas column length and the water temperature.
2. Drain the water jacket and fill it with ice water. *After 5 minutes equalize the mercury levels* and record both the confined-gas column length and the water temperature. Stir the water thoroughly with the thermometer before reading the temperature.
3. Drain out the ice water and fill the jacket again with tap water. Make a steam generator from a 500-mL Florence flask (Fig. 25.4) and use it to heat the water in the jacket with steam. Take out the steam line from time to time and, after stirring the water, measure the temperature. When it has reached about 40°C, wait a few minutes to allow the temperatures in the system to become constant *and then equalize the mercury levels.* Record both the confined-gas column length and the temperature.
4. Again introduce steam to heat the water in the jacket to about 60°C. *After a few minutes equalize the mercury levels* and record both the confined-gas column length and the temperature, stirring the water thoroughly before reading the temperature.
5. Repeat this procedure, this time making measurements at about 90°C.

Proceed to the calculations and plotting of data for Parts A and B of the experiment. Do not put the apparatus away until you are sure that the data are sufficiently good that measurements will not have to be repeated.

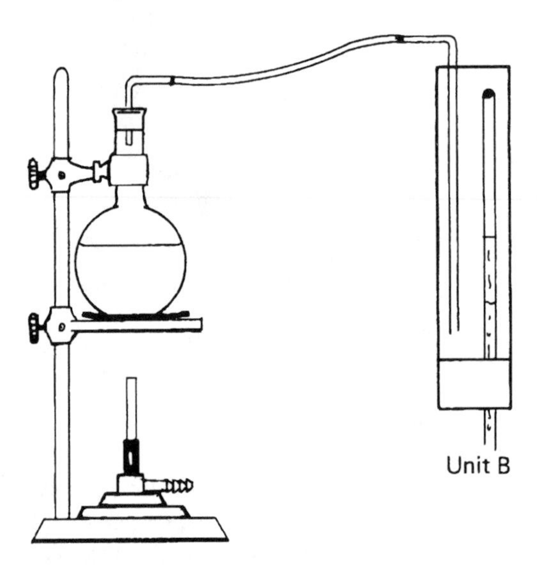

FIGURE 25.4
Apparatus for heating the water in Unit B.

CALCULATIONS

A. Effect of Pressure on the Volume of a Confined Gas at a Constant Temperature

Determine the volume of the confined gas column for each data point using the mathematical statement for the volume of a cylinder.

$$V_{cyl} = \pi r^2 h$$

where $\pi = 3.14$

$r = 0.200$ cm (cylinder radius)

$h =$ length of confined gas

Keep three digits in the volume values. Depending on the measurements made for the gas column length, you may violate significant figure rules. In this case you may need to do so to detect volume differences. Record these experimentally determined volumes in the table on the report sheet.

Using atmospheric pressure as P_1 and the volume experimentally determined at atmospheric pressure as V_1, use Boyle's law (Equation 1) to calculate the volume that the confined gas should ideally occupy at the highest pressure that you measured. Determine the percent error between the gas-law calculated volume and the experimentally determined volume.

For each of the pressure-volume measurements, calculate and record in the last column of the pressure-volume table, the inverse of the pressure (1/P).

Prepare a graph that illustrates the relationship between pressure and volume of a confined gas at constant temperature. Put pressure on the x-axis and volume on the y-axis. You do not need to start each axis at zero, but the scale must include the range of your pressure and volume numbers. The scale for each axis should be expanded so that the graph of your data points occupies most of the page. Label each axis with the name of the variable plotted and the units of measure for that variable. For example, "Pressure (cm Hg)." Plot each of the pressure-volume data points and circle each point. Draw a smooth curve through the data points. If you have followed all the correct graphing techniques and your data points do not lie on a continuous smooth curve, this suggests either measurement errors or calculation errors (usually in calculation of gas pressure). If this occurs, first double-check all calculated values. If you find no mistake, you may want to remeasure some data points.

Mathematical equations to describe curves are more complex than equations for straight lines, which are simple statements of proportions. Therefore, when a graph of original data is not a straight line, transformations of original measurements are sometimes made and these transformations are plotted in an attempt to find a straight-line relationship.

Prepare a second graph to illustrate the relationship between volume and the inverse of the pressure (1/P). Put volume on the y-axis and 1/P on the x-axis. The same specifications for each scale described previously apply to this graph. Plot each of the V-1/P data points and circle each point. Draw a straight line through these points to illustrate the proportional relationship between volume and inverse of the pressure of a gas at constant temperature.

B. Effect of Temperature on the Volume of a Confined Gas at a Constant Pressure

As with the pressure-volume data, determine the volume of the confined gas column for each measured temperature using the mathematical statement for the volume of a cylinder ($V_{cyl} = \pi r^2 L$). Using room temperature as T_1 and the volume experimentally determined at room temperature as V_1, use Charles' law (Equation 2) to calculate the volume that the confined gas should have occupied at the highest temperature that you measured. Determine the percent error between the gas-law calculated volume and the experimentally determined volume.

Prepare a graph that illustrates the relationship between temperature and volume of a confined gas. Put temperature on the x-axis and volume on the y-axis. Again, establish each scale to include the range of measurements for the variable. The scale for each axis should be expanded so that the graph of the data points occupies most of the page. Label each axis with the appropriate variable and unit of measure. Plot each of the temperature-volume data points and circle each point. Draw a straight line through the data points to illustrate the proportional relationship between temperature and volume of a gas at constant pressure. If your data points are somewhat scattered, you have made measurement errors, and you may want to go back and remeasure some data points.

Prepare another graph from which you will obtain a numerical value in degrees Celsius for absolute zero. The axes for this graph must be set up in a certain way. Use the 11-in. length of the paper for the x-axis and put temperature (°C) on this axis. Prepare a temperature scale to range from -400°C to +100°C. Use the 8-1/2-in. length of the paper for the y-axis and plot volume on this axis. Prepare a scale to range from 0 cm^3 to your largest calculated value. Plot each temperature-volume data point and circle each point. Draw a straight line through the data points. Extrapolate this line (mark with a dashed line) to the x-axis. The point at which the extrapolated line crosses the x-axis is called absolute zero. Record this value on the report sheet. Calculate the percent error between this experimental value and the accepted value, if the accepted value for absolute zero is -273°C.

Experiment 25
The Gas Laws

Name____

Date____

REVIEW QUESTIONS

1. The confined gas trapped in the apparatus can be compressed by raising the mercury reservoir (Unit A). If enough pressure had been used, could the volume have been reduced to zero? Explain.

2. If the confined gas had been cooled to absolute zero, would the volume of the confined gas have become zero? Explain.

3. If a droplet of water was accidentally trapped in the confined-gas region of the apparatus, how would it affect the volume versus temperature measurements? (Hint: Remember that water has significant vapor pressure that varies as the temperature changes.)

4. Are pressure and volume of a confined sample of gas (T = constant) directly or inversely proportional to each other?

5. Are temperature and volume of a confined sample of gas (P = constant) directly or inversely proportional to each other?

6. If the volume of a sample of confined gas is increased at constant temperature, what will happen to the pressure of the confined gas?

7. If a sealed glass bottle containing air is placed in a refrigerator, what will be observed if the bottle is removed from the refrigerator and the seal is immediately removed?

8. If the temperature of a sample of confined gas is increased but the pressure is decreased, what will happen to the volume of the confined gas?

9. Why do air bubbles from a scuba diver get larger as the bubbles get closer to the surface of the water?

10. Explain why a cake will "fall" if it is removed from the oven before the batter has had a chance to "set up."

Experiment 25
The Gas Laws

Name______________________________

Date______________________________

REPORT SHEET

A. Effect of Pressure on the Volume of a Confined Gas at a Constant Temperature

How can you cause the pressure on the confined gas to be above atmospheric pressure with this apparatus?

How can you cause the pressure on the confined gas to be below atmospheric pressure with this apparatus?

Constant (room) temperature:____________ °C   Atmospheric pressure:____________ cm Hg

Distance between mercury levels (cm Hg)	Pressure on confined gas (cm Hg)	Length of confined gas column (cm)	Volume of confined gas (cm^3)	1/P (1/cm Hg)

The distance between the mercury levels is used to calculate the pressure on the confined gas sample, but it is not equal to the pressure on the confined gas. Explain why the distance between the mercury levels does not directly measure the pressure on the gas.

Using atmospheric pressure as P_1 and the volume measured at atmospheric pressure as V_1, use Boyle's law (Equation 1) to calculate the volume that the confined gas should have occupied at the highest pressure that you measured. Determine the percent error between the calculated volume and the measured volume.

B. Effect of Temperature on the Volume of a Confined Gas at a Constant Pressure

Atmospheric (constant) pressure: ____________ cm Hg

Temperature of confined gas (°C)	Length of confined gas column (cm)	Volume of confined gas (cm^3)

Why are the mercury levels in unit A and unit B adjusted to the same height each time a set of data is obtained?

Using room temperature as T_1 and the volume measured at atmospheric pressure as V_1, use Charles' law (Equation 2) to calculate the volume that the confined gas should have occupied at the highest temperature that you measured. Determine the percent error between the calculated volume and the measured volume.

Using the appropriate graph, determine the temperature at which the volume of the gas would be zero (the x-intercept).

Experimental absolute zero: ____________ °C

Calculate the percent error between this experimental value and the accepted value.

Attach both plots to the report sheet.

EXPERIMENT 26

Charles' Law

Purpose: To test the validity of Charles' Law in the relationship between volume and temperature for a given quantity of gas.

Materials: Mineral oil.

Apparatus: Small ruler, thermometer, 150-mm test tube, 100-mm capillary tube, ring stand, clamp, split one-hole rubber stopper for holding thermometer, Bunsen burner.

Safety Precautions: You will be heating oil. It may reach temperatures as high as 200° C. Remember that this is twice as hot as boiling water! Be careful with the hot oil and avoid burns! You will be using a mercury thermometer. Be very careful in handling it. It is expensive and the mercury is difficult to clean up and toxic if spilled! As always, safety glasses are required.

Hazardous Waste Disposal: Return the mineral oil to the appropriate container. Wash the test tube and the thermometer carefully to remove all traces of the oil. Discard the used capillary tube in the container for broken glass (or other receptacle) as instructed.

INTRODUCTION

The first quantitative measurements of physical phenomena involved accurate weighings. Beyond that there was not much early chemists could do with solids. Pure liquids are few in number, so chemists turned to gases for their investigations.

Before much progress could be possible, two inventions were necessary: the *thermometer* and the *barometer.* The thermometer is so familiar to us that it is difficult to imagine a time when people did not routinely think in terms of *temperature*—there was hot and there was cold, but fine distinctions between the two were subject to personal interpretation. The first practical thermometer was developed by Gabriel Fahrenheit (1686–1736), who used liquid mercury in a sealed evacuated tube to quantify changes in temperature. Some modern thermometers use liquids like alcohols, colored with red dyes, in the place of more expensive and toxic mercury, but the basic mercury thermometer is still widely used.

Curiously, the first barometers also used mercury. It is a liquid and, therefore, flows evenly. But more importantly, it is 13.5 times more dense than water. Evangelista Torricelli (1608–1647) constructed the mercury barometer by inverting an evacuated tube over a pool of liquid mercury. He observed that the height of the column of mercury that rose up in the tube varied from one day to the next. We now understand that he was measuring atmospheric pressure—the weight of all the layers of air pressing down on the surface of the earth. Physical scientists quickly learned that the barometer could be used to measure pressures of gases in closed containers.

The third parameter for investigating the physical behavior of gases is volume. This presented no difficulty since volume is a simple geometric concept and all educated people were well-versed in Euclidean geometry.

Thus the stage was set for the discovery of the Gas Laws. In summary, they are:

$P \times V = \text{constant}$	(Robert Boyle)
$V/T = \text{constant}$	(Jacques Charles)
$PV = nRT$	(Ideal Gas Law)

There are, of course, others as well. The first two rely on all other variables remaining constant. The Ideal Gas Law incorporates all possible variables and relates them to the universal gas constant R. (You may be surprised to learn how often R turns up in other important equations, many of which don't appear to have anything to do with gases.)

In this experiment, you will attempt to verify Charles' Law by making several volume measurements of a trapped gas at various temperatures. If V/T is a constant, then it follows that as V gets bigger, T must get bigger, and conversely, that V decreases as T decreases. A plot (graph) of V vs. T should give a straight line. If you make careful measurements, your resulting plot should give a line straight enough to confirm Charles' Law.

PROCEDURE

A. Heat the Oil

You will need to bring a small ruler with you to lab. At least one side of the ruler must be marked in millimeters.

Obtain a thermometer, a 150-mm test tube, a 100-mm capillary tube, and enough mineral oil to fill the test tube about two-thirds full. (Insert the thermometer to the bottom. There should not be so much oil that it overflows.)

Clamp the test tube in place about 1 inch above the top of a Bunsen burner and begin heating the oil. Do this carefully! Small amounts of water in the oil may tend to pop and spatter. Use a stirring rod to move the oil around, especially from top to bottom. Notice the swirling patterns in the oil. This happens because layers of different temperatures have different densities.

The most important and most difficult part of this experiment is ensuring an even temperature of the oil. Since you are heating the bottom, the lower parts of the oil will tend to be hotter than the upper levels. Stir the hot oil with long, gentle up and down movements.

B. Insert the Capillary Tube

Check the temperature of the oil occasionally. When it reaches 180°–220° C, discontinue heating. Stir thoroughly for a few seconds and drop the capillary tube, *open end down*, into the oil. Insert the thermometer. You may wish to clamp it in place, using a split, one-hole rubber stopper. If it does not seem dangerously top-heavy, just allow it to rest on the bottom of the test tube.

Watch the capillary tube closely. A pocket of air is trapped inside the tube and as the oil (and air) cools, the volume of the air will decrease and oil will be sucked back into the tube. Watch for the first movement of oil into the capillary tube and then make your first set of measurements. For each data point you will record temperature (to the nearest whole degree Celsius) and the height of the *column of air* (not the height of the oil) in the capillary (to the nearest millimeter).

It is essential to keep mixing the oil gently throughout the experiment, but do not agitate the capillary tube. If it is broken, the experiment must be restarted. If your stirring rod will not fit into the test tube alongside the thermometer, it is acceptable to stir with the thermometer. Just remember that the thermometer is a delicate and expensive instrument.

Take subsequent readings as the temperature falls. (Perhaps at 1 minute, then 2 minutes.) Since the air has a very different density from the oil, it should be easy to distinguish. Hold your ruler next to the test tube and estimate the length of the air column to the nearest millimeter.

Make 10 to 15 total readings as the temperature approaches room temperature.

For one final reading, support an ice water bath around the test tube and stir. Allow 4 to 5 minutes (with gentle stirring of the oil) to reach a thermal equilibrium. Do not assume that the oil will easily reach 0° C. A temperature of 5°–10° C is okay, but just be certain the oil is the same temperature throughout. Record this data pair. (Note that it is necessary to remove the ice bath and wipe away the condensation to complete this measurement.)

Return the oil to the container. Avoid getting any water in it. Wash the test tube with soap and water. Return the *cleaned* thermometer.

CALCULATIONS

Make a graph of temperature (°C) vs. the height of column of air (mm). (We are making the assumption that the height of the column is directly proportional to the volume.) Be sure the plot is *space-filling*. Check the remarks in the appendices of this laboratory manual for instructions on procedures for correct graphing technique. Draw the best-fitting straight line through the points. If you know how, or you have a calculator that will do it for you, calculate the *correlation coefficient* (optional). Write this on the graph. Enter on your report sheet your comments and conclusions from the graph; i.e., you set out to prove Charles' Law, how well have you done that?

Now make a second graph. First convert all of your temperatures to degrees Kelvin. Make a plot of temperature (K) vs. height of the air column (mm), but this time start both axes at zero. On this graph all of the points will be bunched up in the upper-right-hand corner. Now draw in the best-fitting straight line. This is not easy to do unless you realize in advance what this is intended to demonstrate. Specifically, you are looking for the *y*- intercept; that is, what is the temperature when volume reaches zero? (Again, a calculator that does x,y data analysis can calculate the intercept for you.) Comment on this plot also on your report form. Turn in the two graphs with your report sheet.

Experiment 26
Charles' Law

Name____

Date____

REVIEW QUESTIONS

1. What is the volume of air represented by your first data (≈ 100 mm)? You need to know that the interior *diameter* of the capillary tube is 1.1 mm. (If you do not know the formula for the *volume of a cylinder* you will have to look it up.) Express your answer in mm^3 (three significant figures).

2. Convert your answer from question 1 to units of liters. (Hint: you might go through cubic centimeters.)

3. Using the volume from question 2 and assuming 1.00 atm of pressure, how many moles of air molecules were inside the capillary tube? If "air" has an average molecular weight of 28.88 g/mol, how much would the trapped air weigh?

4. The experiment assumes that the height of the column of air is directly proportional to the volume of air inside the capillary tube. Explain why this is a valid assumption.

5. Jacques Charles (1746–1823) was one of those adventurous Frenchmen who constructed and ascended in balloons. In August, 1783, Charles decided to employ hydrogen instead of hot air. He used 1,000 lbs. of iron in excess acid to produce the gas via

$$2\ Fe + 6\ H^+ \longrightarrow 2\ Fe^{+3} + 3\ H_2$$

Assuming 1.00 atm pressure and 30.0° C (warm summer afternoon), how big (what volume in L) was his balloon?

Experiment 26
Charles' Law

Name_______________

Date_______________

REPORT SHEET

Temp (°C)	Ht (mm)	Temp (°C)	Ht (mm)
________	________	________	________
________	________	________	________
________	________	________	________
________	________	________	________
________	________	________	________
________	________	________	________
________	________	________	________
________	________	________	________

Comments/conclusions on Graph I:

Comments/conclusions on Graph II:

EXPERIMENT 27

Percent By Volume of Alcohol in Wine

Purpose: To determine the alcohol content of wine or beer by distillation and measurement of the density of the distillate; to provide a practical application of laboratory skills and measurement techniques introduced in earlier experiments.

Materials: Apparatus illustrated in Figure 27.1, including distilling flask with side arm, condenser, and thermometer. Wine or beer of known alcohol content and an unknown prepared by either diluting wine or adding additional ethanol to produce a concentration between 2% and 20% alcohol.

Waste Disposal: No hazardous materials are involved.

INTRODUCTION

Quantitative analysis is a process of analytical chemistry used extensively in many areas besides chemistry laboratories. It is a process used by industries in quality control and research and development activities. It is used in medicine to measure various components of the body. The main goal of quantitative analysis is to determine the proportion of constituents in a given substance. In contrast, the goal of qualitative analysis is to determine the identity of the constituents of a given substance. Very often these processes overlap in actual analytical work.

The analysis of wine in this activity is quantitative in that you will determine the percent by volume of alcohol in wine. Wine is a mixture of water, alcohol (also known as ethyl alcohol or ethanol), sugar, and flavor-producing ingredients. If wine were a simple mixture of water and ethanol, it would be easy to determine the percent of alcohol in the mixture simply by measuring the density of the alcohol-water mixture. A standard table such as Table 27.1 would help determine the percent composition of alcohol corresponding to the measured density. However, the presence of sugar and other ingredients causes the density of wine to be considerably different from that of a simple alcohol-water mixture.

The analysis of wine incorporates the methods of distillation to separate the components of a mixture and of density measurements to determine the percentage of one of the components in the mixture. Distillation will separate the alcohol (along with some water) from the nonvolatile ingredients in wine. This alcohol-water mixture is called the distillate. The density of the distillate will be used to determine the percent by volume of alcohol in the original wine sample.

You will work with two concentration units for solutions as you complete this activity. One will be percent by weight of alcohol in the distillate, an alcohol-water mixture. A value of 52% alcohol by weight in the distillate is equivalent to the ratio 52 g alcohol/100 g distillate. The other unit is percent by volume of alcohol in wine and is based on the ratio of alcohol volume to wine volume. For example, if a 30-mL sample of wine contains 1.2 mL of alcohol, the percent by volume of alcohol in the wine can be determined.

$$\frac{1.2 \text{ ml alcohol}}{30 \text{ mL wine}} \times 100 = 4.0\% \text{ by volume alcohol in wine}$$

This 4.0% by volume is then equivalent to the ratio 4.0 mL alcohol/100 mL wine.

TABLE 27.1 Weight percent and density of ethanol-water mixtures

Percent by Weight of Ethanol in Mixture	Density of Mixture (g/mL)	Percent by Weight of Ethanol in Mixture	Density of Mixture (g/mL)
0.50	0.997	24.0	0.963
1.00	0.996	28.0	0.957
2.00	0.995	32.0	0.950
3.00	0.993	36.0	0.943
4.00	0.991	40.0	0.935
5.00	0.989	44.0	0.927
6.00	0.988	48.0	0.918
7.00	0.986	52.0	0.910
8.00	0.985	56.0	0.900
9.00	0.983	60.0	0.891
10.0	0.982	64.0	0.882
12.0	0.979	68.0	0.872
14.0	0.977	72.0	0.863
16.0	0.974	76.0	0.853
18.0	0.971	80.0	0.844
20.0	0.969	100.0	0.789

PROCEDURE

Set up a simple distillation apparatus as illustrated in Figure 27.1. Use a tared (empty weight previously determined to the nearest 0.001 g) 150-mL beaker as the collection vessel. Record the empty weight of this beaker.

Obtain exactly 50.0 mL of wine in a 100-mL graduated cylinder and record this volume on the report sheet. Put the wine sample into the distillation flask (sidearm flask), making sure that no wine runs into the side arm. (If some wine is lost this way, rinse and dry all the equipment and start over.) Make sure that all the connections are tight and that water is running through the outside jacket of the condenser properly before you heat the wine.

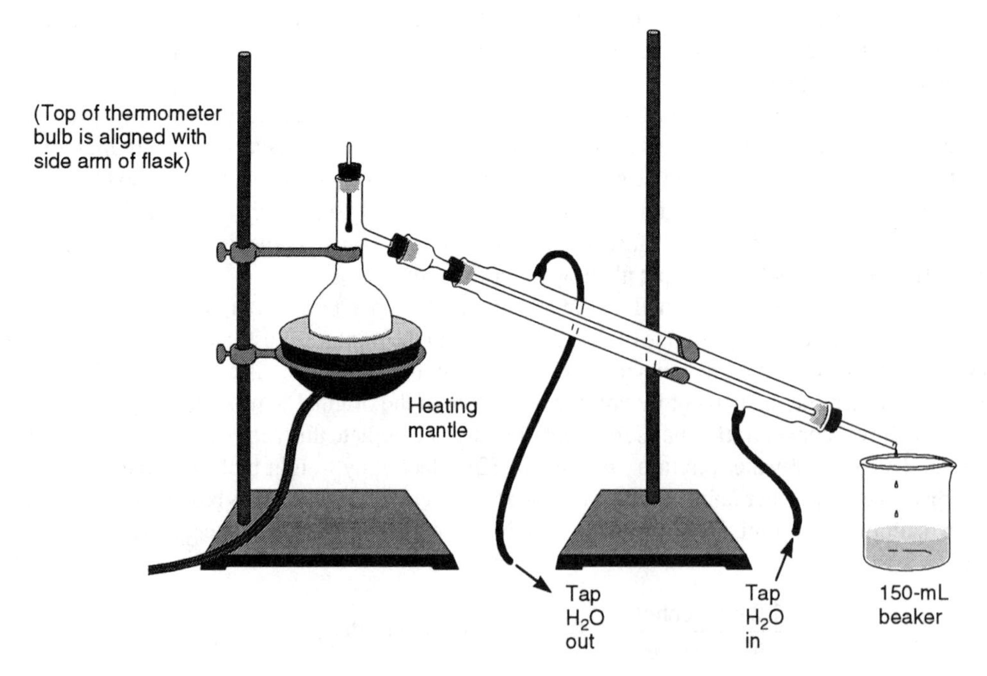

FIGURE 27.1
Distillation apparatus.

Slowly heat the wine to a *gentle* boil with a Bunsen burner or electric heating mantle. Perform the distillation *slowly*. While gently heating the wine, observe the temperature of the vapor produced. Initially the temperature will be about 80°C because alcohol boils at a lower temperature than water. As the mixture continues to boil, the mixture in the distillation flask will contain less alcohol and the temperature will rise slowly to the boiling point of water (99–100°C). Continue to boil the wine gently until a constant temperature close to 100°C is reached. When this temperature has been reached, it is assumed that all of the alcohol and some of the water in the original wine sample has been distilled into the 150-mL beaker. Generally, the amount of distillate obtained will be 25 mL or less.

When the distillation of the alcohol is complete, remove the heat and let all the distillate come through the condenser. Weigh the beaker plus distillate on the milligram balance. Record this value and subtract the tare weight of the beaker to determine the total number of grams of distillate obtained. To determine the density of the distillate, a 25-mL graduated cylinder should be weighed to the nearest 0.001 g, and some of the distillate (10 to 20 mL) added to the graduated cylinder. Weigh the graduated cylinder and its distillate contents to the nearest 0.001 g, record this weight, and subtract the empty cylinder weight to find the weight of distillate in the graduate cylinder. **Note:** *This weight will be less than the total weight of distillate collected in the beaker because not all of the liquid will have been transferred to the graduated cylinder.* Place the graduated cylinder on a level desk surface, and **carefully read the volume of distillate in the cylinder to a precision of at least 0.1 mL. This is an important step. Make sure the volume is read correctly.** Record this volume.

Before repeating the procedure with a new sample of the wine (trial 2), complete the calculations for the sample of the wine just distilled.

CALCULATIONS

A. Density of Distillate

Use the weight of the distillate in the graduated cylinder and its volume to determine the density of the distillate to the appropriate number of significant figures.

B. Percent by Weight of Alcohol in Distillate

Using Table 27.1, locate the density of distillate. Record the corresponding weight percent of alcohol in the distillate (g alcohol/100 g distillate) on the report sheet.

If the density value you calculated for the distillate lies between two numbers on the chart, you should use the method of interpolation to obtain a more accurate percent-by-weight value. For example, if you have an experimental density of 0.940 g/mL, it lies in the interval of 0.943 to 0.935 g/mL on Table 27.1. This corresponds to the interval 36.0 to 40.0% by weight. To determine the percent by weight corresponding to this experimental density, determine the difference between your experimental density and the density at the "top" of the interval. Then divide this by the difference between the density at the "top" and that at the "bottom" of the interval.

$$\frac{(0.943-0.940)}{(0.943-0.935)}=\frac{0.003}{0.008}=\frac{3}{8}$$

Multiply this fraction by the interval value in the percent-by-weight column.

$$3/8 \times (40.0 - 36.0) = 1.5$$

Add this number to the "top" number in the percent-by-weight column to determine the percent by weight corresponding to a density of 0.940 g/mL.

$$(36.0 + 1.5)\% = 37.5\%$$

C. Percent by Volume of Alcohol in Wine

Up to this point you have made calculations concerning alcohol in the distillate. Since all the alcohol in the distillate originated in the wine, you make some calculations concerning the original wine sample. For this section on the report sheet, show all the setups of numbers with units that will lead you to an answer, and remember to apply significant figure rules.

1. **Weight of alcohol in distillate.** To calculate the actual weight of alcohol in the *total* distillate, multiply the decimal equivalent of the percent by weight of alcohol in the distillate by the *total* weight of the distillate.
2. **Volume of alcohol in distillate.** Calculate this value using the weight of alcohol in the distillate and the density of pure alcohol (0.789 g/mL). Divide the weight of alcohol in the distillate by the density of pure alcohol.
3. **Percent by volume of alcohol in wine.** Since all the alcohol in the distillate originally came from the wine sample, determine the percent by volume of alcohol in the wine by using the following relationship.

$$\text{vol\%} = \frac{\text{volume of alcohol}}{\text{volume of wine}} \times 100$$

Repeat the entire procedure and calculations with a different sample of the same wine or with another "unknown" wine sample as instructed. Usually the first time a person performs a new procedure, the results are not as accurate as they could be. Going through the procedure one more time will help perfect lab technique and perhaps lead to more accurate results.

Experiment 27

Percent By Volume of Alcohol in Wine

Name____________________________________

Date_____________________________________

REVIEW QUESTIONS

1. An alcohol-water mixture has a weight of 22.125 g. If a 21.1-mL sample of this mixture has a weight of 18.120 g,

a. Determine the density of this mixture.

b. Determine the percent by weight of alcohol in this mixture.

c. Determine the weight of alcohol in the mixture.

d. If the density of pure alcohol is 0.789 g/mL, how many milliliters of alcohol are present in the mixture?

2. If a 250-g alcohol-water mixture is 26% alcohol by weight, determine the weight of alcohol in the mixture.

3. If a sample of an alcohol-water mixture contains 4.703 g of alcohol, determine the volume of alcohol in the sample (density pure alcohol = 0.789 g/mL).

4. What is the other major ingredient in wine besides alcohol and water?

5. Why was it necessary to distill the wine in this experiment? Could a person simply determine the density of a wine sample directly and then use Table 27.1 to get the weight-percent alcohol in the wine?

Experiment 27
Percent By Volume of Alcohol in Wine

Name______________________________

Date______________________________

REPORT SHEET

Data from Distillation

Identity of wine sample used____________________ Volume of wine used____________________

Weight of beaker plus distillate =

Tare weight of beaker =

Total weight of distillate =

Weight of graduated cylinder plus part of distillate =

Tare weight of graduated cylinder =

Weight of distillate in graduated cylinder =

Volume of distillate in graduated cylinder =

Calculations (For each calculation, show setup with correct units and rounding as dictated by significant figure rules.)

A. Density of Distillate

B. Percent by Weight of Alcohol in Distillate (From Table 27.1; interpolate if necessary)

C. Percent by Volume of Alcohol in Wine

1. **Weight of alcohol in distillate**

2. **Volume of alcohol in distillate**

3. **Percent by volume of alcohol in wine**

Second Wine Sample or Unknown

Data from Distillation

Identity of wine sample used ______________________ Volume of wine used ______________________

Weight of beaker plus distillate =

Tare weight of beaker =

____

Total weight of distillate =

Weight of graduated cylinder plus part of distillate =

Tare weight of graduated cylinder =

____

Weight of distillate in graduated cylinder =

Volume of distillate in graduated cylinder =

Calculations (For each calculation, show setup with correct units and rounding as dictated by significant figure rules.)

A. Density of Distillate

B. Percent by Weight of Alcohol in Distillate (From Table 27.1; interpolate if necessary)

C. Percent by Volume of Alcohol in Wine

1. **Weight of alcohol in distillate**

2. **Volume of alcohol in distillate**

3. **Percent by volume of alcohol in wine**

EXPERIMENT 28

Molecular Weight of Oxygen

Purpose: To experimentally determine the molecular weight of a common gas.

Materials: $KClO_3$, MnO_2, glass wool.

Waste Disposal: When the test tube has cooled, the residue may be removed by adding water and allowing the contents to begin dissolving. This will loosen the caked material and allow its removal. Dispose of the solid in the solid waste receptacles located throughout the lab.

Safety Precautions: Potassium chlorate is a strong oxidizing agent that is capable of supporting the combustion of paper and other solids. Do not pour hot molten potassium chlorate into waste containers.

INTRODUCTION

Under standard conditions of temperature and pressure, one mole of gas occupies a volume of about 22.4 liters, subject to Gas Law errors. Measurement of the molecular weight of an unknown gas is, therefore, simply a matter of finding the weight of 22.4 liters at STP, where standard temperature is 273°K (0°C) and standard pressure is 760 mm.

It is usually impossible to make the actual measurements at STP, and in any event, 22.4 liters is too great a volume for convenient weighing. Therefore, the weight of a smaller volume is determined under room conditions. A gas calculation then gives the volume at STP, and since a weight-volume relationship is known, a simple proportion gives the weight of 22.4 liters. This principle provides a means of measuring the approximate molecular weight of a gas, or any substance that can be readily converted into a gas.

In this experiment, the practical difficulties of weighing a small amount of gas are avoided by weighing a solid substance, $KClO_3$, from which the gas can be generated and collected. The residue is then weighed and the weight of the gas evolved is obtained by subtraction.

The weight of one mole of gas can also be calculated by use of the Ideal Gas Law, $PV = nRT$. The number of moles of gas in the sample, n, is then $n = PV/RT$. If the pressure P is in atmospheres and the V is in liters, then $R = 0.0821\ L \cdot atm \cdot °K^{-1} \cdot mol^{-1}$. The weight of a mole is found by the relationship $M = g/n$, where g is the weight of the sample.

PROCEDURE

(Be sure to wear safety glasses.)

Weigh approximately 4 grams of $KClO_3$ into a crucible, add one-fourth spatula of MnO_2, and mix without grinding. Put the mixture into a *large*, dry Pyrex test tube, insert a loose plug of glass wool, and weigh the tube and contents accurately (to the nearest milligram if possible).

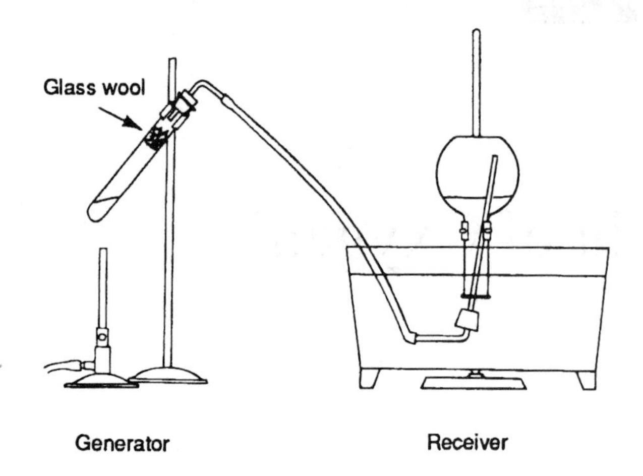

FIGURE 28.1
Apparatus for generation and collection of oxygen.

Set up the apparatus shown in Figure 28.1. The receiving flask is a 1,000-mL Florence flask fitted with a *two-hole* stopper carrying a glass tube or a polyethylene tube (to guard against glass cuts) with a right-angle bend. The tube must reach almost to the bottom of the flask when the stopper is in place. A foot of rubber tubing is attached to this glass tube. (The use of a small trap at the lower end of the delivery tube reduces chance of water entering the generator.)

Fill the flask with water, insert the tubing and stopper, and invert into a pneumatic trough. Support the flask with a ring stand and clamp.

Attach the generating tube to the apparatus and heat the generator gently. Before heating, remove the stopper from the neck of the receiving flask to permit easier exit for water. Heat slowly to prevent mechanical carryover of fine solid particles or undue vaporization of KCl.

Do not let molten $KClO_3$ come into contact with the rubber stopper. Continue to generate oxygen until the receiver is about three-fourths full. Make sure that the end of the delivery tube is well above water level, then allow the apparatus to cool to room temperature *without disconnecting.* After the tube has partially cooled, it may be immersed in water to hasten the cooling.

When the generator has cooled, insert the stopper in the receiving flask without disconnecting the generator, remove the flask from the ring stand, and lay it on its side in the pneumatic trough, keeping the outlet hole of the stopper under water. Hold the flask in such a position that the levels of the water inside and outside are the same. This equalizes the gas pressure with atmospheric pressure. With the flask in this position place a finger over the open outlet hole and lift the flask from the water. Set it upright, mark the position of the bottom of the stopper with a gummed label, and disconnect the generator. Take the temperature of the water in the flask, which will be the temperature of the gas, and read the barometer.

The receiving flask has some water in it at this point. Add additional water, *carefully keeping track of the volume added,* with a graduated cylinder to fill the flask to the level of the gummed label which was used to mark the bottom of the stopper. Measure as accurately as possible the volume of water required to fill the receiving flask to this level. This volume equals the volume of wet oxygen which was generated. Weigh the generator and contents accurately. The loss in weight is the weight of oxygen evolved.

From the data you now can correct your oxygen volume to STP and calculate the weight of 22.4 liters of oxygen.

Experiment 28

Molecular Weight of Oxygen

Name______________________________

Date______________________________

REVIEW QUESTIONS

1. Discuss the errors, other than careless manipulation or measurement, to which this experiment is subject.

2. Would your calculated molecular weight have been too high or too low in each of the following situations?

a. A large glass bubble was present in the flask before heating.

b. The generator was not allowed to cool before measuring the gas volume.

c. Some moisture was present in the generator when the first weighing was made.

d. Heating was so rapid that solid KCl was swept out of the generator.

e. The water vapor correction was not made.

f. The measured gas volume was not corrected to STP.

g. The water levels inside and outside the receiver were not equalized.

3. What is the purpose of the glass wool plug in the apparatus?

4. Why must the glass tube in the receiver be so long?

5. What improvement in technique can you suggest? (The author would be glad to be informed of these.)

6. Why is MnO_2 used? What other substance could have been used?

7. Calculate the percent error inherent in the weight of oxygen produced. You will need to estimate the weighing error for the balance used in your laboratory. How could the accuracy be improved?

Experiment 28

Molecular Weight of Oxygen

Name______________________________

Date_______________________________

REPORT SHEET

	I		II	
Initial weight of generator and contents =		g		g
Final weight of generator and contents =	____________	g	____________	g
Weight of oxygen =		g		g
Temperature of gas =	____________	°C	____________	°C
Barometric pressure =		mm		mm
Pressure of water vapor present =	____________	mm	____________	mm
Pressure of oxygen =		mm		mm
Uncorrected volume of oxygen (measured) =	____________	mL	____________	mL

Volume of oxygen corrected to STP

Calculation:

Molecular weight of oxygen (weight of 22.4 L at STP)

Calculation:

Percent error

Calculation:

EXPERIMENT 29

Equivalent Weight and Oxidation Number

Purpose: To measure the equivalent weight of a metal and to determine its oxidation number.

Materials: Al wire, 6 N HCl solution. Unknowns: zinc, zinc or aluminum alloys. (Magnesium can be used but reacts with dangerous speed.)

Waste Disposal: Transfer the reacted metal and hydrochloric acid from the H_2 generator to the heavy metal waste container.

INTRODUCTION

The equivalent weight (combining weight) of a substance is the weight that will gain or lose one mole (Avogadro's number) of electrons in a specific redox reaction or will take part in Avogadro's number of electrovalent bonds in an ion combination reaction. It is, in effect, the weight of a metal that will liberate 1.008 g of H_2 from an acid, since one mole of H^+ ions (1.008 g) will gain one mole of electrons.

In this experiment a measured amount of H_2 is liberated from an acid by a known weight of metal, and the amount of metal needed to liberate 1.01 g of H_2 is calculated.

The oxidation number of the metal is determined by comparing the measured equivalent weight with the atomic weight read from a periodic chart. The atomic weight is the product of the equivalent weight and the oxidation number, using the fact that a mole of metal will lose a whole number of moles of electrons.

PROCEDURE

Assemble the apparatus shown in Figure 29.1. The generator is a 250-mL Erlenmeyer flask fitted with a *one-hole* stopper carrying a short right-angle glass or polyethylene tube. Attach a two-inch piece stopper. The receiver is a 500-mL Florence flask.

A. Equivalent Weight of Aluminum

Obtain a piece of aluminum metal weighing about 0.25 g and clean off the oxide coating with steel wool. Weigh it accurately (to 0.001 g if possible) and wrap it around the generator outlet tube as shown. Put 30 mL of dilute HCl in the flask, insert the stopper firmly, and attach a foot of rubber tubing. The delivery tube in the receiver can also be of polyethylene to eliminate danger of glass cuts. It should go through a two-hole rubber stopper and should reach nearly to the bottom of the flask when the stopper is in place. Fill the receiver with water, insert the stopper, and invert in a water-filled trough. Support it with a clamp and remove the stopper. There should be no bubbles.

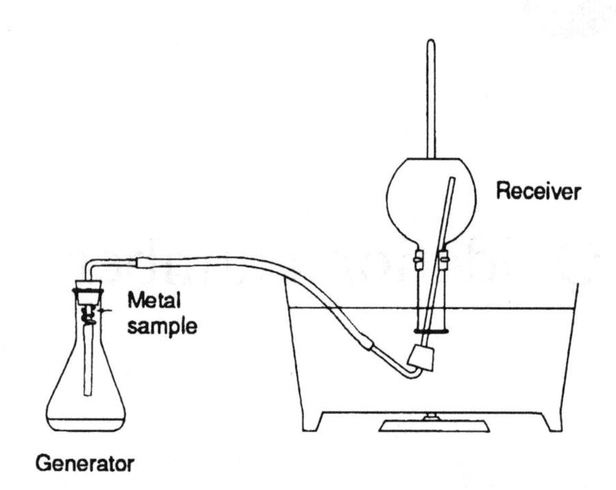

FIGURE 29.1
Apparatus for quantitative collection of hydrogen.

When the whole system is connected and has been checked, wrap the generator in a towel and invert it to bring the acid in contact with the metal. When all the metal has reacted, cool the generator to room temperature without disconnecting. The cooling can be hastened by putting the generator into water.

With generator and receiver still connected, insert the stopper in the receiver and lay it on its side in the trough, keeping the outlet under water. Raise or lower the flask to equalize water levels within and without, put your finger over the hole in the stopper, and set the receiver upright. Mark the position of the bottom of the stopper with a gummed label, remove the tubing, and measure as accurately as possible the water needed to fill the flask. This represents the volume of H_2 generated under room conditions (plus water vapor).

Correct the volume of H_2 to STP, remembering that you have a wet gas. Then calculate the weight of this gas from the weight and volume of one mole of H_2. From the weight of the metal used and weight of H_2 it liberates, calculate the equivalent weight of the metal. Then calculate the oxidation number (charge of the metal ion) from your equivalent weight and the atomic weight.

B. Equivalent Weight of an Unknown

Obtain a piece of unknown metal weighing approximately 1 gram, record its number, and determine its equivalent weight by the procedure used for aluminum. The oxidation number cannot be determined unless the metal is known and is a pure sample.

Experiment 29

Equivalent Weight and Oxidation Number

Name____________________________________

Date_____________________________________

REVIEW QUESTIONS

1. To what errors, other than carelessness, is this experiment subject? How might these errors be reduced or eliminated?

2. Would your calculated equivalent weight have been too high or too low if acid had contacted the metal before the generator was attached?

3. Why does the air already in the generator not introduce an error?

4. Would your equivalent weight have been too high or too low if the oxide coating of the metal had not been removed?

5. What is the relationship between the equivalent weight and the mole of a substance? Explain why this is true.

6. Which would liberate the most hydrogen from an excess of acid: 10 g of Na, 10 g of Mg, 10 g of Zn, 10 g of Al, or 10 g of Fe? Which would liberate the least H_2? Explain the reasoning behind your answers.

7. Could you have measured the equivalent weight of copper by the method of this experiment? Explain. Devise an experiment by which this could be done.

Experiment 29
Equivalent Weight and Oxidation Number

Name________________________________

Date________________________________

REPORT SHEET

Number of unknown ___________

	Aluminum	Unknown
Weight of metal =		
Temperature of gas (water temperature) =		
Barometric pressure =		
Water vapor pressure =	________________________	________________________
H_2 pressure =		
Measured H_2 volume =		
Volume of dry H_2 at STP =		
Weight of H_2 =		
Equivalent weight of metal =		

Calculation of volume of dry H_2 at STP:

Calculation of weight of H_2:

Calculation of equivalent of metal:

Value from literature ________________________

Percent error ________________________

Calculation of oxidation number of metal:

1. H_2 and Cl_2 react according to the equation $H_2 + Cl_2 \rightarrow 2\ HCl$. Using the equivalent weight of H_2 given and the atomic weight of Cl_2 from the periodic chart, calculate the equivalent weight of chlorine.

2. Chlorine gas and iron react to form iron(III) chloride. Write a balanced equation and use your calculated equivalent weight of chlorine to find the equivalent weight of iron.

3. The following half-reaction shows the pattern of electron loss that iron undergoes in the reaction of question 2: $Fe \rightarrow Fe^{3+} + 3e^-$. Again calculate the equivalent weight of iron, this time using the known atomic weight and the number of electrons lost.

4. With weaker oxidizing agents, iron reacts according to the following equation:

$$Fe \rightarrow Fe^{2+} + 2e^-.$$

Calculate the equivalent weight of iron for this half-reaction.

5. Calculate the equivalent weight of each substance from the half-reaction equation:

a. S in the reaction $S + 2e^- \rightarrow S^{2-}$

b. S in the reaction $S + O_2 \rightarrow SO_2$

c. Ca in the reaction $Ca \rightarrow Ca^{2+} + 2e^-$

d. $KMnO_4$ in the reaction $KMnO_4 + 8H^+ + 5e^- \rightarrow + Mn^{2+} + 4H_2O$

e. $H_2C_2O_4$ in the reaction $H_2C_2O_4 \rightarrow 2H_2 + 2CO_2 + 2e^-$

f. O_2 in the reaction $O_2 + 4e^- \rightarrow 2O^{2-}$

6. Calculate the equivalent weight of each for ion combination reactions:

$CuSO_4$ $Al_2(SO_4)_3$ Na_3PO_4

$NaOH$ $AlCl_3$ KNO_3

$BaCl_2$ H_2SO_4

7. In an experiment 0.435 g of Mg gave 457 mL of moist H_2 at 23 °C and 740 mm pressure. Calculate the equivalent weight and oxidation number of Mg.

8. A sample of Al weighing 3.26 g reacted with excess acid to produce 0.360 g of H_2. Calculate the equivalent weight of Al. Look up the atomic weight and calculate the oxidation number.

9. In a certain reaction 4.25 g of Fe reacted with oxygen to produce 6.08 g of Fe_2O_3. Calculate the equivalent weight of Fe.

10. By use of the equivalent weights previously found, calculate how much:

a. Fe would react with 10.0 g of O_2 (forming Fe_2O_3)

b. Fe would react with 7.00 g S (forming FeS)

c. $KMnO_4$ would react with 25.0 g $H_2C_2O_4$

d. $AlCl_3$ would react with 10.0 g NaOH

e. $BaCl_2$ would react with 50.0 g $Al_2(SO_4)_3$

f. H_2SO_4 would react with 75.0 g of $CaCO_3$

EXPERIMENT 30

Redox Titration of Iron

Purpose: To calculate the %Fe in an unknown by titrating the Fe^{+2} in an oxidation-reduction reaction with $KMnO_4$.

Materials: Solid unknown FAS (primary reagent for Fe^{+2}), concentrated H_2SO_4, *standardized* 0.01 M $KMnO_4$ solution, R–Z Reagent .

(Reinhardt–Zimmermann, "R–Z Reagent" is a catalyst for the redox reaction: dissolve 70 g $MnSO_4 \cdot 4H_2O$ in 500 mL of water, add 125 mL of concentrated H_2SO_4, then 125 mL of concentrated (85%) H_3PO_4, dilute to 1 L).

Apparatus: 100-mL volumetric flask, 10-mL graduated cylinder, 50-mL buret, 20-mL pipet, pipet bulb, 500-mL Erlenmeyer flask.

Safety Precautions: Wear your safety glasses at all times during this experiment. The solutions will contain concentrated sulfuric and phosphoric acids. These can cause serious skin burns. Avoid contact. Potassium permanganate ($KMnO_4$) is a strong oxidizing agent and may be violently reactive with other chemicals. It is notorious for staining everything including glassware and clothing. Rinse all glassware just as soon as you finish using it.

Hazardous Waste Disposal: The solutions may be rinsed down the drain with a very large volume of water. The acidic solutions must be highly diluted to protect plumbing.

INTRODUCTION

The essentials of oxidation-reduction reactions have been discussed previously. You should review the appropriate sections in your text before beginning this experiment.

Iron in iron ore is typically determined by redox titration with $KMnO_4$. The ore is first crushed and then dissolved by heating in concentrated HCl. This converts all Fe to Fe^{+3}. Next, the Fe is pre-reduced to Fe^{+2} by the use of stannous chloride (another redox reaction). Finally, the Fe^{+2} is titrated with $KMnO_4$. The *unbalanced* reaction is

$$H^+ + Fe^{+2} + MnO_4^- \longrightarrow Mn^{+2} + Fe^{+3} + H_2O$$

You must balance this reaction. The reaction is catalyzed (speeded up) by Mn^{+2}, which also happens to be a product of the reaction. Thus, the reaction is *autocatalytic*. You will add some Mn^{+2} in the beginning to get things started, otherwise the reaction is frustratingly slow. The catalyst also contains high concentrations of acid, H^+, and H_3PO_4, which serve to de-colorize the Fe^{+3} product. (Fe^{+3} is normally yellow.)

Your unknown is already in the ferrous ion state so it will not require pre-reduction with stannous chloride.

Your instructor may require that you standardize the $KMnO_4$ yourself with a known amount of Fe^{+2}. If so, carry out the titrations in the same manner.

PROCEDURE

1. Your unknown is often called FAS. Weigh approximately 2.35 g out to the nearest 0.001 g, or 0.0001 g if you have access to an analytical balance. Record the weight as measured.
2. Transfer quantitatively to a 100-mL volumetric flask. Dissolve in about 50 mL of water, then add 5 mL of concentrated sulfuric acid. *CAUTION!* Make up to volume; that is, dilute with distilled water up to the etched mark for 100 mL, stopper, and *mix thoroughly*.
3. Pipet 20 mL of your FAS solution into a 500-mL Erlenmeyer flask. Add 10 mL of R–Z Reagent (contains Mn^{+2} and acid) and enough distilled water to make *about* 300 mL total volume.
4. Take no more than 150 mL of standardized $KMnO_4$ solution. Record the concentration. Rinse and fill a buret. Cover the remaining solution.
5. In this titration read the buret to the nearest 0.05 mL. You will discover that the intense color makes this difficult. Make an initial reading and record.
6. The titrant here is a self-indicator. As quickly as it enters the solution, it disappears. The product Mn^{+2} is virtually colorless. The Fe^{+3} may be slightly yellow. When the solution is completely titrated, the *next drop* of $KMnO_4$ has nothing to react with so you can see it. The end point is the first pale pink that persists. The appearance will be nearly the same as for phenolphthalein but it is only a coincidence that they are both pink. They are NOT RELATED.
7. Do not add more than 1 mL at a time, especially at the beginning. The reaction will be slow at first. Unlike acid-base, redox is not instantaneous. If you proceed too fast, brown MnO_2 will begin to precipitate out and the solution stoichiometry will be wrong. Read the end point volume to the nearest 0.05 mL.
8. The solution may be discarded down the sink. Flush with plenty of water.
9. Perform two more trials.
10. Rinse all glassware thoroughly. Rinse the buret first with tap water, then with distilled water. Hang upside down, stopcocks open to dry.

CALCULATIONS

1. Average the three titration volumes (assuming the same amount of FAS, 20 mL, was used each time).
2. Write the balanced equation between Fe^{+2} and MnO_4^-.
3. Calculate moles MnO_4^-.
4. Convert to moles Fe^{+2}.
5. Convert to grams of Fe.
6. Since you took a 20-mL aliquot each time from a solution that had a total volume of 100 mL, you were only titrating 20/100 or 1/5 of the total FAS. Therefore, multiply by 5 to get total grams Fe.
7. Calculate the %Fe.

$$\%Fe = \frac{\text{grams Fe}}{\text{grams FAS}} \times 100\%$$

8. Show all your calculations on your report sheet.

Experiment 30

Redox Titration of Iron

Name________________________________

Date________________________________

REVIEW QUESTIONS

1. Rewrite the iron-permanganate reaction and indicate: (a) oxidizing agent, (b) reducing agent, (c) the species that gets oxidized, and (d) the species that gets reduced.

2. An acidic solution of permanganate ion reacts with oxalate ion to form carbon dioxide and Mn(II). Write a balanced equation for the reaction.

3. Referring to the reaction in #2, if 38.4 mL of 0.150 M $KMnO_4$ solution is required to titrate 25.2 mL of sodium oxalate solution, what is the concentration of oxalate ion?

4. Suggest one other way to determine Fe quantitatively.

5. What is the principal use of manganese?

6. In the balanced Fe^{+2}/MnO_4^- equation, how many moles of electrons are lost? How many moles of electrons are gained?

7. In performing the experiment, one student failed to mix her FAS solution in the volumetric flask before taking the 20 mL aliquot. In this case would the %Fe she calculated be too high, too low, or indeterminate?

Experiment 30

Redox Titration of Iron

Name______________________________

Date______________________________

REPORT SHEET

Concentration of $KMnO_4$ ____________________ *M*

Weight of FAS ____________________ g dissolved in ____________________ mL

Titration data ____________________ mL FAS used

Trial 1	Trial 2	Trial 3
mL	mL	mL
– mL	– mL	– mL
mL	mL	mL

Average mL $KMnO_4$ ____________________ mL

Balanced equation

Calculations–calculate %Fe by weight in FAS

EXPERIMENT 31

The Atmosphere

Purpose: To note the presence of water vapor and CO_2 in the air and to estimate the amount of oxygen present; to observe noble gas spectra.

Materials: $CaCl_2$ (anhydrous), NaOH pellets, saturated solution $Ba(OH)_2$, 6 N NaOH solution, 30% pyrogallol solution.

Waste Disposal: All solutions may be rinsed down the drain with water.

INTRODUCTION

The atmosphere is a mixture of gases that rests on the earth's surface with a weight of about 14.7 pounds per square inch at sea level. It becomes less dense the greater the distance from the surface of the earth and gradually blends into the almost perfect vacuum of interstellar space.

The general composition of the atmosphere is remarkably constant from place to place over the globe, with only the water vapor content varying widely. The carbon dioxide content varies slightly. The average composition by volume calculated to a dry gas basis is as follows:

N_2 .....78.03% | Ne 0.0012%
O_2 ..... 20.99% | He 0.0004%
A ..... 0.94% | Kr 0.0001%
CO_2..... 0.03% | Xe 0.000008%

The water vapor content may vary from a small fraction of one percent to four or five percent. Other gases may be present locally in small amounts. Collodial particles of dust and smoke, as well as water droplets, may also be found.

The noble gases in the atmosphere cannot be detected chemically because of their inactivity and low concentration. The spectroscope, however, reveals them. If an atom is excited by heat or an electrical discharge, its electrons absorb energy and assume higher energy levels. As the atom moves out of the high-energy area its electrons drop back to thier normal energy levels, giving off the excess energy in the form of light. Since an electron can only occupy definite energy levels, the released energy has definite values that can be measured in terms of the wavelength of the light emitted. Also, since no two elements have exactly the same electron pattern, no two elements give off exactly the same wavelengths of light. The spectroscope separates light into its component wavelengths and provides for measuring them. Hence the spectrum (light pattern) of an element serves to identify it.

PROCEDURE

A. Water Vapor

1. Place on a watch glass some pieces of anhydrous $CaCl_2$ and on another watch glass a few pellets of NaOH. Weigh each one quickly, allow to stand for an hour, and weigh again. Record your observations.
2. By means of a wet and dry bulb hygrometer or a sling pyschrometer, determine the relative humidity of the room. Make sure that the wick of the wet bulb thermometer is wet, then gently fan air over the instrument for one minute, or, if a sling psychrometer is used, swing it for one minute. Be sure you have plenty of space for swinging. Read and record the temperatures and determine the difference between wet and dry bulb readings. From Table 31.1 read the relative humidity, where d-w is the difference between the dry and wet bulb thermometer readings, and r.h. is the relative humidity.

B. Carbon Dioxide

1. Pour some clear $Ba(OH)_2$ solution on a watch glass and allow to stand in the air for one hour. Observe and report.
2. Put some fresh $Ba(OH)_2$ solution in a test tube and by means of a glass tube bubble your breath through the solution for a short time. Observe and record your results.

TABLE 31.1 Humidity from Wet and Dry Bulb Readings.

	d-w = 0°C	d-w = 1°C	d-w = 2°C	d-w = 3°C	d-w = 4°C	d-w = 5°C
Temp °C	*r.h. %*	*r.h. %*	*r.h. %*	*r.h. %*	*r.h. %*	*r.h. %*
15	100	90	80	71	61	52
16	100	90	81	71	62	54
17	100	90	81	72	64	55
18	100	91	82	73	65	56
19	100	91	82	74	65	58
20	100	91	83	74	66	59
21	100	91	83	75	67	60
22	100	92	83	76	68	61
23	100	92	84	77	69	61
25	100	92	84	77	70	63
30	100	93	86	79	73	67

	d-w = 6°C	d-w = 7°C	d-w = 8°C	d-w = 9°C	d-w = 10°C	d-w = 11°C
Temp °C	*r.h. %*	*r.h. %*	*r.h. %*	*r.h. %*	*r.h. %*	*r.h. %*
15	44	36	27	20	12	5
16	46	37	32	24	15	8
17	47	39	32	24	17	10
18	49	41	34	27	20	13
19	50	43	35	29	22	15
20	51	44	37	30	24	18
21	52	46	39	32	26	20
22	54	47	40	34	28	22
23	55	48	42	36	30	24
24	56	49	43	37	31	26
25	57	50	43	37	31	26
26	58	51	46	40	34	29
28	59	53	48	42	37	32
30	61	55	50	44	39	34

C. Estimation of Oxygen Content

Fit a 500-mL Florence flask with a tight-fitting, one-hole rubber stopper. Into the hole put a short piece of fire-polished glass tubing fitted with a rubber tube and pinch clamp as shown in Figure 31.1. Be sure that the clamp closes the tube tightly. Mark the position of the bottom of the stopper with a gummed label. Then insert an empty test tube in the flask and fill the flask to the mark on the neck with water, measuring the amount required. This establishes the volume of the flask.

Empty the flask but do not dry it. Put 20 mL of 6 N NaOH into it and put 5 mL of a 30% pyrogallol solution in the test tube. (Be careful! Pyrogallol solution will stain clothing.) Insert the test tube in the flask and stopper the flask tightly, making sure that the rubber tube is closed.

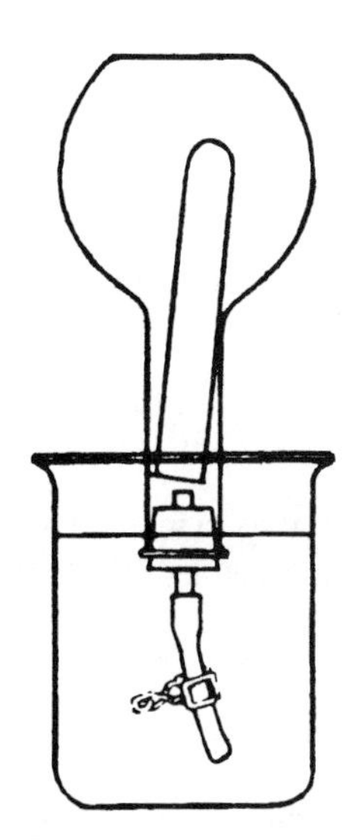

FIGURE 31.1
Apparatus of oxygen-content estimation.

Tip and rotate the flask so that the solutions mix and come in contact with the entire inner surface of the flask. The pyrogallol solution absorbs oxygen from the trapped air, creating a partial vacuum in the flask. Quickly place the flask in a beaker with the stopper and outlet tube under water, as shown in the diagram. This reduces leakage of air into the flask. Frequently, over a period of 15 minutes, remove the flask briefly and gently shake or rotate it to redistribute the solution over the inner surface.

Hold the outlet tube under the surface of the water in the pneumatic trough and release the pinch clamp. After the water flow has ceased, lower the flask in the water to equalize the levels inside and out, keeping the outlet under water, and then close the clamp. Remove the flask and measure the volume of solution it now contains.

Calculate the percent by volume of oxygen in the air from the volume of air originally trapped and the volume of oxygen absorbed by the pyrogallol solution.

D. Spectroscopic Examination of Noble Gases (Demonstration)

The instructor will set up the spectroscope with spectrum tubes of one after another of the noble gases of the atmosphere. An illuminated scale is seen superimposed on the spectrum lines in the eyepiece. The instructor will have various students read the positions of the lines on the scale (as many as possible). Record this data and draw the lines in their proper positions on the spectrum diagram of the report sheet. If the eyepiece scale is properly positioned according to the directions with the instrument, the actual wavelengths can be determined.

Name________________________________

Date________________________________

REVIEW QUESTIONS

1. To what errors is the method for estimating oxygen in the air subject? How might these be reduced or eliminated?

2. Look up or devise a method for measuring quantitatively the amount of CO_2 in the air. If your method shows promise, the instructor may let you try it.

3. Calibration curves are generally supplied with spectroscopes, which relate scale division to actual wavelengths. If such a curve is available, determine the wavelengths of some of the more prominent spectral lines in your diagrams.

4. Helium, the first noble gas to be discovered, was actually first observed in the sun. Find out why the sun's light gives a continuous spectrum, what the origin of the dark (Frauenhofer) lines is, and how they relate to "bright line" spectra of the elements. Even a small spectroscope will usually separate a great many Frauenhofer lines from sunlight.

5. Recently the noble gases have been found to be not completely inert. Look up and report on compounds that have been made from them. Discuss possible electron sharing arrangements in these compounds.

6. Assuming an air pressure of 14.7 lb per square inch, calculate the pressure on a square mile of the earth's surface.

7. Using the following table, estimate your body area and calculate the total air pressure on your body at sea level.

Height	Weight (lb)	Area (sq. in.)	Height	Weight	Area
5'0"	120	2,320	5'6"	150	2,740
5'2"	130	2,470	5'8"	160	2,880
5'4"	140	2,600	6'0"	180	3,160

Experiment 31
The Atmosphere

Name____________________
Date____________________

REPORT SHEET

A. Water Vapor

1. What change does the $CaCl_2$ and the NaOH undergo? ____________________

____

	$CaCl_2$	NaOH
Final weight of glass and contents =		
Initial weight of glass and contents =	______	______
Gain in weight =		

Explain: ____________________

____

What is the process called? ____________ Under what conditions does it take place? Be specific.

____

What uses for $CaCl_2$ are based on this phenomenon?

(a) ____________________

(b) ____________________

Define relative humidity: ____________________

____

2. Measurement of relative humidity:

Dry bulb temperature = $^\circ$

Wet bulb temperature = ______ $^\circ$

Difference = $^\circ$

Relative humidity (from table) = %

B. Carbon Dioxide

1. What is seen in the $Ba(OH)_2$ solution? ____________________

What is the precipitate? ____________________

What does the test show? ______________________________

Write the equation: ______________________________

2. What does the test show regarding the composition of exhaled air from the lungs? ______________________________

____

C. Estimation of Oxygen Content

	I	II
Total volume of flask =		
Volume of NaOH and pyrogallol =	______________	______________
Volume of H_2O which entered =		

$$\text{Percent } O_2 = \frac{[\text{Volume } H_2O \text{ that entered } (O_2\text{volume})]}{[\text{Volume of air trapped}]} \times 100$$

$$\frac{[\qquad\qquad]}{[\qquad\qquad]} \times 100\% = \qquad\qquad \%$$

What is the gas remaining in the flask after the reaction (principal component)? ______________________________

____

D. Spectroscopic Examination of the Noble Gases

Helium

1	2	3	4	5	6	7	8	9	10

Neon

1	2	3	4	5	6	7	8	9	10

Argon

1	2	3	4	5	6	7	8	9	10

Krypton

1	2	3	4	5	6	7	8	9	10

Xenon

1	2	3	4	5	6	7	8	9	10

EXPERIMENT 32

Oxygen and Some Oxides

Purpose: To prepare oxygen by several methods; to study behavior in combustion; to observe certain properties of its oxides; to prepare ozone and observe some of its properties.

Materials: $KClO_3$, MnO_2, Fe_2O_3, PbO_2, CaO, $NaNO_3$, MgO, P_2O_5, C (charcoal), S, glass wool, steel wool, Mg ribbon, Fe wire, 0.2 N $AgNO_3$ solution, 6 N HNO_3, 6 N H_2SO_4.

Apparatus: Water electrolysis apparatus, battery charger, induction coil, ozonizer, deflagrating spoon that is easy to clean or can be discarded (can be made from a 6" × 1/2" strip cut from a steel can; indent one end with a ball-peen hammer over a tapered hole bored in a steel block, then bend a cup at one end and a handle at the other at right angles to the shaft).

Waste Disposal: Dispose of mixtures containing lead or silver in the heavy metal waste container. The electrolysis apparatus contains a sulfuric acid solution and should be treated carefully because the glass apparatus is easily broken, and the sulfuric acid that spills would need immediate attention. Neutralize any spilled sulfuric acid with powdered sodium bicarbonate (baking soda).

INTRODUCTION

Oxygen can be readily obtained in the laboratory by decomposing a number of compounds containing it. However, many compounds that contain oxygen do not decompose to yield it. Industrially, O_2 is obtained from the air by first liquefying and then fractionally distilling it. Relatively small amounts are obtained by the electrolysis of water.

One of the most practical laboratory preparations involves the catalytic decomposition of $KClO_3$ according to the reaction:

$$2\ KClO_3 \rightarrow 2\ KCl + 3\ O_2$$

Oxidation reactions with most fuels are, of course, more intense in pure O_2 than in air, which is only about 21% O_2 by volume. Oxides of the elements in the fuel are produced in general. Metallic oxides, if they show any appreciable reaction with water at all, produce a basic solution and are called basic anhydrides. Nonmetallic oxides that react appreciably with water produce acidic solutions and are called acid anhydrides. In this experiment iron and magnesium are used as examples of metals. Sulfur is an example of a nonmetal.

Ozone, O_3, a high-energy allotrope of oxygen, can be formed by several endoenergetic (endothermic) reactions. Most of its reactions are similar in kind but are more intense than those of O_2.

PROCEDURE

A. Preparation of Oxygen

1. **Electrolysis of water (demonstration).** At the beginning of the laboratory period the instructor will put into operation an apparatus for decomposing H_2O into its elements. A direct electric current is passed through H_2O slightly acidified with H_2SO_4 (for better conductivity). The electrodes are arranged so that the two gases evolved are collected separately. At the end of the period these gases are tested. Record the identity of each gas, the electrode at which it was liberated, and the relative amount of each. Write the balanced equation for the reaction.
2. **Decomposition of compounds containing oxygen.** *Be sure to wear your safety glasses throughout this experiment!* Prepare five clean, dry test tubes and label them respectively Fe_2O_3, MnO_2, PbO_2, $NaNO_3$, MgO. Put about one-half inch of the named substance into each of them. Heat each strongly, testing for evolved O_2 frequently with a glowing splint. *Be careful not to drop the splint into the molten material. A strong oxidizing agent may react explosively.* Report your results. Do all substances containing oxygen evolve O_2?

B. Collection of Oxygen

1. **Investigation of $KClO_3$ reaction.** Put a spatula of Fe_2O_3 and a loose plug of glass wool into a large (25 × 150 mm), clean test tube. Weigh tube and contents to the nearest 0.01 gram. Roughly weigh out about five grams of $KClO_3$, remove the glass wool plug, put the $KClO_3$ into the test tube, and at once reinsert the plug. Be sure that no glass wool is lost. The purpose of the plug is to prevent molten $KClO_3$ from contacting the rubber stopper and also to reduce loss of KCl by sublimation.

   Weigh tube and contents accurately after shaking to mix $KClO_3$ and catalyst. Find the exact weight of $KClO_3$ by difference. Set up the apparatus shown in Figure 32.1. Be sure to clamp the generator test tube near the stopper and slant it toward the closed end. Fill three gas bottles with water and invert them in a water-filled pneumatic trough.

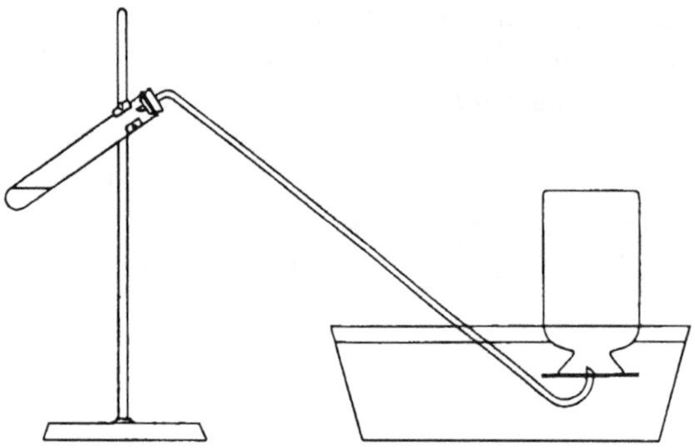

   FIGURE 32.1
   Apparatus for oxygen collection.

   Start heating the generating tube but let the first few bubbles, which are heat-expanded air, escape. Then move the gas bottles one by one over the delivery tube outlet and collect three bottles of O_2 gas. Slip glass plates over the mouths of the bottles before removing from the water and set the covered bottles on the desk for later use.

   Disconnect the delivery tube and continue to heat the generator strongly as long as there is any activity. *Do not let any part of the material get red-hot.* Then let the generator cool and weigh it accurately. While it is cooling you may proceed with Part C. Determine the weight of KCl remaining and compare it with the theoretical amount from the weight of $KClO_3$ used. Explain the difference.

   Remove the glass wool plug, fill the test tube almost full of water, insert a stopper, and shake as long as dissolving is occurring. Filter the solution to get rid of the catalyst. Put about 10 mL of the filtrate into a clean test tube, add a few drops of HNO_3, and about 1 mL of $AgNO_3$ solution. The white precipitate is AgCl, and shows that the chloride ion Cl^- was present. Write the ionic equation. Put a spatula of $KClO_3$ in a test tube, add 10 mL of distilled water, and make the $AgNO_3$ test. Is Cl^- present? Repeat the test with a similar amount of KCl. What conclusion can you draw?

C. Formation of Some Oxides and Their Reactions with Water

1. Obtain a deflagrating spoon and make sure it is clean. Put into it a pea-sized piece of charcoal, ignite it, and lower it into one of the bottles of O_2, replacing the cover as fully as possible. When the flame has expired slip the cover aside and add one-half inch of distilled water. Replace the cover and shake. Finally, test the water with litmus paper and record your results.
2. Clean the deflagrating spoon and repeat the experiment with one-half spatula of sulfur. Add distilled water to the bottle, test with litmus as before, and report your results.

3. Slip the cover of the last bottle of oxygen aside enough to add one inch of water and then replace it. Fasten a small wad of steel wool to a 6-inch piece of iron wire. Holding it by the wire, heat it until it glows then lower it *quickly* into the O_2. Do not drop it. Detecting acidity or basicity with litmus is not possible because iron oxide is essentially insoluble in water.
4. Hold a piece of magnesium ribbon in the tongs, ignite it, and let it burn in the air. Catch the oxide on a watch glass. Do not look directly into the flame. Mg burns so easily with O_2 that even the dilute concentration of it in the air permits rapid reaction. Add a few drops of water to the oxide of Mg and make the litmus test. The reaction with water is slow.

D. Unknown Oxides

Obtain three unidentified oxides—CaO, P_2O_5, and MgO. Determine by the test used in this experiment which behave like metal oxides and which behave as nonmetal oxides.

E. Ozone (Demonstration)

The instructor will set up an ozone generator (Figure 32.2) which consists essentially of a tube situated between two electrodes so that a high voltage discharge can be passed through oxygen. A $KClO_3$ oxygen generator, a source of bottled O_2, or an air line is connected to the ozone generator and a gentle stream of gas is passed through the discharge. Ozone can be detected by its odor or by the color change in a piece of wet starch-iodide paper held near the outlet.

To demonstrate the effect of ozone on rubber, the instructor will test the elasticity of the rubber tubing connected to the ozone generator outlet. Note the result and explain.

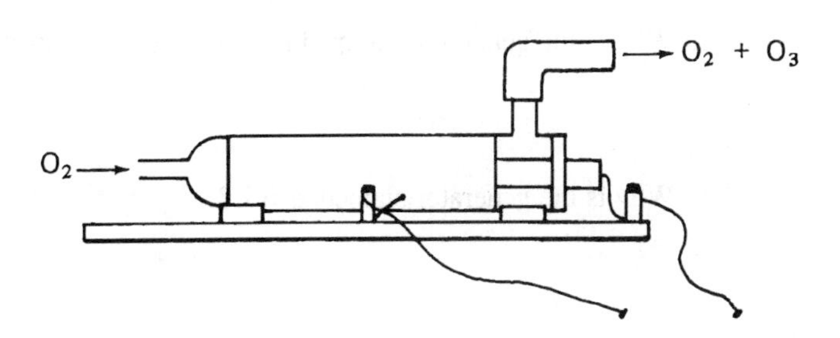

FIGURE 32.2
Apparatus for ozone generation.

Experiment 32

Oxygen and Some Oxides

Name______________________________

Date______________________________

REVIEW QUESTIONS

1. a. Why was Fe_2O_3 used in this experiment? (Part B1)

   b. Refer to a previous experiment and name several other substances that could have been used in place of Fe_2O_3.

2. Why is the generator clamped near the rubber stopper?

3. Why is the generator slanted as it is?

4. What would happen if a glowing splint fell into molten $KClO_3$, or if the molten $KClO_3$ touched rubber? Explain.

5. From your collection of O_2 in contact with H_2O and your observation of the air in contact with rivers, lakes, and oceans what can you assume regarding the solubility of O_2 in H_2O? Look up facts and report.

6. What is the literal meaning of the term "basic anhydride"?

7. For each element in the following list write the formula of the oxide and predict whether the oxide is basic, neither, or both.

   Na, Ca, Cl, Al, Fr, Sb

8. What reactions occur in starch-iodide paper in contact with O_3? In general starch-iodide paper is a test for what kind of substance?

9. What commercial use could be made of the behavior of O_3 toward rubber?

Experiment 32

Oxygen and Some Oxides

Name________________________________

Date________________________________

REPORT SHEET

A. Preparation of Oxygen

1. Electrolysis of water

	Positive electrode	Negative electrode
Volume of gas produced	________________	________________
Identity of gas	________________	________________

Balanced equation:____

2. Decomposition of compounds containing oxygen

Oxide	Observation	Equation
Fe_2O_3		
MnO_2		
PbO_2		
$NaNO_3$		
MgO		

B. Collection of Oxygen

1. Investigation of $KClO_3$ reaction

Weight of tube + Fe_2O_3 + $KClO_3$ =

Weight of tube + Fe_2O_3 = ________________

Weight of $KClO_3$ =

Weight of tube and contents after heating =

Weight of tube + Fe_2O_3 = ________________

Weight of product (KCl) =

Calculation of theoretical weight of KCl:

Account for any significant difference.

2. Write ionic equation for Cl^- test ______________________________

Does $KClO_3$ give this test? ____________________ Does KCl give it? ____________________

Tentative conclusion: ______________________________

C. Formation of Some Oxides

Element	Metal or nonmetal	Observation: Acidic or basic on burning	Equation for Combustion
Carbon (charcoal)			
Sulfur			
Iron			
Magnesium			

Equation for reaction of each oxide and water, if observable:

1.

2.

3.

4.

D. Ozone

Equation for formation of ozone: ____________________

Explain action of ozone on starch-iodide paper: ______________________________

Explain action of ozone on rubber: ______________________________

EXPERIMENT 33

Colligative Properties

Purpose: To test the colligative property of freezing point depression. To examine the effect of the number of ions per formula weight on freezing point depression.

Materials: NaCl, $CaCl_2 \cdot H_2O$, $C_{12}H_{22}O_{11}$ (sucrose), $Fe(NO_3)_3 \cdot 9H_2O$, for each student $Ce(NH_4)_2(NO_3)_6$ for demonstration.

Apparatus: 150-mL beakers, thermometers accurately scaled in the –20 to +20° C range.

Safety Precautions: Wear your safety glasses at all times in the lab. Use caution when handling the solids. Be careful when handling the thermometer. Do not drop it or allow it to roll off the table. The mercury inside is particularly difficult to clean up.

Hazardous Waste Disposal: All the solutions may be washed down the drain with plenty of water.

INTRODUCTION

Colligative properties are physical properties that depend on the *quantity* of discrete solute particles and not (very much) on the identity of the solute. Examples of colligative properties include boiling point, freezing point, osmotic pressure, and vapor pressure.

Pure water freezes at 0° C (273 K), boils at 100° C (373 K), and exerts a vapor pressure of 23.76 mm Hg at 25° C (298 K). It is possible to estimate new physical properties which will be observed upon the addition of a solute. (**Note:** Our calculations are *only* estimations, the validity of the equations breaks down at higher concentrations.)

Let's take an example. Dissolve 10g of ethanol, C_2H_5OH, in 100g of pure water. What will the new boiling point be? Freezing point?

The concentration of solute particles is usually expressed in *molality* for colligative properties (the exception is osmotic pressure). Recall that molality is defined as "moles of solute per kilogram of solvent." Therefore, we must first calculate the molality of the solution.

$$10 \text{ g } C_2H_5OH \times \frac{1 \text{ mole}}{46.069 \text{ g}} = 0.217 \text{ mole ethanol}$$

$$\text{molality} = \frac{\text{moles ethanol}}{\text{kg water}} = \frac{0.217 \text{ mol}}{0.100 \text{ kg}} = 2.17\ m$$

The equations relating boiling point elevation and freezing point depression to molality are

$$\Delta T_b = k_b \times m \times i$$
$$\Delta T_f = k_f \times m \times i$$

where ΔT is the boiling point or freezing point change, m is the molality, k is a constant for the particular solvent, and i is a multiplier we shall discuss shortly.

For water the constants have the values $k_b = 0.52°\ C/m$ and $k_f = 1.86°\ C/m$. So for this theoretical ethanol solution $\Delta T_b = 0.52°\ C/m \times 2.17\ m \times 1 = 1.13°\ C$, or the boiling point is *increased* by 1.13° C to 100° C + 1.13° C = 101.13° C. Likewise, $\Delta T_f = 1.86\ °\ C/m \times 2.17m \times 1 = 4.04°\ C$ and the freezing point is *decreased* by 4.04° C to 0° C – 4.04° C = –4.04° C.
Do not confuse freezing point and freezing point depression!

Would our calculations be any different if we had chosen anything else to dissolve in the water? Maybe. Remember that substances usually come in two basic classifications: ionic and covalent, or more precisely, *electrolytes* and *nonelectrolytes*. Electrolytes dissolve to give *ions* (charged particles) in solution while nonelectrolytes dissolve to give *molecules*. In our example ethanol is a nonelectrolyte. A mole of ethanol dissolves to give a mole of ethanol molecules. The same number of moles of any other nonelectrolyte (e.g., 13.47g $C_2H_6O_2$ or 25.82g of $SOCl_2$) would give the *same* molality and therefore the same boiling point elevation and freezing point depression.

On the other hand, an electrolyte would give a value of ΔT that is (theoretically) twice, three times, four times, or more as much as a molar equivalent of nonelectrolyte. The multiplying factor, i, adjusts the equations to compensate for this. One mole of ethanol yields one mole of ethanol molecules in solution. However, one mole of common table salt (NaCl) when dissolved in water yields *two* moles of *ions* in solution.

$$NaCl_{(s)} \xrightarrow{H_2O} Na^+{}_{(aq)} + Cl^-{}_{(aq)} \qquad \text{2 moles}$$

Other electrolytes yield more than 2.

$$(NH_4)_2SO_4 \xrightarrow{H_2O} 2\ NH_4^+{}_{(aq)} + SO_4^{-2}{}_{(aq)} \qquad \text{3 moles}$$
$$Fe(NO_3)_3 \xrightarrow{H_2O} Fe^{+3}{}_{(aq)} + 3\ NO_3^-{}_{(aq)} \qquad \text{4 moles}$$
$$Al_2(CO_3)_3 \xrightarrow{H_2O} 2\ Al^{+3}{}_{(aq)} + 3\ CO_3^{-2}{}_{(aq)} \qquad \text{5 moles}$$

The values of i are NaCl (i=2), $(NH_4)_2SO_4$ (i=3), $Fe(NO_3)_3$ (i=4), and $Al_2(CO_3)_3$ (i=5). In order to analyze i fully, you must be able to recognize and interpret the formulas for ionic compounds, and especially the formulas and charges of the common polyatomic anions.

Sometimes a substance may be covalent in nature, but possesses a slight tendency to ionize in aqueous solution. The ionization almost always occurs as a result of the formation of H^+, the H specifically having been attached to O or N (note the same qualifications as for hydrogen bonding).

One example is the organic class of carboxylic acids, general formula R–COOH, where R is any carbon chain. The sharp taste of vinegar is caused by the presence of acetic acid, CH_3COOH.

```
   H   O              H   O⁻
   |   ||             |   ||
H—C—C—OH   ⇌   H—C—C—O + H⁺
   |                  |
   H                  H
```

The double-headed arrow indicates that the process is an *equilibrium*. An equilibrium constant, K_a, indicates the extent to which the written reaction occurs. K_a for acetic acid is very small (1.8×10^{-5}), implying that only about four out of every thousand acetic acid molecules undergo this reaction. It is reasonable then to assume that the value of i for acetic acid is approximately one. If acetic acid were completely dissociated i would be equal to 2. For most compounds like this, the dissociation is quite small and the ionization may be ignored for the purpose of colligative property calculations.

In this experiment you will investigate several solutes to determine their effect on the freezing point of aqueous solutions. You should be able to observe how changes in the value of i cause changes in ΔT and therefore the freezing point.

PROCEDURE

1. Work in pairs.

   Each pair will need a thermometer, two 150-mL beakers, a plastic weigh boat, a stirring rod, a large beaker of crushed ice, and some distilled water.

2. First, check the accuracy of your thermometer. Fill one 150-mL beaker nearly full with crushed ice. Add approximately 50 mL of distilled water. Stir for about 30 seconds or until the ice bath reaches thermal equilibrium. Be certain that your stirring rod is clean. Check the temperature. Record the temperature of your pure ice bath on your report form. The thermometer should read 0.0° C. If it does not, repeat this step making certain no contamination is possible. If you still have difficulty, consult the instructor or a TA.
3. In one clean, dry 150-mL beaker weigh out 50.0 g of water. In the other, weigh out 50.0 g of ice (nearly a full beaker of ice). Dry off the outside of the beakers. Try to get close to 50.0 g, but do not waste too much time attempting to get exactly 50.00 g.

   In the weighing dish, weigh out approximately the recommended amount of the solute. Record the exact amount used. The amounts are

2.91g	$NaCl$	17.11g	$C_{12}H_{22}O_{11}$ (sucrose)
7.35g	$CaCl_2 \cdot 2H_2O$	20.19g	$Fe(NO_3)_3 \cdot 9H_2O$

   Again, the amounts do not have to be exactly the same, just reasonably close.

   To ease competition for resources, you may do the solids in any order.
4. Transfer the weighed solid to the beaker of distilled water. Try to insure a quantitative transfer, but do NOT use any more water.

   Stir with the stirring rod until *all* the solid has dissolved. Do not get impatient here—all the solid must be dissolved before you can go on to the next step.
5. Pour the solution into the 150-mL beaker with the ice. CAREFULLY AND GENTLY stir with the thermometer. This will help the thermometer come to thermal equilibrium.

   After about 30 seconds of gentle stirring, most but NOT ALL, of the ice will have melted. Check the temperature. Since solid ice is present, the solution is by definition at its freezing point. After another 30 seconds or so check the temperature again. If it has not changed, record the temperature to the nearest 0.1° or 0.2° C, as allowed by your thermometer. If the temperature is still falling, wait another 30 seconds and check again. (Record the coldest stable temperature.)
6. Dispose of solution by pouring it down the sink and flushing it with water. Rinse *and dry* both beakers, the weigh boat, the stirring rod, and the thermometer.
7. Repeat for the other solutes.
8. If a demonstration solution is set out, record the data for that.
9. If time allows and you are skeptical of a measurement, repeat it. There is much source for error. Remember also that the theoretical calculations are just estimations. In order to attain measurable changes, very large amounts of solid are used. As a result, the high concentrations cause the basic assumptions of the equations to break down. Nevertheless, these substances should exhibit a clear trend.

   Return the clean, dry thermometer.

CALCULATIONS

Calculate the molality of each solution. Convert grams solute into moles solute, being careful to include any water of hydration into the molar mass. Next, it is permissible to assume that the ice plus the water is the total solvent, so in each case the solvent weighed 100.0 g or 0.100 kg. Divide each number of moles by 0.100 kg to get the molality. Record this value to three significant figures.

Examine the formula to determine the value of *i*. For these solutes all values of *i* will be small whole numbers (no partial ionization here).

For each solution calculate the theoretical freezing point in °C. Recall that the equation for freezing point depression is $\Delta T_f = k_f \times m \times i$ and that for water, $k_f = 1.86°C/m$. Don't confuse freezing point with freezing point depression. It is only a coincidence that the solvent water freezes at 0.0°C. Therefore, the theoretical freezing point for each solution is $0.0 - \Delta T_f$, or simply $-\Delta T_f$.

Finally, comment on your report sheet how the experiment shows the relationship between freezing point and number of particles in solution.

Experiment 33
Colligative Properties

Name______________________________

Date______________________________

REVIEW QUESTIONS

1. Calcium chloride, rather than sodium chloride, is often used by homeowners to melt snow and ice on sidewalks. Give at least two good reasons for the preference.

2. What is the boiling point of a solution made of 200 g of benzene (solvent) to which 10.25 g of naphthalene, $C_{10}H_8$ (a nonelectrolyte) has been added? (You will need the constants for benzene from your text or the *CRC Handbook of Chemistry and Physics*.) Show all of your work.

3. We assumed that the ice was pure H_2O—that it had no contaminating solutes present in the solid ice. Is this a reasonable assumption given that the laboratory ice machine uses tap water?

4. Permanent antifreeze is also called "summer-winter" antifreeze. Why is the solution useful for *both* winter and summer driving?

5. Explain why molarity and molality are nearly equal for dilute aqueous solutions.

6. In an actual experiment, the boiling point of a mixture of ethanol and water turned out to be very different from that calculated from colligative properties. Suggest a reason why.

7. Aside from seasoning, why is salt added to a pot of boiling water to prepare foods?

Experiment 33
Colligative Properties

Name ______________________________

Date ______________________________

REPORT SHEET

100.0 g of pure ice water freezing point ____________________ °C

Solute	mass(g)	Observed f.p. (°C)	*m*	*i*	Theoretical f.p. (°C)
NaCl					
$CaCl_2 \cdot 2H_2O$					
$C_{12}H_{22}O_{11}$					
$Fe(NO_3)_3 \cdot 9H_2O$					
$Ce(NH_4)_2(NO_3)_6$ (or formula of other demonstration)					

Conclusions:

EXPERIMENT 34

Molecular Weight by Freezing Point Depression

Purpose: To measure the molecular weight of a substance by its effect on lowering the freezing point of a solvent.

Materials: Naphthalene, 1,4-dichlorobenzene; 1,4-dibromobenzene; 1,4-nitrochlorobenzene; 1-hexadecanol; 4-nitrotoluene; 2-chloro-4,6-dinitrophenol.

Waste Disposal: Dispose of naphthalene, mixtures of naphthalene, and other substances in the waste disposal container.

INTRODUCTION

One mole of any nonionized, nonvolatile solute dissolved in 1,000 g of a solvent lowers the freezing point of the solvent by a fixed amount. This quantity, called the molal freezing point depression constant, is a colligative property of the solvent, because it depends on the *number* of solute particles in a given amount of solvent rather than on their nature. For the solvent naphthalene, used in this experiment, the molal freezing point depression constant is 6.85° Celsius.

The weight of a mole of solute can be determined by measuring the lowered freezing point, which is produced by a known weight of the solute dissolving in a known weight of naphthalene. The weight of solute per 1,000 g of solvent is easily calculated. The number of moles this represents is determined by comparing the freezing point lowering it produces with that produced by one mole. From this the weight of one mole is calculated.

PROCEDURE

A. Freezing Point of Pure Naphthalene

Since ordinary laboratory thermometers are often in error by a degree or so, the melting point of naphthalene should be measured with your thermometer. The apparatus consists of a 25- × 150-mm test tube, clamped to a ring stand, so that the bottom can be heated gently with a Bunsen burner. Weigh the clean, dry test tube and small beaker in which it is held upright as accurately as possible. Then weigh out roughly 15 grams of naphthalene, put it in the tube, insert the thermometer, and carefully melt the sample. When the temperature has reached about 95° C, remove the flame and let the naphthalene cool while stirring. The thermometer should read to the nearest 0.1° C at 1-minute intervals and the readings recorded. At the freezing point the temperature will remain virtually constant until solidification is nearly complete. Why? Continue recording the temperature until it has been constant for at least 5 minutes. Remelt the naphthalene, remove the thermometer and clean it, and weigh the tube and contents accurately while mounted in the same beaker. The naphthalene is weighed after the melting point has been taken because some material is lost by vaporization. Plot the cooling curve for naphthalene on the graph and record the exact melting point.

B. Molecular Weight of 1,4-dichlorobenzene

On a piece of paper weigh out roughly 2 grams of 1,4-dichlorobenzene and pour it carefully into the naphthalene through a wide-stemmed funnel. This keeps it off the walls of the test tube. Weigh the test tube and contents accurately again. The increase in weight is the exact weight of 1,4-dichlorobenzene added. Mount the test tube on the ring stand and melt the sample while stirring gently with the thermometer. When the temperature has reached 95° C remove the flame and allow the sample to cool while stirring. Read and record the temperature to the nearest 0.1° C at 1-minute intervals. At the freezing point there is an interval of several minutes when no temperature occurs, and crystals form. Consider this first break in cooling rate to be the freezing point of the solution. Record this temperature and determine the freezing point lowering.

From the concentration of your solution calculate the quantity of solute which would have been present in 1,000 g of naphthalene. From your freezing point lowering compared with the lowering of a 1-molal solution (6.85°C), determine the molal concentration of your solution. The information you now have enables you to calculate the grams per mole of solute.

C. Molecular Weight of an Unknown

Obtain an unknown, which may be one of the substances listed among the materials, or one selected by your instructor. Start with a clean, dry test tube and weigh it accurately as before. Add about 15 g naphthalene, weigh accurately again, and record the weights. Record the number of the unknown and add about 2 grams of it to the naphthalene, keeping it off the walls of the test tube. Weigh accurately and record the weight. Carefully melt the sample with the thermometer in place. When the temperature has reached 95° C remove the flame and let the sample cool while stirring. Record the temperature to every 0.1° C every minute. Determine the freezing point lowering, using the freezing point of pure naphthalene you determined in part A. Calculate the molecular weight of the unknown. The instructor will supply the name of the unknown, and you can calculate your percent error.

All test tubes should be cleaned by melting the contents and pouring them into the waste jar provided, *not* into the sink. Final cleaning can be done by benzene under the fume hood, followed by acetone and distilled water.

Experiment 34

Molecular Weight by Freezing Point Depression

Name______________________________________

Date_______________________________________

REVIEW QUESTIONS

1. To what errors, other than carelessness, is this experiment subject? How might these errors have been eliminated or reduced?

2. If the freezing point of the solution had been read 0.2° C lower than correct, what would have been the percent error?

3. Why does the temperature remain constant for several minutes during the solidification of the pure naphthalene?

4. Why does the temperature of the solution during solidification not remain as constant after the addition of 1,4-dichlorobenzene as it was with pure naphthalene? Why take as the freezing point the first break in the cooling curve?

5. Would the melting point have been too high, too low, or unaffected if:
   (a) some 1,4-dichlorobenzene had remained stuck to the weighing paper or the upper part of the test tube;

   (b) some rubber from a stopper had gotten into the naphthalene;

   (c) the sample had not been completely dissolved;

   (d) some unmelted naphthalene had stuck to the upper part of the tube;

   (e) the naphthalene had been impure to begin with;

   (f) the thermometer had read 0.3° C too high in both the pure naphthalene and the solution;

   (g) the solute had ionized to an appreciable extent? Explain your answers.

6. What other method, different in procedure but very similar in principle, might have been used for molecular weight determination?

7. How many molecules of 1,4-dichlorobenzene were present in the samples you used? How many molecules of naphthalene were present?

8. The molal boiling point elevation constant of naphthalene is 5.65° C. If the boiling point of pure naphthalene is 218° C at 1 atmosphere, what would have been the boiling point of your 1,4-dichlorobenzene solution?

Experiment 34

Molecular Weight by Freezing Point Depression

Name____________________

Date____________________

REPORT SHEET

A. Freezing Point of Naphthalene

Weight of tube + beaker + naphthalene =

Weight of tube + beaker =

Weight of naphthalene =

Temperature Readings

1.________	6.________	11.________
2.________	7.________	12.________
3.________	8.________	13.________
4.________	9.________	14.________

B. Molecular Weight of 1,4-dichlorobenzene

Weight of tube + beaker + naphthalene =

Weight of tube + beaker =

Weight of naphthalene =

Weight of tube + beaker + naphthalene + sample =

Weight of tube + beaker + naphthalene =

Weight of sample =

Temperature Readings

1.________	8.________
2.________	9.________
3.________	10.________
4.________	11.________
5.________	12.________
6.________	13.________
7.________	14.________

Freezing point of solution ________ Freezing point lowering ________

Calculation of molecular weight:

Percent error ____________

C. Molecular Weight of an Unknown

Number of unknown ______________________

Temperature Readings

Weight of tube + beaker + naphthalene	=	1. __________	8. __________
Weight of tube + beaker	=	2. __________	9. __________
Weight of naphthalene	=	3. __________	10. __________
		4. __________	11. __________
Weight of tube + beaker + naphthalene + sample	=	5. __________	12. __________
Weight of tube + beaker + naphthalene	=	6. __________	13. __________
Weight of sample	=	7. __________	14. __________

Freezing point of solution __________ Freezing point lowering __________

Calculation of molecular weight:

Identity of unknown __________

True molecular weight __________

Percent error __________

EXPERIMENT 35

Catalysis

Purpose: To observe the behavior of catalysts in several reactions.

Materials: $KClO_3$, MnO_2, Fe_2O_3, Al_2O_3, Pb_3O_4, CuO, PbO_2, SiO_2, ascorbic acid (vitamin C) tablets; raw potato; 3% H_2O_2; 6% H_2O_2, 6 N H_2SO_4; $KMnO_4$ (8 g/l); $H_2C_2O_4$ (40 g/l); 0.2 N solutions of $Pb(NO_3)_2$, $BaCl_2$, and $Cu(NO_3)_2$.

Equipment: 110° C laboratory oven.

Waste Disposal: Dispose of all mixtures containing lead, barium, or copper compounds in the heavy metal waste container.

INTRODUCTION

Catalysts change the rate of chemical reactions without themselves being permanently changed. They are usually used to increase reaction rates, but may be introduced to retard reactions, in which case they are negative catalysts or inhibitors.

Catalysts apparently work by making possible reaction paths that require less activation energy. In some cases a reacting substance forms an intermediate compound with the catalyst. This intermediate substance decomposes very readily to form the final products, and liberates the catalyst for another cycle ($KClO_3$ decomposition). In other cases a reactant is adsorbed to the surface of a catalyst, providing a high concentration or an activated form of the reactant (Pt in H_2 gas reactions). Inhibitors (catalysts that slow reactions) apparently act by destroying positive catalysts or by interfering in some manner with the normal reaction mechanism.

The complex reactions that occur in any life process depend on catalysts called enzymes.

PROCEDURE

NOTE: The reaction of Part E requires an entire laboratory period. Set it up first, put it aside to work, and proceed with the rest of the experiment.

A. Catalysis in $KClO_3$ Decomposition

Label six clean, dry test tubes as follows: MnO_2, Fe_2O_3, Al_2O_3, Pb_3O_4, CuO, and SiO_2. Put about one-fourth inch $KClO_3$ into each test tube. Have ready on separate weighing papers (properly labeled) a half-spatula of each substance to be tested for catalytic action.

Heat the $KClO_3$ in the first test tube until it is just molten, then quickly pour in the MnO_2 from the first paper. Note if there is gas evolution and quickly test the gas with a glowing splint if active evolution occurs. Repeat the process with each of the other substances being tested. Record your results. Write the equation for the reaction.

B. Recovery of Catalyst

Weigh a large, dry test tube to the nearest 0.01 gram. Add to it about 0.5 gram of CuO and weigh accurately again. Get the exact weight of CuO by difference. Next add about 5 grams of $KClO_3$, roughly weighed, and insert a loose plug of glass wool in the mouth of the tube. Heat the material until all gas evolution ceases, then let it cool. Meanwhile, weigh a filter paper accurately and set up a filter. When the reaction tube is cool, fill it almost full of demineralized water, insert a stopper, and shake until dissolving appears to have stopped. Pour the solution carefully through the filter and again add water to the test tube if any undissolved material remains. Repeat until the test tube is clean, pouring all the liquid through the filter.

When the filter has drained, fill it again with distilled water, let it drain, and dry the paper in an oven at 110° C. Finally weigh it accurately and determine the weight of CuO recovered. How does it compare with the amount of catalyst originally used?

C. Catalysis in H_2O_2 Decomposition

Prepare seven clean test tubes and label them respectively: no catalyst, MnO_2, Fe_2O_3, Al_2O_3, CuO, PbO_2, and SiO_2. Mix 10 mL of 3% H_2O_2 solution and 30 mL of distilled water and put about 5 mL into each test tube. Put into each test tube about one-half spatula of the appropriate substance being tested for catalytic action and note if there is any sign of gas evolution. Compare results with the sample that contains no catalyst and report. Write the equation for the decomposition of H_2O_2. In Experiment 14 this decomposition was catalyzed by NH_4OH and $MnCl_2$.

D. Catalysis in Oxidation by $KMnO_4$

Into a 25- × 150-mm test tube put 5 mL each of $H_2C_2O_4$ and H_2SO_4 solutions. Measure out 5 mL of $KMnO_4$ solution, *quickly* add it to the $H_2C_2O_4$-H_2SO_4 mixture and measure the time it takes for the mixture to clear. Record your observations.

Prepare four separate mixtures of 5 mL each of $H_2C_2O_4$ and H_2SO_4. To one add 2 drops of $Pb(NO_3)_2$ solution, to another 2 drops of $BaCl_2$, to the third add 2 drops of $Cu(NO_3)_2$, and to the fourth add 2 drops of 6% H_2O_2. Add 5 mL of $KMnO_4$ solution to each and time the reaction. Write the equation for the reaction.

E. Action of an Inhibitor

Certain fruits and vegetables such as bananas, apples, peaches, and potatoes darken in color when cut and exposed to the air. The reaction is an air oxidation catalyzed by the enzyme tyrosinase. Various substances can inhibit this reaction, presumably either by combining with the enzyme and destroying its action or by uniting with the substrate (substance which darkens) and preventing its oxidation.

Ascorbic acid (vitamin C) functions as an oxidation inhibitor. Crush a 100-mg tablet in 50 mL of distilled water and stir until maximum solution has occurred. Cut two thin slices from a potato just before the experiment, dip one into the ascorbic acid solution and the other into plain water and expose both to air on a paper towel. After about 15 minutes pour a few more drops of ascorbic acid on the treated slice. Lay both slices aside until the end of the period and observe. Record your observations.

Experiment 35
Catalysis

Name________________________________

Date________________________________

REVIEW QUESTIONS

1. Can a catalyst initiate a reaction that otherwise would not occur at all?

2. Does the quantity of a catalyst present have any significant effect on a reaction?

3. Can a catalyst cause a normally reversible reaction to go more nearly to completion? Explain.

4. What is meant by autocatalysis?

5. Look up some industrial reactions that make use of catalysts and name the catalyst.

6. Discuss practical situations other than the reaction in the experiment in which inhibitors are used.

7. Refer to textbooks or reference books on physiology, biochemistry, or botany and describe some specific applications of catalysis to life processes.

8. Many reactions require the application of activation energy in order to start them. Discuss the effect of a catalyst on activation energy.

Experiment 35
Catalysis

Name____

Date____

REPORT SHEET

A. Catalysis in $KClO_3$ Decomposition

Rate each substance A (good action), B (some action), or C (no action)

Substance Tested	Observation	Rating
MnO_2		
Fe_2O_3		
Al_2O_3		
Pb_3O_4		
CuO		
SiO_2		

Write the equation for the reaction:

B. Recovery of a Catalyst

Weight of tube and catalyst =

Weight of tube only =

Weight of catalyst used =

Weight of filter paper and catalyst =

Weight of filter paper only =

Weight of catalyst recovered =

How does the amount of catalyst recovered compare with that originally used?

Account for any difference:

C. Catalysis in H_2O_2 Decomposition

Rate each substance A (good action), B (some action), or C (no action)

Substance Tested	Observation	Rating
No catalyst		
MnO_2		
Fe_2O_3		
Al_2O_3		
CuO		
PbO_2		
SiO_2		

D. Catalysis in Oxidation by $KMnO_4$

Rate each substance A (good action), B (some action), or C (no action)

Substance Tested	Observation	Rating
$Pb(NO_3)_2$		
$BaCl_2$		
$Cu(NO_3)_2$		
H_2O_2		

E. Action of an Inhibitor

What differences are observed between the two samples?

What practical application has the observed behavior?

EXPERIMENT 36

Hydrogen

Purpose: To prepare hydrogen, H_2, by a practical laboratory method; to observe some of the physical and chemical properties of H_2.

Materials: Mossy Zn, 6 NH_2SO_4, CuO, $CaCl_2$.

Apparatus: Drying tubes, unglazed porcelain cup, gas explosion apparatus.

Hazard: Because of the danger of explosion, the instructor may prefer to do some or all of this experiment by demonstration.

INTRODUCTION

Hydrogen gas, the least dense material on the face of the earth, is the most plentiful element in the universe and among the most plentiful on earth. It does not occur as a free element in the earth's atmosphere in appreciable amounts because it readily combines with oxygen to form water, and because a fraction of its molecules are always moving rapidly enough to exceed escape velocity such that the earth's gravitational field cannot hold it. The sun and other stars are composed primarily of elemental hydrogen, and hydrogen's nuclear transformation (fusion) to form helium produces the energy radiated from the sun and stars.

The gas has some extremely important industrial applications and it is prepared commercially by electrolysis of water or as the by-product of a number of petroleum refining operations such as catalytic cracking, hydroforming, and alkylation. It is easily prepared in the laboratory by the reaction of a moderately active metal with a mineral acid, but an element of danger is involved unless certain precautions are observed. The generator must be wrapped in a towel and there must be no open flame within 3 feet of the generator unless it is free of air. In this experiment you will prepare the gas and make some observations on its solubility, density, ability to diffuse, reducing power, and capacity for explosion.

PROCEDURE

A. Laboratory Preparation of H_2

Assemble the apparatus diagrammed in Figure 36.1. The generator is a 250-mL thick-walled glass bottle fitted with a two-hole stopper carrying a thistle tube and a delivery tube. The thistle tube should extend to one-fourth inch from the bottom of the bottle when the stopper is firmly in place. Put into the bottle 20 mL of distilled water and about 20 g of mossy zinc. Wrap the body of the generator in a towel and fasten the towel with rubber bands.

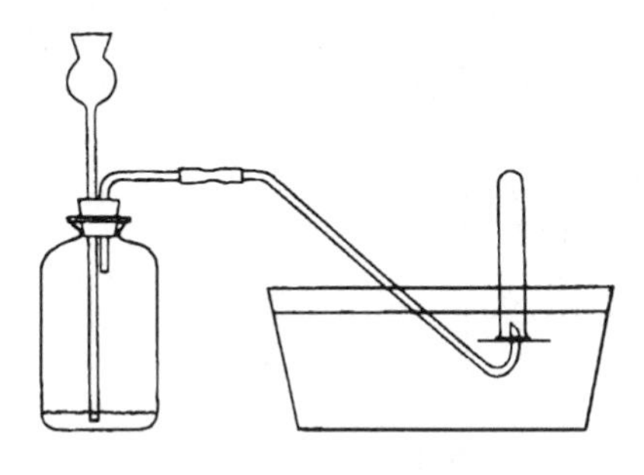

FIGURE 36.1
Hydrogen Generator.

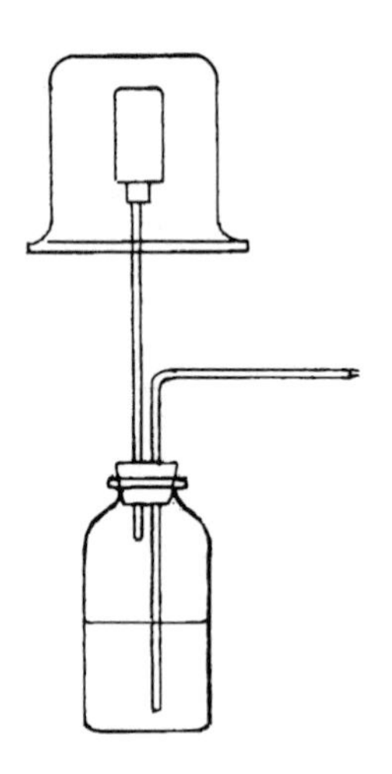

FIGURE 36.2
Diffusion Apparatus.

Before starting to generate H_2, some other apparatus must be prepared. Assemble the diffusion apparatus pictured in Figure 36.2. The upper vessel is an unglazed porcelain cup fitted with a one-hole rubber stopper and a 16-inch glass tube inserted into another bottle that is half-filled with water. It has a right-angle tube drawn to a tip and reaching almost to the bottom of the bottle. All stoppers must fit tightly.

Stopper a dry 500-mL Florence flask and weigh it to the nearest 0.01 gram or the nearest 0.001 gram if possible. Record the weight and set the flask aside.

Have your instructor inspect all apparatus. When it has been approved, measure out 50 mL of dilute (6 N) H_2SO_4 and pour about 15 mL of it into the generator through the thistle tube. Let the evolved gas escape for a few minutes, then collect some in a small inverted test tube filled with water. When the tube is full of gas put your thumb over the mouth, carry the tube still inverted to a lighted Bunsen burner (at least 3 feet from the nearest generator), and ignite it. A whistling explosion tells you that the gas is still mixed with air. Repeat the test until the H_2 burns quietly. At this point the tube can be laid in the trough and you can see the gas burn as water displaces it. *Do not try to explode a glass vessel of H_2 any larger than a test tube. The explosion may hurl broken glass.*

B. Density of Hydrogen

When all the air has been expelled from the generator, invert your weighed flask, clamp it to a stand, and insert a glass tube all the way to the top. Connect the tube to the generator through a $CaCl_2$ drying tube and let H_2 pass through until the air has been swept out. Put the stopper in place and weigh the flask to the nearest 0.01 (or 0.001) gram. Record the weight. Does it weigh more or less than the air-filled flask? Explain.

C. Diffusion of Hydrogen

Fill a 250-mL beaker with water and invert it in the trough over the outlet of the generator. When it is full of H_2 slip a watch glass under it, carry it to the diffusion apparatus, and lower the inverted beaker over the porous cup. Report your observation and explain it. After a few minutes remove the beaker and observe the apparatus. Explain what happens.

D. Reduction of CuO

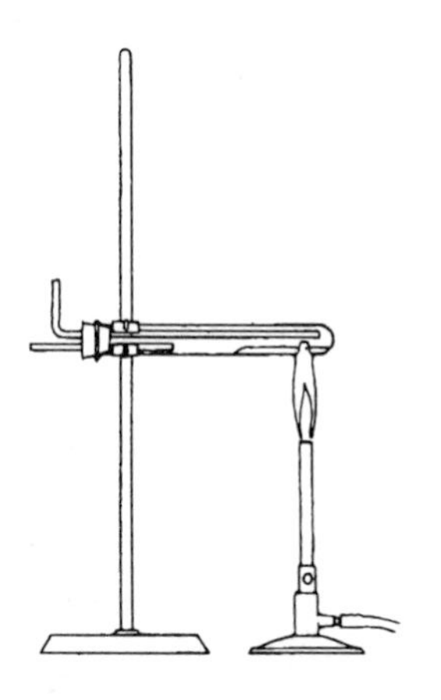

FIGURE 36.3
Reduction Apparatus.

Assemble the reduction apparatus illustrated in Figure 36.3, and prepare to attach it to the generator through a drying tube containing granular $CaCl_2$. The empty reduction tube is first weighed as accurately as possible (nearest 0.001 g). About one gram of CuO is added and the system is weighed accurately again. The actual weight of CuO is determined by difference.

The CuO is spread out thinly before the apparatus is attached to the generator. To the right-angled outlet tube of the reduction apparatus, connect a length of rubber tubing leading to the pneumatic trough. Again wrap the generator in a towel, add the remainder of the H_2SO_4, and let the gas sweep through the entire system. From time to time test for complete removal of air from the apparatus. When the system is free of air, *but not before*, remove the outlet tube from the pneumatic trough and heat the CuO in the reduction apparatus with a Bunsen burner. Report what happens and explain.

Continue to heat the tube and contents in a stream of H_2 for 10 minutes after there is no further change in appearance. Then disconnect the apparatus and allow it to cool. After cooling, weigh it accurately again. Calculate the weight of copper you theoretically should have received and compare it with the actual amount. Explain any difference. Calculate the percent error.

The instructor may direct you to measure the amount of water also produced. To do this, a second drying tube is filled with fresh $CaCl_2$, closed with a wad of cotton, and accurately weighed. It is attached to the outlet by a short rubber connector and the experiment is carried out. Finally, the drying tube is again weighed to determine the amount of H_2O collected. Compare this amount with the amount theoretically expected.

E. Explosive Mixture (Demonstration)

The instructor will provide a 1-liter or 1-quart steel can that has the valve stem from an automobile tire soldered to a hole in the bottom, as shown in Figure 36.4. The opening in the end *must* be as wide as the can. The valve core is removed, but a well-fitting valve cap is provided. The instructor will displace the air from an H_2 generator, collect a can of gas by water displacement, slip a glass plate over the bottom, and arrange the can on three rubber stoppers, first removing the plate. He or she will then unscrew the valve cap, use a pipe cleaner to remove any water from it, and ignite the gas as it comes out. For a time it will burn quietly. Then it explodes violently. Explain.

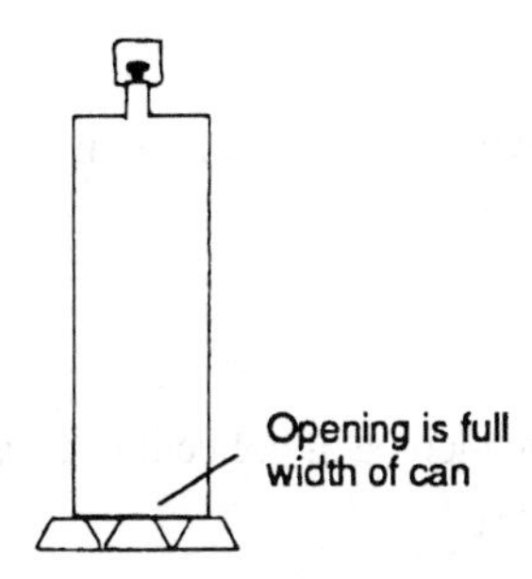

FIGURE 36.4
Gas Explosion Apparatus.

Experiment 36
Hydrogen

Name____________________________________

Date_____________________________________

REVIEW QUESTIONS

1. Research and discuss several ways by which H_2 is prepared commercially.

2. Hydrogen gas is an essential raw material for a number of important industries. Write a report on one or more of these.

3. Does the method of collecting H_2 used here give you any clues regarding the solubility in water? Explain. Look up its solubility.

4. Look up the melting point, boiling point, and critical temperature of H_2. Do these values suggest why H_2 used to be called a "permanent gas"?

5. Why is the gas issuing from the generator explosive at first but not later? Would it have exploded if you had opened the generator?

6. Look up the explosive ranges of hydrogen and of other common gases.

7. What metal oxides are commercially reduced by H_2?

8. Research and report on the uses to which differences in diffusion rates have been utilized in the preparation of nuclear explosives and fuels.

Experiment 36
Hydrogen

Name ____________________
Date ____________________

REPORT SHEET

A. Laboratory Preparation of H_2

Why should the thistle tube extend almost to the bottom of the flask? ____________________

____

Why is the generator wrapped in a towel? ____________________

Write the equation for the reaction in the generator: ____________________

____

B. Density of H_2

Weight of flask filled with air =

Weight of flask filled with H_2 = ____________________

Difference =

Explain the difference in weight: ____________________

____

C. Diffusion of H_2

Why will an inverted beaker retain most of the H_2 in it? ____________________

____

What happens when the beaker of H_2 is lowered over the porous cup? ____________________

____

Explain: ____________________

____

What happens when the beaker is removed? ____________________

____

Explain: ____________________

____

D. Reduction of CuO

Weight of tube and CuO =

Weight of empty tube =

Weight of CuO =

Weight of tube and Cu =

Weight of empty tube =

Weight of Cu =

Calculation of theoretical weight of Cu:

Calculation of percent error:

If the actual weight of Cu differed from the theoretical weight what was the probable reason? ______

______

______

Why is the drying tube put between the generator and the reduction apparatus? ______

______

What is observed in the reduction tube? ______

______

Write the equation for the reduction reaction: ______

______

E. Explosive Mixture

What is the color of the H_2 flame? ______

Why does the gas at first burn quietly and then explode? ______

______

Look up the explosive ranges of the following flammable gases and vapors:

Gas	Explosive Range	Vapor	Explosive Range
H_2		Gasoline	
Natural Gas, CH_4		Ethanol	
CO		Acetone	
C_3H_8		Diethyl Ether	

EXPERIMENT 37

Molecular Models

Purpose: To learn more about the geometry of covalent molecules based on their Lewis Dot Structures and the VSEPR Model.

Apparatus: Geometric model kit with centers for 2, 3, 4, 5, and 6-coordinate geometry, springs for making double bonds, pliers to remove stuck bonds.

Safety Precautions: None

INTRODUCTION

Structural chemistry is that branch of chemistry concerned with how molecules are put together. Special analytical techniques, notably x-ray crystallography, can unambiguously establish the precise geometry of a molecule.

From early data, it became clear that it was possible to *predict* the structure of a molecule without doing the analytical experiment. The predictions are based on the *V*alence *S*hell *E*lectron *P*air *R*epulsion model—the VSEPR model (usually pronounced "vesper"). First proposed by Sidgwick and Powell in 1940, and later expanded by Nyholm and Gillespie, VSEPR predicts that electron clouds (usually pairs) will be oriented around an atom in such a way as to minimize interaction and, therefore, place the electron pairs as far apart from each other as possible. It is one of the most useful theories in chemistry—it is almost never wrong.

Before beginning this exercise, you should review the pertinent material in your textbook on Lewis Dot Structures and VSEPR geometries.

The purpose of this exercise is for you to experience the geometrical shapes of molecules by constructing models of them. Pictures in a book or on a blackboard serve a purpose, but they remain two-dimensional images. It is hoped that this exercise will enhance your thinking processes for spatial relationships, and that you will begin to "see" these more clearly in your mind.

Most model kits contain balls of different colors to represent each of the several common elements. Some of the work has been done for you in that the holes drilled in the balls are placed in the correct orientation for geometries common to those elements. Common colors include (but may not apply to your model kit):

Black = carbon (4 holes, 4 bonds)
Yellow = hydrogen (1 hole, 1 bond)
Red = oxygen (2 holes, 2 bonds)
Green = Cl
Brown = I } Halogens (1 hole, 1 bond)
Orange = Br
Blue = nitrogen - (There are 5 holes, but use only the 3 pyramidal holes. Avoid the holes that go straight through.)

Long and short sticks are single bonds. Length is not relevant here. Springs are used for multiple bonds; e.g., two springs equal a double bond, three a triple bond.

PROCEDURE

Assemble yourselves into groups of three to six in order to share the equipment. If you lack a piece for one construction, improvise or borrow from your neighbor. Please share generously. This is a learning experience. Take your time.

Construct the molecules indicated on your report sheet, or any other molecules you may be assigned. Follow the directions on the report sheet. Use pencil for your drawings so you can erase easily.

If you do not have time to complete this exercise, or if you wish to use the models in your homework, they may be available to be checked out from the stockroom or directly from your instructor.

Experiment 37
Molecular Models

Name______________________________
Date______________________________

REVIEW QUESTIONS

1. What is the difference between isomers and resonance structures?

2. Draw all the resonance structures for nitrate ion, NO_3^-.

3. Draw the two resonance structures for the cyclic molecule benzene, C_6H_6.

4. Draw the Lewis Dot and geometric structures of SF_4. Indicate the bond angles and specify if they are non-ideal.

5. What is the atomic hybridization of Cl in ClF_3? Is the molecule polar? Explain.

6. On some periodic tables, H is placed in Group 17 (or Group VIIA), along with the halogens. Based on this exercise, explain why.

7. What is special about 5- and 6-coordinate geometries?

8. Why is it difficult to construct a Lewis Dot Structure for NO? What kind of molecule is this?

9. The geometry that chemists call "octahedral" is sometimes called "square bipyramid" by mineralogists. Explain how this is logical.

Experiment 37
Molecular Models

Name____

Date____

REPORT SHEET

Geometry. Draw the Lewis Dot Structures for the molecules listed and then draw the structures, indicating the geometry as accurately as you can and describing the correct geometry (e.g., tetrahedral, bent, etc.).

Molecule	Lewis Dot	Structure & Name of Geometry
NH_3		
CO_2		
CF_4		
O_3^{-2}		
OCl_2		
H_2O		

Isomers. Draw structures for the formulas. The number in parentheses indicates the number of isomers.

Formula	Structures
$C_2H_2Cl_2$ (3) (contains C = C)	
C_2H_6O (2)	

CHClBrI (2) These are *optical isomers.* They are nonsuperimposable mirror images.

Expanded Octets. You must use the extra model kits that have pieces for more than four electron pairs on the central atom.

Molecule	Lewis Dot	Structure & Name of Geometry
PCl_5		
SF_6		

ClF_3

ClF_5

XeF_4

EXPERIMENT 38

Phosphate Contamination in Water

Purpose: To illustrate a spectrophotometric method of quantitative analysis; to measure the phosphate content of water samples.

Materials: 100 ppm phosphate standard (0.4395 g of KH_2PO_4 made up to a total volume of 1.00 liter with distilled water); Vanadate/molybdate solution [add 1.25 g of ammonium metavanadate, NH_4VO_3, to 300 mL of water with stirring; add 330 mL 12 M HCl and continue to stir until all solid has dissolved; cool to room temperature, add 25 g of ammonium molybdate, $(NH_4)_6Mo_7O_{24} \cdot 4\,H_2O$ and stir until dissolved; dilute to 1 L]; samples of water (free of suspended materials) from different sources such as local lakes or streams, tap water, water from a private well, melted snow, rainwater, etc.

Apparatus: Bausch and Lomb Spectronic Twenty spectrophotometer or similar instrument (zero and light intensity controls labeled; tray set up with waste container, tissues, and cuvette for each instrument); 5-mL measuring pipet.

Waste Disposal: All standard solutions and mixtures used for spectrophotometric analysis may be rinsed down the sink. Unused portions of ammonium molybdate reagent or ammonium metavanadate reagent should be poured into the heavy metal waste container.

INTRODUCTION

Quantitative measurement is an important tool in chemistry. You are already familiar with balances used in comparing the masses of objects. This experiment is designed to introduce additional techniques for quantitative measurement. These will be used to measure the amount of phosphates in a water sample. Phosphates, which are essential nutrients for plant life, are one of several substances that contribute to the growth of algae and premature aging or "eutrophication" of lakes. Among the sources of phosphates in water are household detergents and fertilizers. Although it is possible that other factors such as nitrate content or dissolved carbon dioxide are the limiting factors in the growth of algae, ecologists advise lowering the amounts of phosphates in surface water.

The analytical method to be used has been selected because of its sensitivity which makes it possible to detect minute quantities of phosphate. Under acidic conditions, orthophosphate ion, PO_4^{3-}, reacts with ammonium molybdate to form molybdophosphoric acid. A yellow compound, vanadomolybdophosphoric acid, is formed in the presence of vanadium. The color intensity produced depends on the amount of phosphate present, and the acidity of the solution. By maintaining the same acidity in all solutions and by adding the same large excess of vanadate/molybdate reagent to each solution, we can insure that the amount of phosphate present controls the intensity of color developed.

It is possible to get a rough idea of the relative amounts of phosphate in the solutions by visually comparing their colors. By use of the spectrophotometer, however, we can obtain a more accurate quantitative measurement. This instrument measures the amount of light transmitted by a sample of the solution. Light from the source is dispersed by a diffraction grating into a spectrum. Light of a narrow band of wavelengths (monochromatic) is selected for illuminating the sample by means of an exit slit.

The amount of light absorbed depends on:

1. The length of the path through the solution;
2. The intensity of color in the absorbing substance;
3. The concentration of the solution.

Since we keep factor 1 constant by using a cuvette of fixed path length, and since the absorbing substance, molybdophosphoric acid, is the same in the water sample as in the standards, the *absorbance* is directly proportional to the concentration of the molybdophosphoric acid.

PROCEDURE

Use hot tap water to clean your 100-mL and 10-mL graduated cylinders and a 5-mL pipet. Phosphate-containing detergents **must not be used.** Phosphate has a tendency to be absorbed on glass surfaces. Rinsing with 3 M hydrochloric acid will help remove absorbed phosphate. Rinse with distilled water at least five times. Do *not* dry the glassware with a towel.

A. Preparation of Five Dilutions of Standard Solution

1. Obtain a sample of the standard phosphate solution in a clean beaker or flask. If the beaker or flask is wet inside, it should be rinsed with a little of the phosphate solution with the rinse mixture being discarded. Rinse the pipet with the standard solution twice, discarding the rinses. Rinsing the flask and/or the pipet avoids dilution of the standard solution if the glassware is wet. Pipet a 1-mL aliquot (a measured amount) of the standard solution into the 100-mL graduated cylinder. To use a pipet, fill it above the top line by drawing the standard solution into it using a rubber squeeze bulb. Quickly place your finger over the upper end of the pipet, and allow the liquid level (the meniscus) to drop to the 0.0 mark. Then allow the pipet to empty slowly, by easing finger pressure on the top of the pipet, until the solution reaches the appropriate mark (1, 2, 3, 4, or 5 mL), reading at the bottom of the meniscus. Stop the flow at the appropriate mark on the pipet by applying more pressure on the upper end with the finger. (You may need to practice pipetting a few times to become proficient.)
2. Add distilled water to fill the cylinder to the 50-mL mark. (Add the last few drops of water with a dropping pipet to avoid overfilling.) Transfer the solution to a clean, dry beaker or flask and label the container. Repeat this process, pipetting 2 mL, 3 mL, 4 mL, and finally 5 mL of the standard solution. Be sure to rinse the graduated cylinder carefully with distilled water between uses. In addition to the five standards, a "*blank,*" consisting of 50 mL of distilled water, should be transferred to a labeled container. Finally, measure 50-mL aliquots of the unknown water samples to be tested into three more clean, dry beakers or flasks. Calculate the phosphate content of each of the dilutions of the standard solution and record them on the report sheet.
3. With the 10-mL graduated cylinder, add 10 mL of the vanadate/molybdate reagent to each of the nine solutions (including the "blank"). Swirl each solution to achieve complete mixing.

B. Use of the Spectrophotometer

1. Check the wavelength setting; it should be 420 nanometers. Adjust the zero knob so that the meter reads "infinite" absorbance or "0.00" % transmittance. Some instruments, such as the Spectronic 21, do not require this step.
2. Fill the cuvette with the "blank" solution (half-full is enough); wipe the fingerprints from the outside with the tissues provided, and insert in the sample compartment. Adjust the knob labeled "light" so that the meter reads "0.00" absorbance or "100" % transmittance.
3. Remove the cuvette. For instruments such as the Spectronic 20 that have an internal shutter to block the light beam when the cuvette is removed, check to make sure the reading returns to infinite absorbance or 0.00 % transmittance.
4. Empty the blank solution from the cuvette (a waste container is provided) and rinse twice with the most dilute of your standard solutions. Half-fill the cuvette with the solution; wipe the outside of the cuvette and insert it in the sample compartment. Read, and record the *absorbance* on the report sheet. Repeat for the other standards you have prepared and for your water samples. Rinse the cuvette carefully with distilled water.
5. On the graph paper (provided at the back of the appendices) make a graph showing the absorbance of each of the standard solutions as a function of its concentration. Use the graph to find the concentration of phosphates in your water samples. Add your measurements to a chart showing the amount of phosphates found in other samples by your classmates, as directed by your instructor.

Experiment 38

Phosphate Contamination in Water

Name__________________________________

Date___________________________________

REVIEW QUESTIONS

1. Write a balanced net ionic equation for the reaction used in this analysis.

2. If H^+ ion, VO_3^- ion, and $Mo_7O_{24}^{6-}$ ion, as well as PO_4^{3-} ion, are required to produce the yellow color, why does the analysis measure phosphate ion content of the water sample?

3. Why is distilled water used for diluting the standard solutions in this experiment?

4. If a given water sample contains 10 parts per million phosphorous, how many grams of phosphorous would a liter (1,000 g) contain?

5. The following terms used in this experiment may be unfamiliar to you. Look up definitions of any terms you do not recognize: spectrophotometric, eutrophication, ecologist, sensitivity, diffraction, monochromatic light, wavelength, nanometer, cuvette.

6. Why are phosphates used in household detergents? What are some of the substitutes that have been suggested for use instead of phosphates for cleaning purposes?

Experiment 38

Phosphate Contamination in Water

Name____________________

Date____________________

REPORT SHEET

Sample	mL of 100 ppm phosphate	Concentration of solution $V_iC_i = V_fC_f$	Absorbance
Blank	0 mL		
1	1 mL		
2	2 mL		
3	3 mL		
4	4 mL		
5	5 mL		
Water Sample #1	XXXXXX	XXXXXXXXXX	
Water Sample #2	XXXXXX	XXXXXXXXXX	
Water Sample #3	XXXXXX	XXXXXXXXXX	

Identity of water samples analyzed:

1. ____________________
2. ____________________
3. ____________________

Concentration of phosphates in water samples:

1. ____________________
2. ____________________
3. ____________________

Attach a graph showing the absorbance of each of the standard solutions as a function of its concentration. Use the graph to find the concentration of phosphates in your water samples.

EXPERIMENT 39

Physical Properties of Water

Purpose: To correct reference points of a thermometer by comparison with the known freezing and boiling points; to measure the molar heat of fusion and molar heat of vaporization of water.

Materials: Ice cubes, NaCl or $CaCl_2$, styrofoam picnic cups (about 200 mL).

Waste Disposal: All solutions may be rinsed away with tap water in a laboratory sink.

INTRODUCTION

The freezing and boiling temperatures of water are the common reference points for thermometer scales and are often used to correct thermometers. In determining the freezing point, a sample of pure water is placed in a freezing bath and its temperature is recorded at regular time intervals. The temperature falls until freezing begins, remains constant during the time interval over which freezing is occurring, then falls again after freezing is complete. The plateau in the curve represents the freezing point.

The boiling point of water is somewhat more difficult to accurately determine because of possible superheating, and because a barometric pressure correction must be made.

The molar heat of fusion, ΔH_f, represents the difference in potential energy content between liquid water and ice at the same temperature. It is defined as the number of calories required to melt one mole of ice at the melting point. The same number of calories are released when one mole of water freezes.

The molar heat of vaporization ΔH_{vap} represents the difference in potential energy content between liquid water and water vapor at the same temperature. It is defined as the number of calories required to vaporize one mole of water at the boiling point.

PROCEDURE

A. Freezing Point of Water

Put about 10 mL of distilled water into a small test tube, insert a thermometer, and immerse the test tube in a freezing mixture of crushed ice and 25 g of salt in a 250-mL beaker. Add water to about three-fourths full. Stir the water in the test tube actively with the thermometer while watching the temperature. When it reaches 5° C start taking readings at 1-minute intervals, reading the thermometer to the nearest 0.1° C, and continuing until the water is frozen. Record results and plot a graph with temperature on the vertical axis and time on the horizontal axis. The horizontal position of the curve represents the release of the heat of fusion, which keeps the temperature constant so long as the freezing continues. The difference between the freezing point recorded on your thermometer and 0.0° C represents the thermometer error.

B. Boiling Point of Water

Put 100 mL of distilled water into an Erlenmeyer flask and add a few capillary tubes to promote smooth boiling. Mount a thermometer and a short right-angle glass bend in a two-hole rubber stopper in the flask so that the bulb is 1 inch above water level. Heat the flask rapidly until near the boiling point, then slowly with a small flame so that the water boils gently and steadily. After 5 minutes of steady boiling, read the thermometer to the nearest 0.1° C and record it. Also read and record the barometric pressure. Calculate the correct boiling temperature for the existing pressure using a 1° C boiling point depression for each 30 mm of pressure, below 760 mm. Use this value to determine your thermometer error.

C. Molar Heat of Fusion of Water (ΔH_f)

Nest two empty styrofoam picnic cups for a calorimeter and weigh to the nearest 0.1 gram. Heat 100 mL of water to about 50° C, put it in the calorimeter, and weigh again. Measure the water temperature to the nearest 0.1° C. Blot one or two ice cubes (a total of about 40 g) that have been out of the freezer long enough to start melting (Why?) and drop them at once into the warm water. Weigh the calorimeter and contents to the nearest 0.1 g, then stir with a thermometer until all the ice has melted. Take the temperature to the nearest 0.1° C.

Calculate the number of calories each gram of ice absorbed in melting. Develop the formula by following the method of question 4 in the Review Questions for this chapter. Remember that all of the heat lost by the water initially present was gained by the ice in melting and by the water from the melted ice as its temperature rose to the equilibrium point. A calorimeter and thermometer correction of 2 cal/degree should be applied. To calculate the molar heat of fusion, multiply the calories/gram by 18.0 grams/mole.

D. Molar Heat of Vaporization of Water (ΔH_{vap})

Obtain from the stockroom a glass connector tube made of 8 mm glass tubing and shaped so that any water that condenses in it runs back into the flask. The parallel ends are about 8 inches apart, and the end that dips into the calorimeter is 1.5 inches long from bend to outlet. Assemble the apparatus shown in Figure 39.1. The small glass bend with a capillary tip serves as a safety valve to prevent air pressure from forcing calorimeter water into the flask. The two nested styrofoam cups used in Part C again serve as a calorimeter. They should be protected from the direct heat of the flame by two ceramic squares.

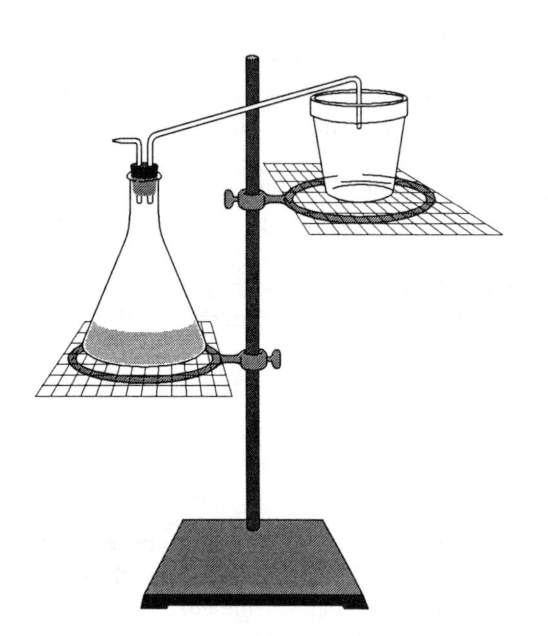

FIGURE 39.1
Heat of vaporization apparatus.

Put about 75 mL of water in the Erlenmeyer flask, add a few boiling chips, and begin heating. Weigh the calorimeter to the nearest 0.1 g, add about 120 mL of ice water (no ice cubes), and weigh again. When the flask is boiling, take the temperature of the calorimeter water to the nearest 0.1° C and mount the calorimeter as shown, making sure that the ceramic squares are in place and that the steam outlet of the connector tube dips just below the level of the calorimeter water.

Pass steam into the calorimeter gently until the calorimeter water has risen in temperature to about 45° C. Remove the calorimeter, stir for a few minutes with the thermometer, and take the temperature to the nearest 0.1° C. Weigh the calorimeter and contents to the nearest 0.1 gram.

Calculate the number of calories absorbed by the water in condensing each gram of steam. The gain in weight of the calorimeter and water represents the condensed steam. Develop your formula by the method in questions 4 and 5 in the Review Questions for this chapter. Remember that the calorimeter water gained all the heat lost by the condensing steam and the cooling of the condensate to equilibrium temperature. Correct the boiling point for barometric pressure and apply the calorimeter and thermometer correction. To calculate the molar heat of vaporization, multiply the calories/gram by 18.0 grams/mole.

Experiment 39
Physical Properties of Water

Name______________________________________

Date_______________________________________

REVIEW QUESTIONS

1. Parts C and D of this experiment are particularly subject to certain errors. What are they, and how could they be reduced?

2. Frequently a time-temperature curve such as plotted in Part A dips below the plateau and then rises again when freezing starts. Explain.

3. Define the boiling point and explain why a change in gas pressure above a liquid changes its boiling point.

4. In an experiment 43.1 grams of ice were placed in a calorimeter containing 100 g of H_2O at 46.5° C. After the ice had melted the water temperature was 8.4° C. Develop a formula for calculating the molar heat of fusion ΔH_f of water from the above data, using the following steps:

a. An expression for the heat gained by the ice in melting.

b. An expression for the heat gained by the water from the melted ice as it rose to equilibrium temperature.

c. An expression for the heat lost by the water originally in the calorimeter in cooling to equilibrium temperature.

d. An expression for the calorimeter and thermometer correction, using 2 cal/degree as the total heat capacity.

e. Remembering that the total heat gained must equal the total heat lost, put these expressions together into a formula for calculating the molar heat of fusion ΔH_f and solve the problem.

5. Using a method similar to that of Question 4, develop a formula for calculating the molar heat of vaporization ΔH_{vap} of water and find its value from the following data:

   Steam was passed into 119.1 g of H_2O at 3.4° C contained in a calorimeter until there was an increase of 8.3 g in weight. The equilibrium temperature was 44.8° C.

Experiment 39
Physical Properties of Water

Name______________________________
Date______________________________

REPORT SHEET

A. Freezing Point of Water

1 min ________ °C	6 min ________ °C	11 min ________ °C	16 min ________ °C
2 min ________ °C	7 min ________ °C	12 min ________ °C	17 min ________ °C
3 min ________ °C	8 min ________ °C	13 min ________ °C	18 min ________ °C
4 min ________ °C	9 min ________ °C	14 min ________ °C	19 min ________ °C
5 min ________ °C	10 min ________ °C	15 min ________ °C	20 min ________ °C

Plot the time-temperature curve:

Temperature

Time

Freezing point (your thermometer) ________________

Thermometer error ________________

B. Boiling Point of Water

Observed boiling point (your thermometer) ________________

Barometric pressure ________________

Calculated BP (from barometric pressure) ________________

Thermometer error ________________

C. Molar Heat of Fusion of Water (ΔH_f)

Weight of calorimeter + water = ____________________

Initial temperature water = ____________________

Weight of calorimeter = ____________________

Equilibrium temperature water = ____________________

Weight of water = ____________________

ΔT calorimeter water = ____________________

Weight of calorimeter + water + ice = ____________________

Equilibrium temperature of water = ____________________

Weight of calorimeter + water = ____________________

Melting temperature of ice = ____________________

Weight of ice = ____________________

ΔT ice water = ____________________

Formula for ΔH_f and calculation:

Value of ΔH_f (from literature) ____________________ Percent error ____________________

D. Molar Heat of Vaporization of Water (ΔH_{vap})

Weight of calorimeter and water = ____________________

Weight of calorimeter = ____________________

Weight of water in calorimeter = ____________________

Weight of calorimeter, water, and condensate = ____________________

Weight of calorimeter and water = ____________________

Weight of condensate = ____________________

Final temperature of water = ____________________

ΔT condensate = ____________________

Formula for ΔH_{vap} Calculation:

Value of ΔH_{vap} (from literature) ________________________ Percent error ________________________

EXPERIMENT 40

Analysis of Vinegar by Titration

Purpose: To illustrate the principles and technique of a volumetric quantitative analysis.

Materials: NaOH pellets, white vinegar, potassium hydrogen phthalate (KHP), phenolphthalein.

Apparatus: 50 mL burets.

INTRODUCTION

Vinegar is essentially a dilute solution of acetic acid in water. The acid concentration of a known volume of vinegar can be measured by adding just enough basic solution of known concentration to exactly react with it.

The experiment is divided into two parts. In Part A an NaOH solution is prepared and standardized; that is, its concentration is accurately measured. In Part B, the analysis proper, this standard solution is used to measure the acid concentration of vinegar. The calculation in both parts is based on the fact that when one substance exactly reacts with another, so that no excess of either remains, the same number of equivalent weights of each reactant is used.

$$\text{number of equivalent weights of base} = \text{number of equivalent weights of acid} \qquad \text{Equation 1}$$

(Actually, it is more convenient to express both acid and base in milliequivalents (MEW), but the equation is still valid.)

In a solution the number of MEWs is the product of the normality (EWs per liter, or MEWs per milliliter) times the volume in milliliters. The number of MEWs present in a weighed solid sample can be found by dividing the weight present by the number of grams in 1 MEW.

For use in Part A, equation 1 reduces to

$$N_B \times V_B = N_A \times V_A \qquad \text{Equation 2}$$

The equation is solved for the normality of the basic solution

$$N_B = \frac{N_A \times V_A}{V_B} \qquad \text{Equation 3}$$

For use in Part B, equation 1 takes the form

$$N_A \times V_A = N_B \times V_B \qquad \text{Equation 4}$$

and is solved for the normality of vinegar, N_A. From this the percent of acid in the vinegar is calculated.

$$N_A = \frac{N_B \times V_B}{V_A} \qquad \text{Equation 5}$$

All volume measurements are made by burets, which can be read to 0.01 mL. The indicator phenolphthalein, which is colorless in acid and pink in base, marks the end point of the reaction. One drop of excess base turns the entire solution a faint pink.

PROCEDURE

A. Standardization of NaOH Solution

Weigh out roughly 6 g of NaOH pellets and place in a 500-mL Florence flask. Be sure to clean up any pellets that may have been spilled, and avoid contact of NaOH with skin or clothing. Fill the flask to the base of the neck with distilled water, stopper, and mix thoroughly. Continue shaking for a minute after all pellets have dissolved to ensure uniformity of solution. Label the flask.

Two samples of the weak acid, potassium hydrogen phthalate (generally called KHP), weighing approximately 1 g each are weighed out as accurately as possible, to the nearest 0.001 or 0.0001 gram. This may be done by first weighing the sample roughly on a platform balance to get approximately 1 gram. The precise weight is determined by weighing two 150-mL or 250-mL Erlenmeyer flasks individually on a milligram or analytical balance. Add one of the samples to each, and again weigh each exactly. The weight of the acid is then obtained by subtraction. Number each flask on the ground glass circle before weighing. Do not waste time trying to get exactly 1 gram, as no advantage is gained. You only need to know exactly how much acid you do have. In quantitative analysis two or more samples are usually run simultaneously to obtain an average value, which is statistically more reliable than a single measurement.

Add 50 mL of distilled water to each sample and swirl until all the acid is dissolved. Use swirling rather than a stirring rod to prevent loss of sample on the stirring rod. In an experiment in which quantities are determined, the loss of even a few drops of solution would introduce a serious error.

Rinse a buret twice with about 10 mL of the NaOH solution to be used in it, making sure that the solution comes in contact with the entire inner surface of the buret. Let the solution run out of the tip. This assures that the standard solution will not be diluted by distilled water clinging to the buret walls. Fill the buret to the zero mark, making sure that no bubbles remain in the tip. A quick squirt of solution will usually sweep them out.

Add 3 drops of phenolphthalein indicator to each KHP sample. Place the first flask on a piece of white paper under the buret and allow the base solution to run in slowly with swirling. Pink color will appear where there is a high local base concentration, but it will vanish quickly as the solution is swirled. Near the end point, the addition should be done a drop at a time with swirling between drops. As soon as the color shows any tendency to persist, the rate of addition should be slowed. Do not add too much base. At the end point 1 drop of base produces a permanent pale pink color. Record the volume of the base used, estimating to the nearest 0.01 mL, then titrate the other two samples in the same way. Calculate the normality of the base from each sample. If the two normalities do not agree to within 0.002, run a third sample. Label your NaOH solution with the average normality obtained and save it for Part B. Keep it stoppered.

B. Analysis of Vinegar

Use another buret to measure exactly 10.00 mL of vinegar into each of two Erlenmeyer flasks. Put 3 drops of phenolphthalein indicator into each flask and swirl.

Rinse your buret with the NaOH solution that is to be used in it, letting it run out of the tip. Then fill the buret to the zero mark, making sure that there are no bubbles in the tip. Titrate each solution with care. If the end point is overshot, add an exactly measured amount of vinegar and again add base carefully to a new end point.

Calculate the normality of the vinegar from the volume of vinegar, the volume of base used, and the normality of the base. The values for the two samples should not differ more than 0.002. If they do, a third sample is necessary. Then, remembering that the formula of acetic acid is $HC_2H_3O_2$, and that the normality gives the number of equivalent weights of acetic acid per liter of vinegar, calculate the number of grams of acetic acid per liter. From the number of grams of acetic acid per liter, calculate the number of grams of acetic acid per 100 mL of vinegar. Assuming that the density of vinegar is the same as the density of water (1.00 g/mL), calculate the percent of acetic acid in the vinegar sample.

Experiment 40

Analysis of Vinegar by Titration

Name______________________________

Date______________________________

REVIEW QUESTIONS

1. Why was the standard NaOH solution not prepared by calculating the amount of solid NaOH needed for 500 mL of solution, weighing it accurately, and making it up to exactly 500 mL of total volume?

2. Find out from a quantitative analysis text why a glass-stoppered buret is never used for the NaOH solution.

3. Why not simply rinse the buret with distilled water rather than the solution to be used in it?

4. Calculate what the percent error would have been in a titration that used 26.65 mL of a solution if a bubble with a volume of 0.30 mL had been swept out of the tip during the titration. Would the normality found for the base have been too high or too low?

5. On the theory that "if a little red is good, a lot is better," a student titrated to a deep-red color, and in so doing, added 3 drops of excess base. If each drop has a volume of 0.05 mL, what percent of error was introduced in a 25.00 mL titration? Was the normality found for the base too high or too low?

6. A student obtained a buret reading of 32.34 mL, but carelessly rounded off the reading to 32.3 mL. What percent of error did the student introduce? Was the normality found for the base too high or too low?

7. Why does the volume of water added to the potassium hydrogen phthalate not have to be measured carefully?

8. Instead of measuring vinegar with a buret, a student used a 25-mL graduated cylinder and lost 0.3 mL which adhered to the walls on pouring. What was the percent error?

9. When $HC_2H_3O_2$ is titrated with NaOH until an exactly equivalent quantity of NaOH has been added, the solution is not neutral. Why? Is the solution acidic or basic? Show why with the help of ionic equations.

10. If NH_4OH had been titrated with standard HCl, the solution at the exact equivalence point would not have been neutral. Show why with the help of ionic equations. Is the solution acidic or basic?

11. If a solution is diluted, the same number of equivalent weights of reagent are still present. Hence, it is possible to calculate a new volume or a new normality from a known greater normality. The equation is:

$$N_1V_1 = N_2V_2$$

If you wish to make 500 mL of a 20.0 N solution of a reagent from 1.5 N solution, how much of the latter solution would you use?

12. Suggest a way of determining the end point if a wine vinegar, which is red in color, had been analyzed.

Experiment 40
Analysis of Vinegar by Titration

Name________________________________
Date_________________________________

REPORT SHEET

A. Standardization

		I	II	III
Weight of beaker + KHP	=	g	g	g
Weight of beaker	=	________ g	________ g	________ g
Weight of sample of acid	=	g	g	g
Volume of NaOH used (V_B)	=	mL	mL	mL
Normality of NaOH(N_B)	=	________ N	________ N	________ N

The two calculations for normality should not differ more than one unit in the thousandths place. If they do, a third sample must be run. Get the instructor's OK before throwing away any solutions.

KHP has the formula $KHC_8H_4O_4$. One hydrogen per molecule reacts with the base. The equation for the standardization reaction is: $KHC_8H_4O_4 + NaOH \longrightarrow NaKC_8H_4O_4 + H_2O$. Calculate its equivalent weight. Also write the milliequivalent weight (MEW).

Equivalent weight of KHP =

Milliequivalent weight of KHP =

Show calculations for normality of the base (N_B):

B. Analysis

	I	II	III
Volume of vinegar used =	mL	mL	mL
Volume of NaOH used =	mL	mL	mL
Average normality of NaOH(N_B) from part A =	N	N	N

I. N_A = = ______________________ N

II. N_A = = ______________________ N

III. N_A = = ______________________ N

Equivalent weight of acetic acid, $HC_2H_3O_2$ = ______________________

Grams of $HC_2H_3O_2$ per liter of vinegar = ______________________

Grams of $HC_2H_3O_2$ per 100 mL (100 g) of vinegar = ______________________

(Assume that vinegar has the same density as water)

Percent acetic acid in vinegar = ____________ %

Equation for analysis reaction:

EXPERIMENT 41

Synthesis of an Iron Complex

Purpose: To synthesize a compound with a complex ion, $[Fe(C_2O_4)_3]^{-3}$.

Materials: Solid $(NH_4)_2Fe(SO_4)_2 \cdot 6H_2O$, 6M H_2SO_4, 1M $H_2C_2O_4$ (oxalic acid), 2M $K_2C_2O_4$ (potassium oxalate), 6% H_2O_2, ethanol, acetone.

Apparatus: 250-mL beaker, 50- or 100-mL graduated cylinder, ring stand, wire gauze, Bunsen burner, thermometer, one 50-mL buret per lab for H_2O_2, Büchner funnel and filtration flask, filter paper.

Safety Precautions: Wear your safety glasses at all times during this experiment. Hydrogen peroxide is a strong oxidizer—it can cause severe skin burns. Oxalic acid and the salt potassium oxalate are moderately toxic. Avoid skin contact. Use caution with concentrated sulfuric acid. Wash your hands before you leave the lab.

INTRODUCTION

Simple salts are ionic compounds that are made up of simple cations and simple anions. When either or both ions are separately, *covalently* bonded to other molecules, the formation is called a *complex ion* or a *coordination compound*. The fundamentals of coordination chemistry should be reviewed in your textbook before beginning this experiment.

In this experiment you will prepare a complex ion composed of iron (III) and the oxalate anion, $C_2O_4^{-2}$. Despite the positive charge on the iron, the iron-oxalate complex ion is the anion of the compound. The formula of the compound is $K_3[Fe(C_2O_4)_3] \cdot 3H_2O$. Since there are three oxalate anions, each at a charge of –2, the sum of the charges in the brackets is [+3 + 3(–2)] or –3. This then accounts for the three potassium +1 ions to balance the charge. When the compound is isolated from aqueous solution it crystallizes as a three hydrate.

PROCEDURE

A. Synthesis of Fe(III) Complex

1. Measure 30 mL of distilled water into a clean 250-mL beaker. Add 6–8 drops of 6 M H_2SO_4.
2. Weigh out 10 g of ferrous ammonium sulfate hexahydrate, $(NH_4)_2Fe(SO_4)_2 \cdot 6H_2O$. Record the amount actually used and dissolve in the acidic water. You may need to warm the solution gently to get all of the solid to dissolve.
3. Measure out 50 mL of 1 M oxalic acid ($H_2C_2O_4$) in a graduated cylinder. Add this to the iron solution slowly and with stirring. A yellow precipitate of iron (II) oxalate will form.

4. Assemble a ring stand, ring, wire gauze, and Bunsen burner. Heat the solution gently to 80° C while stirring constantly. Do NOT heat over 80° C. CAUTION! This mixture will tend to bump and may splatter. Continue stirring constantly. After a few minutes discontinue heating.
5. Allow the solution to settle. While still very warm decant the supernatant solution, saving the precipitate.
6. Heat 60 mL of distilled water to 50°–60° C. Use half of the distilled water to wash the solid and then decant. Wash with the second 30-mL hot water portion and decant.
7. Add 18 mL of 2 M potassium oxalate to the yellow solid. Heat the solution just to 40° C and then discontinue heat.
8. A buret of 6% hydrogen peroxide will be set up in the lab. CAUTION! From the buret slowly add 17 mL of the H_2O_2 solution while stirring. A rust-brown precipitate of $Fe(OH)_3$ may form. Continue.
9. Heat the mixture to boiling. While boiling, add a total of 13 mL of 1 M oxalic acid, the first 9 mL all at once, and the last 4 mL dropwise.
10. At this point any rust-brown solid should have dissolved; if any remains add more oxalic acid dropwise until it is completely dissolved. Discontinue heating.
11. Transfer the solution to a clean 150-mL beaker and add 15 mL of ethanol. Layer this gently over the top. Do not stir.
12. Cover the beaker with a watch glass. Leave in your drawer until next week.

B. Isolation of Fe(III) Complex

Second Week

Filter the crystals with a Büchner funnel. You may wish to wash with a small amount of acetone to dry. Weigh dried product.

CALCULATIONS

Assuming iron is the limiting reagent, calculate the theoretical yield and the % yield.
Describe the product.

Experiment 41

Synthesis of an Iron Complex

Name________________________________

Date_________________________________

REVIEW QUESTIONS

1. How is iron (II) converted to iron (III)? Write a balanced ionic equation to show this step.

2. Why does the addition of ethanol cause the precipitate to form?

3. Oxalate is a special kind of ligand. Explain. Draw the structure of the iron-oxalate complex ion.

4. What is the formal name for the compound you have just synthesized?

5. Why is it useful to wait a week before isolating the product?

Experiment 41
Synthesis of an Iron Complex

Name ______________________________

Date ______________________________

REPORT SHEET

Weight of $(NH_4)_2Fe(SO_4)_2\ 6H_2O$ used ____________________ g

Weight of $K_3[Fe(C_2O_4)_3]\ 3H_2O$ obtained ____________________ g

Product description:

Calculations:

Theoretical yield ____________________ g

% yield ____________________

(show all calculations)

EXPERIMENT 42

Corrosion

Purpose: To study some of the factors that influence the corrosion of metals and to note some protective measures.

Materials: Agar, phenolphthalein, 6 N HCl, 3 M H_2SO_4, 1 N NaOH, 1% $K_3Fe(CN)_6$ (fresh), 3% H_2O_2, 3% NaCl in H_2O, 3% NaCl in 3% H_2O_2, 0.2 N NaOH, saturated CO_2 solution, NaCl. Sheet zinc, copper foil or gauze, fine copper wire, galvanized iron sheet, tin plate, finishing nails, paper clips, 4 x 4 inch sheets of carbon steel, 12% chrome steel, and 18-8 stainless steel (18% Cr-8% Ni).

Waste Disposal: Iron, zinc, and copper solids should be disposed of in separate marked containers. Jelled agar may be disposed of in a waste basket. All solutions may be rinsed down the sink drain with water.

INTRODUCTION

Voltaic cell action plays an important role in the corrosion of metals. In one type of cell, anodic areas and cathodic areas develop when the metal or combination of metals is in contact with an electrolyte. The metal goes into solution at the anodic area (corrodes), with liberation of electrons. These electrons flow through the metal to the cathodic area, where they serve to inhibit corrosion either by tending to reverse the electron loss of metals in this area or by discharging H^+ ions to produce a protective layer of H_2, or both. The following are some of the conditions responsible for anodic and cathodic areas:

1. When a more active metal is in contact with a less active one, the more active one becomes anodic and corrodes; the less active one becomes cathodic and is protected.
2. When an area of strained metal is in contact with unstrained metal, the strained one corrodes and the unstrained one is protected.
3. When a clean metal area is in contact with an oxide-coated one (adherent oxide), the clean area corrodes.

The electrolyte may be an adsorbed film of moisture containing traces of dissolved salts, CO_2, and O_2; it may be a liquid solution in contact with the metal surface; or it may be soil moisture touching buried metal. These three cases represent nonuniform metal surfaces in contact with the same solution. Voltaic cells that promote corrosion may also be set up if a uniform metal is in contact with a nonuniform solution, as when different solution areas contain different concentrations of dissolved O_2 or electrolyte.

If iron is in contact with copper in an electrolyte, corrosion of the iron is speeded up, whereas the copper is protected. Iron, the more active metal, becomes anodic and the copper becomes cathodic. The equations can be represented as shown on the following page.

The cathodic area represented in the equations below becomes slightly alkaline because of excess OH^- from water.

Anodic Area

Fe $Fe^{2+} + 2e^-$ Fe $\longrightarrow Fe^{2+} + 2e^-$

Presence of O_2 oxidizes Fe^{2+}

$Fe^{2+} \xrightarrow{O_2} Fe^{3+}$

This prevents reversal of the first reaction and promotes further dissolving.

$Fe^{3+} + 3\,OH^- \longrightarrow Fe(OH)$

(from H_2O)($Fe_2O_3 \cdot \times H_2O$)

Cathodic Area

$2\,HOH \rightleftarrows 2\,H^+ + 2\,OH^-$

$2\,H^+ + 2e^- \longrightarrow H_2$ (protective layer over the copper)

Since Cu^{2+} holds electrons more tightly than Fe^{2+}, electrons from Fe would tend to prevent dissolving of copper.

$Cu^{2+} + 2e^- \longrightarrow Cu$

If iron and zinc are in contact (e.g., galvanized iron) in an electrolyte, the more active zinc becomes anodic and its corrosion protects the cathodic iron. If sufficient oxygen is present at the cathode area, the protecting film of H_2 may be removed by oxidation and corrosion may occur.

Certain metals such as aluminum, magnesium, chromium, and nickel form thin but tight coatings in air that protect them from further oxidation (the metal becomes "passive"). Chromium and nickel in stainless steel impart this same ability to the alloy. Zinc protects itself in air by forming basic zinc carbonate $ZnCO_3 \cdot Zn(OH)_2$ on its surface.

Iron seems to have only a slight tendency to become passive. The normal iron rust $Fe_2O_3 \cdot X\,H_2O$ is too loose and scaly to protect, but Fe_3O_4 can be artificially produced, as in bluing of steel. Iron in contact with solutions with different O_2 content seems to become passive where the O_2 concentration is greatest and to corrode where it is least.

PROCEDURE

A. Cell Action in Corrosion

Weigh out 1 gram of agar and soften it in a few mL of cold distilled water. Bring 75 mL of distilled water to a boil, add it to the agar in a beaker, and heat gently until all the agar has dissolved. Add 10 drops of phenolphthalein and then add 1 N NaOH from a dropper until the solution just becomes pink. Discharge the pink color by touching a stirring rod lightly to the surface of 6 N HCl solution and stirring the mixture with it. Add 3 mL of fresh 1% $K_3Fe(CN)_6$ solution.

Immerse three paper clips in 3 M H_2SO_4 and warm until H_2 evolution starts. This dissolves the thin protective oxide coating on the surface of the metal. In a similar manner, clean half of a bright nail, taking care that the acid does not touch the other half of the nail. Wash off the acid thoroughly from paper clips and nail.

Distribute the agar mixture among three beakers (250-mL) and allow it to cool. Before it sets, put into the solutions the following items, taking care that they do not touch each other:

1. A cleaned paper clip holding a small piece of sheet zinc (iron and a more active metal).
2. A cleaned paper clip holding a three-fourths inch square of copper gauze (iron and a less active metal of relatively large area).
3. A cleaned paper clip holding a 1-inch piece of fine copper wire (iron and a less active metal of relatively small area).
4. A bright nail (the metal will have been strained by the cold work of forming the head and the point).
5. A bright nail, half of which has been cleaned of its passive coating.
6. A piece of galvanized iron and a piece of tinned iron, each with a deep scratch on the surface.

Observe all samples for half an hour and report results. The anodic areas gradually develop a dark-blue color as dissolved iron(II) ions react with ferricyanide ions.

$$3\,Fe^{2+} + 2\,Fe(CN)_6^{3-} \longrightarrow Fe_3(Fe(CN)_6)_2 \text{ dark-blue precipitate}$$

The corresponding precipitate at the anodic zinc area is white. A pink phenolphthalein color develops near cathodic areas where an excess of OH^- ions is produced.

B. Corrosion on Carbon Steel, Chrome Steel, and 18-8 Stainless Steel (Demonstration)

1. **Passivity, oxygen concentration cell.** The instructor will prepare an agar mixture like the one of Part A and will pour a pool about an inch in diameter on the clean surface of each of the following: a sheet of carbon steel, a sheet of chrome steel, and a sheet of 18-8 stainless steel. After a period of time, note which samples corrode and which do not and explain. Note also in the corroded sample which portion of the agar solution is anodic and which is cathodic and explain. (Recall that the outer portion of the agar gel that is in contact with the air contains more dissolved oxygen than the inner portion.) This is a typical concentration cell.

2. **Corrosion in ionic solutions.** To the balance of the still hot agar mixture from Part B1, the instructor will add about 2 g of NaCl and stir to dissolve. Next, pools about 1 inch in diameter are poured on clean portions of the carbon steel, the chrome steel, and the 18-8 stainless steel. Note and report how these three metals react to the salt mixture. Note the effect of ionic solutions on corrosion and also which portion of the agar-coated area is anodic and which is cathodic.

C. Corrosion and Protection Resulting from an Outside Current (Demonstration).

The instructor will prepare another agar mixture as in Part A. While it is cooling he or she will fasten copper wires to two paper clips and will immerse the clips in the mixture. When the agar has set, the instructor will connect the wires to the terminals of a dry cell or a DC power supply. After a few minutes observe and record the results.

D. Corrosive Action of Various Solutions

Obtain six iron nails (8-10 penny) and clean off the passive oxide coating by allowing them to stand in warm H_2SO_4 solution until active H_2 evolution occurs. Wash off the acid thoroughly. Label six test tubes as follows:

1. Tap water
2. 3% NaCl
3. 3% H_2O_2–3% NaCl
4. 3% H_2O_2 (no NaCl)
5. 0.2 N NaOH
6. CO_2 solution

Pour an inch of the appropriate solution into each of the tubes and add to each a cleaned nail, which should protrude above the surface of the liquid. Allow the tubes to stand until the next class period and note evidence of corrosion.

Experiment 42
Corrosion

Name________________________

Date________________________

REVIEW QUESTIONS

1. Why does a large copper area in contact with iron promote corrosion more than a small one?

2. Figure out or look up a reason why a solution with a high ionic strength should be more corrosive than water in which few ions are present.

3. What is meant by a sacrificial metal? Suggest some practical uses.

4. Exactly how does a coat of paint or a film of oil protect metal?

5. What may happen when a paint film starts to break up, and why?

6. Suggest some possible dangers inherent in applying undercoating to a car.

7. What would happen if iron sheeting was fastened with copper rivets? Would you expect corrosion to be fast or slow? Explain.

8. What would happen if copper sheeting was fastened with iron rivets? Would you expect corrosion to be fast or slow? Explain.

9. What bearing does the pH of a solution have on its ability to corrode iron?

10. List as many practical anticorrosion measures as possible.

Experiment 42
Corrosion

Name____________________________________
Date_____________________________________

REPORT SHEET

A. Cell Action in Corrosion

1. Which metal is anodic?____________________ Is corrosion speeded or slowed? ____________________

   What evidence is there of zinc reaction? ____

2. Which metal is anodic?____________________ Iron proctected or corrosion?____________________

3. What bearing does the relative areas of the two metals have on corrosion?____________________

   ____

4. Which part of the nail is anodic? ____________________ Why?______________________________

5. Which part of the nail is cathodic? ____

6. Does the scratch in the galvanized sheet show any dissolved iron?____________________________

   Is the zinc surface anodic or cathodic? ____________________ Explain (with equations) what happens at the zinc surface and at the exposed iron: ____

   ____

   Will zinc continue to protect the iron after the surface is broken?____________________________

   What other metals will similarly protect iron? ____

   Does the scratch in the tinned plate show any dissolved iron?______________________________

   Is the tin surface anodic or cathodic? ____

   How does tin protect iron?____

   Does it continue to protect after the surface is broken?__________________________________

   Give two ways in which a coat of paint protects a surface: a. _______________________________

   b. ____

B. Corrosion on Carbon Steel, Chrome Steel, 18-8 Stainless Steel (Demonstration)

1. Passivity, oxygen concentration cell

   Which steel showed the most corrosion? ____

   Which steel showed the least corrosion? ____

   In the corroded sample, which part of the agar gel showed corrosion and which showed protection? ________________

   ____

How do corroded areas correlate with oxygen concentration? ____________________

Explain ____________________

2. Corrosion in ionic solutions

Did all of the steels show corrosion? ____________ Which showed the most? ____________

Which showed the least? ____________________

How did the amount of corrosion compare with that of Part B1? ____________________

____

Are the same relative areas corroded as in Part B1? ____________________

C. Corrosion and Protection Resulting from an Outside Current

Which piece of metal is protected? ____________ Explain why: ____________

____

How might the principle discussed in this section be used to protect a steel structure under ground? Be specific:

____

D. Corrosive Action of Various Solutions (Observed after 24 Hours)

Sample	Appearance of Sample	Order of Corrosive Attack
#1		
#2		
#3		
#4		
#5		
#6		

Which showed the greater reaction, #1 or #2? ____________ Explain: ____________

____

Explain the extensive reaction of #3. ____________________

____

Is there any evidence of attack on #4? ____________ Explain: ____________

____

Is there evidence of corrosion on #5? ____________________

____

EXPERIMENT 43

Colloids

Purpose: To prepare several colloidal sols of different characteristics and observe some of their properties.

Materials: 0.2 N FeCl3, 0.2 N $AgNO_3$, concentrated NH_4OH, 1% tannic acid, As_2O_3, $(NH_4)_2S$, 95% ethanol, saturated $Ca(C_2H_3O_2)_2$, kerosene, soap solution, concentrated HCl, 14 M NH_4OH, 6 N Na_2SO_4, 1 N $Al_2(SO_4)_3$, 1:200 methylene blue, decolorizing carbon, cotton.

Apparatus: Narrow beam light source, Brownian movement cell, microscope.

Waste Disposal: Arsenic-containing and silver-containing mixtures should be placed in the proper waste container.

INTRODUCTION

Colloidal sols differ from true solutions mainly in the size of their dispersed particles. The sols have particles ranging from 1 to 100 nanometers in diameter, whereas true solution particles are ions or small molecules falling below that size range. Dispersions having larger particles are called suspensions. A colloidal sol, therefore, represents a state of matter rather than a particular kind of matter. Colloids in which the dispersed particles are liquid droplets are called emulsions.

Gels are sols in which the dispersed particles are tangled and interlaced masses of fibers or long crystals that have entrapped relatively large amounts of liquid (the dispersion medium). Aerosols have colloidal solid particles or liquid droplets dispersed in a gas.

The particular size range of colloids gives them their characteristic properties. Although individual particles may be invisible to the eye or microscope, they are large enough to reflect light, so they show the Tyndall effect and appear as tiny pinpoints of light under the ultramicroscope. The particles are too large to have molecular motion but may be moved by the impact of molecules, and hence exhibit the Brownian movement. They pass through ordinary filters but some of them can be stopped by special filters. They cannot pass through parchment, animal membranes, or other semipermeable partitions.

Colloidal particles partake of the nature of more massive particles, but their small size gives them enormous surface area per unit of weight. Therefore they show such surface phenomena as adsorption to a high degree. The highly selective character of their adsorption is shown by the fact that some colloids adsorb only negative ions from the dispersion medium and others adsorb only positive ones. This results in an overall change of one sign on the surface of the particles, which produces a mutual repulsion and prevents coalescence of the particles into lumps. A sol can often be precipitated by adding it to an electrolyte containing high valence ions of opposite charge. These neutralize the adsorbed charges and permit coagulation.

Many colloids are stabilized by adsorbing a covering layer of foreign matter that keeps the particles apart. Films of egg adsorbed to droplets of oil stabilize the colloid mayonnaise. Such covering layers are called protective colloids.

Lyophilic sols (solvent attracting) have a strong tendency to unite with the dispersion medium, and are thereby stabilized. Common examples of this are glues and gels. Lyophobic sols (solvent avoiding) show little or no tendency to interact with the dispersion medium and are more easily converted into larger particles and precipitated. Hydrated ferric oxide, arsenic sulfide, and metallic silver or gold sols are of this type.

Colloidal sols can be prepared by (1) condensation methods in which true solution particles are built up to colloid size by chemical reaction, or (2) dispersion methods in which more massive material is broken down to colloid size. The colloid mill and the Bredig arc are examples of the latter method.

PROCEDURE

Preparation and Properties of Colloidal Sols

1. **Hydrated ferric oxide (condensation method).** Heat 100 mL of distilled water in a small beaker and pour into it 20 mL of freshly prepared $FeCl_3$ solution. A red colloidal sol of hydrated ferric oxide $Fe_2O_3 \cdot X\ H_2O$ is formed by a hydrolysis reaction.

$$2\ Fe^{3+} + (X + 3)H_2O \longrightarrow Fe_2O_3 \cdot X\ H_2O + 6\ H^+$$

Label the sol. Test it for the Tyndall effect and compare with a true solution.

Add 1 drop of Na_2SO_4 solution to 5 mL of the sol in a test tube. The $Fe_2O_3 \cdot X\ H_2O$ sol tends to adsorb positive ions from solution and is stabilized by the mutual repulsion of the positively charged particles.

2. **Metallic silver (condensation method).** Add concentrated NH_4OH dropwise to 2 mL of $AgNO_3$ solution until the brown precipitate that first formed has just disappeared. Pour the solution into 100 mL of distilled water and add 2 drops of a 1% tannic acid solution. The tannic acid acts as a reducing agent and converts silver ions to free silver.

$$Ag^+ + e^- \longrightarrow Ag$$

Label the sol. Make the test for the Tyndall effect.

Add 1 drop of $Al_2(SO_4)_3$ solution to a 5-mL sample of the sol. Add a similar amount of the $Fe_2O_3 \cdot X\ H_2O$ to another sample. Observe and report. Negative ions are adsorbed.

3. **Arsenious sulfide (condensation method).** Put 1 g As_2O_3 **(Caution! Poison)** in 100 mL of distilled water and boil for a few minutes. Set the solution aside to cool, filter it, and add to the filtrate 6 drops $(NH_4)_2S$ solution. An ion combination reaction produces colloidal As_2S_3.

$$2\ As^{3+} + 3\ S^{2-} \longrightarrow As_2S_3$$

Label the sol. Put the filter paper in a waste jar at once and wash the beaker and funnel. Test the sol for the Tyndall effect. Add a few drops of $Al_2(SO_4)_3$ to 5 mL of the sol. Add 2 mL of As_2S_3 sol to 5 mL $Fe_2O_3 \cdot X\ H_2O$ sol. Observe and report.

4. **Gel formation (condensation method).** Quickly add 2 mL of saturated $Ca(C_2H_3O_2)_2$ solution, while stirring, to 10 mL of 95% ethanol in a small beaker. Try igniting the mixture. Observe and report.
5. **Emulsion (dispersion method).** Mix 5 mL water and 5 mL kerosene in a test tube by violent shaking. Observe and report. Add 1 mL soap solution and shake again. Report.
6. **Smoke (aerosol).** Put 2–3 drops of concentrated HCl on a small tuft of absorbent cotton and a similar amount of concentrated NH_4OH on another tuft. Put both tufts of cotton in a large beaker so that they do not touch each other. Cover with a watch glass. Test for the Tyndall effect.
7. **Brownian movement (by demonstration).** Introduce some cigarette smoke into the cell of the Brownian movement apparatus, direct the light source into the cell, and focus the microscope on the visible beam. Best results are obtained if there is not too much smoke in the cell. This is a simple form of ultramicroscope, an instrument much used in colloid research.
8. **Adsorption.** Add 1 mL methylene blue solution to 200 mL of water in a large beaker. Then put in 2–3 g of decolorizing carbon, stir thoroughly, and filter part of the mixture. Observe and report.

Experiment 43
Colloids

Name________________________________
Date_________________________________

REVIEW QUESTIONS

1. To get some idea of the increase in surface area as a given mass of matter becomes more finely divided, calculate the surface area of a cube 1 cm on an edge. Then imagine this cube cut into cubes 1×10^{-6} cm on an edge and calculate the new area. By what factor is the area increased?

2. Considering the intermediate size range of colloidal particles, what two opposite approaches could be used in making them? Give an example of each.

3. Tabulate at least three differences among solutions, colloids, and suspensions.

4. Define the terms peptization, dialysis, hydrophobic, and hydrophilic.

5. Exactly what happens when an emulsion "breaks"?

6. Exactly what happens when a jelly "weeps"?

7. What is done in the process of "homogenizing" milk, and why does this retard cream separation?

8. Look up methods of separating colloids from noncolloids.

9. Make a table showing the dispersed phase and the dispersion medium of each of the following colloids; state whether each component is a solid, a liquid, or a gas; and what the particular name (if any) of that type of colloid is: coal smoke, jello, mayonnaise, pearl, grey hair, foam rubber, paint, a photographic emulsion, milk, whipped cream, fog.

Experiment 43
Colloids

Name____________________
Date____________________

REPORT SHEET

Preparation and Properties of Colloidal Sols

	Sol	Color	Clarity	Tyndall effect	Stabilized by	Ion added	Effect
1	$Fe_2O_3 \cdot X\ H_2O$						
2	Ag						
3	As_2O_3						

Explain the action of each effective ion ____________________

____

4. Gel formation

Describe the gel: ____________________

Does it burn? __________ What commerical product does it resemble? ____________________

What substance produces the interlacing fibers in a fruit jelly? ____________________

Why do some fruit juices such as elderberry and strawberry not gel properly even when boiled down? __________

____

What happens when a jelly "weeps"? ____________________

5. Emulsion

What happens when water and kerosene are shaken? ____________________

What happens on shaking when soap is added? ____________________

Explain: ____________________

What is the difference in principle between this method of colloid formation and the preceding ones? Be specific:

____

6. Smoke (aerosol)

What is seen in the beaker? ____________________

What substance is formed?____________ Write the equation: ____

____

Explain the "whiskers" or surface deposits on laboratory bottles. ____________________________________

____

Does the aerosol show the Tyndall effect? ____

What kind of aerosol particles make a searchlight beam visible at night?______________________________

____

7. Brownian movement

What is seen in the cell? ____

Can you see detail of the particles? ____

What makes them move? ____

Why do they sometimes seem to dissolve? ____

____

8. Adsorption

Is the filtrate colored?____________ Explain: ____

____

Can you suggest a practical use for this procedure? ____

____

EXPERIMENT 44

Electrochemistry

Purpose: To associate the production of an electric current with chemical change; to measure electrode potentials (magnitude of the tendency to lose or gain electrons); and to construct some voltaic cells.

Materials: 18 gauge Cu wire; 1/2 × 3 inch strips of Zn, Cd, Pb, Al, and Mg; carbon rods (flashlight cell electrodes); MnO_2, NH_4Cl; 1 M solutions of $CuSO_4$, $ZnSO_4$, $AgNO_3$, $CdSO_4$, $Pb(NO_3)_2$; 3 M H_2SO_4.

Apparatus: Digital voltmeters (20,000 ohms) with alligator clip leads, D.C. power supply.

Waste Disposal: All solutions containing ions of heavy metals should be poured into the heavy metal waste container provided in the lab. The metal strips should be wiped clean, rinsed, and then placed into appropriately labeled waste beakers on the reagent shelf using care that each metal is put in the correct container. The metal strips will be recycled.

INTRODUCTION

Theoretically a metal crystal consists of ions occupying lattice points with an "atmosphere" of more or less free electrons between them. An electric current is the movement of these electrons caused when an oxidation reaction in an outside source feeds additional electrons into the metal at one point and a reduction reaction removes them at another. The free electrons are repelled from the area of excess and drawn toward the area of deficiency. Such a combination of an oxidation half-reaction and a reduction half-reaction produces a voltaic cell. The electrode at which oxidation occurs (electrons are liberated) is the anode and the one at which reduction occurs (electrons are gained) is the cathode.

The drive behind this electron flow (the cell voltage) depends on the relative amounts of energy liberated by the reduction compared with the energy needed to cause the oxidation. Both the oxidation and the reduction reactions are necessary for the whole process, but they can be measured separately if some arbitrary assumptions are made. The half-reaction, $2\,H^+ + 2(e^-) \rightleftharpoons H_2$, is taken as a standard and arbitrarily assigned a reduction potential (measure of the tendency to accept electrons) of 0.00 volts, while the oxidation potential for the same reaction (in the reverse direction) is also 0.00 volts. Other half-reactions are compared to the hydrogen half-cell. For quantitative measurement of standard oxidation potentials, a hydrogen electrode in a 1 M solution of H^+ ions is used as one half-cell, and the metal being tested in a 1 M solution of its ions is the other. The two half-cells are connected by a salt bridge or a porous membrane such as unglazed porcelain or sintered glass. The voltage produced by the cell is measured. If the metal-metal ion half-cell (M/M^{2+}) loses electrons to H^+ ions, it is assigned a positive oxidation potential (negative reduction potential, E°) equal to the voltage given by the cell, and the electron flow is from this cell to the hydrogen half-cell (H_2/H^+). If the metal half-reaction is able to remove electrons from H_2, a negative oxidation potential (positive reduction potential) is assigned to it, and the electron flow is opposite to the first.

Ion concentrations also affect the strength of oxidation and reduction tendencies, so all such concentrations are 1 M in measuring standard cell potentials. For more precise work the activity, or *effective* concentration, should be used, although it frequently is not known. Other facts such as resistance to electron flow or the difficulty of crossing crystal boundaries also play a role.

The hydrogen reference electrode is impractical for class use, so in this experiment the reference electrode consists of Cu in a 1 M solution of Cu^{2+} ions (Cu/Cu^{2+}), which has an assigned reduction potential E° of 0.34 volts relative to the hydrogen electrode (the oxidation potential is –0.34 volts). Several metal-metal ion half-cells are combined with the reference electrode and the cell voltage of each is measured. From this the standard half-cell potential of the metal is calculated, using the relationship $E°_{ox} + E°_{red}$ = cell voltage. Other combinations of the half-cells measured are made, their voltages are predicted and measured. A dry cell and a lead storage cell are constructed and their voltage is measured.

PROCEDURE

A. Standard Half-Cell Potentials E° of Several Metals

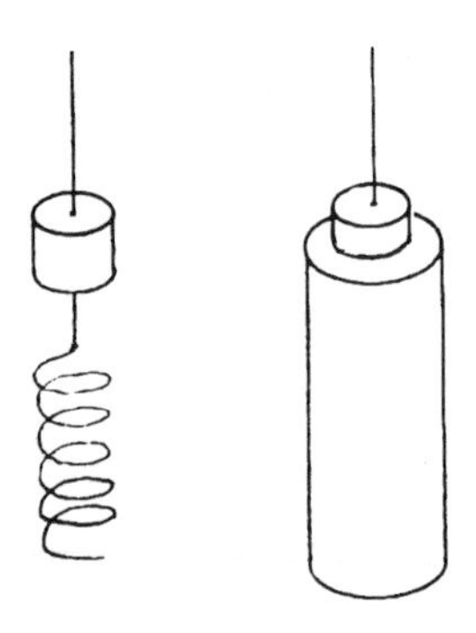

Cu/Cu^{2+} reference cell.

1. **Construction of the reference electrode.** A copper reference cell is made by putting 10 mL of a 1 M $CuSO_4$ solution into a 25 × 75 mm unglazed porcelain cup and inserting a coil of 18 gauge copper wire. The coil is made by wrapping about 2 feet of wire around a 1/2-inch test tube and leaving a 3-inch end projecting for electrical contact. This projecting end is threaded through a one-hole stopper fitted to the porcelain cup. The cell is kept immersed in a beaker of water until needed.
2. **Reduction potential of zinc ($Zn/Zn^{2+}//Cu^{2+}/Cu$).** A zinc-half-cell is made by putting about 15 mL of 1 M ZnO_4 in a 50-mL beaker and inserting a zinc strip freshly cleaned with steel wool. Connect the negative (black) lead from the voltmeter to the copper half-cell and the positive (red) lead to the other metal. Lower the copper reference cell into the $ZnSO_4$ solution in the beaker and read and record the cell voltage. If a negative reading is obtained, the reduction potential of the zinc is less than that of copper and the zinc is being oxidized. Which half-cell is the anode (oxidation)? Discard the Zn^{2+} solution and wash the outside of the porous cup with distilled water. Calculate the reduction potential of zinc from the relationship

$$E°_{ox} + E°_{red} = \text{cell voltage}$$

3. **Reduction potential of silver ($Ag/Ag^{+}//Cu^{2+}/Cu$).** Make a silver half-cell by pouring about 15 mL of 1 M $AgNO_3$ into a 50-mL beaker and by putting in a coil of copper wire with one end projecting. (Keep the solution off hands and clothing to avoid stains.) A layer of silver crystals quickly forms on the surface of the copper wire. Why? Again connect the negative (black) lead from the voltmeter to the copper half-cell, and the positive, (red) lead to the wire from the silver half-cell. Lower the cup into the beaker containing the silver half-cell and read the voltage. A positive voltage reading indicates that the reduction potential of copper is less than that of silver. Silver is being reduced and copper is being oxidized. Discard the Ag^{+} solution and clean the outside of the cup as before. Calculate and record the reduction potential of silver.
4. **Reduction potential of cadmium ($Cd/Cd^{2+}//Cu^{2+}/Cu$).** Prepare a cadmium half-cell by pouring about 15 mL of 1 M $CdSO_4$ into a 50-mL beaker and inserting a strip or rod of cadmium metal. Set up the complete cell with the copper reference cell as before. Again connect the negative (black) lead from the voltmeter to the copper half-cell and record the cell voltage. Calculate the reduction potential of cadmium.
5. **Reduction potential of lead ($Pb/Pb^{2+}//Cu^{2+}/Cu$).** Prepare a $Pb/Pb^{2+}//Cu^{2+}/Cu$ cell in the manner previously described. Again connect the negative (black) lead from the voltmeter to the copper half-cell, and measure the cell voltage. Calculate the reduction potential of lead and record it. Look up and record the standard reduction potentials of Zn, Ag, Cd, and Pb.

B. Cell Voltages

Using the reduction potentials of the metals you measured, design two voltaic cells and predict their voltages. Then construct the cells and measure and record the voltages. Either solution can be used in the porous cup. Compare the observed values with the predicted ones.

C. Zinc Dry Cell

Weigh out about 3 grams of MnO_2 and 1/2 gram NH_4Cl, mix them in your crucible, and add enough water to make a stiff paste. Connect a narrow zinc strip (1 × 1/4 inch) to the negative terminal of the voltmeter and a carbon rod to the positive terminal. Dip the two electrodes into the paste and record the cell voltage. Write equations for the reactions at the electrodes. Read and record the voltage of an ordinary zinc flashlight cell. Design a cell that should have a higher voltage than the cell constructed and check it. Discuss its practicality.

D. Lead Storage Cell

Clean two lead strips with steel wool, connect then to a voltmeter, and dip them into a 3 M H_2SO_4 solution. Does the voltmeter register? Disconnect the voltmeter and connect the electrodes to a direct current source (power supply) at a voltage that barely produces gas evolution. Be sure the electrodes are well separated in the H_2SO_4 solution. After a brown coating (PbO_2) has developed on one electrode (which one?), disconnect the direct current source and connect the electrodes to the voltmeter, with the PbO_2 coated electrode attached to the positive terminal. Record the voltage. Write the equation for the formation of PbO_2. Write equations for the reactions of a Pb storage cell and show direction of charge and discharge.

Experiment 44
Electrochemistry

Name ______________________________

Date ______________________________

REVIEW QUESTIONS

1. What must be the arrangement of the half-cells to get a practical flow of current from a redox reaction?

2. What kind of change in cell voltage would be expected if 0.1 M $CuSO_4$ solution had been used? If a 0.1 M $ZnSO_4$ solution had been used? Explain.

3. Would a zinc "dry cell" work if the contents were really dry? Explain.

4. If a heavy current is drawn from a zinc dry cell for a short time, the flow of current decreases, but recovers after a short rest. Explain why.

5. What changes limit the "shelf life" of a zinc dry cell?

6. Explain what a fuel cell is and give its advantages over other kinds of batteries.

7. Give the chemistry of the mercury cell and list advantages and disadvantages compared with the zinc dry cell.

8. Look up the chemistry of the nickel cadmium storage cell and compare this cell with the ordinary lead storage cell.

Experiment 44
Electrochemistry

Name______________________________

Date______________________________

REPORT SHEET

A. Standard Reduction Potentials E^o of Several Metals

	Metal	Complete cell formula	Half-cell reaction negative electrode (anode)	Half-cell reaction positive electrode (cathode)	Calculation of E^o	E^o from literature
2	Zn					
3	Ag					
4	Cd					
5	Pb					

Why does Ag deposit on the Cu electrode? ______________________________

B. Voltages of Two Cells

Complete cell formula	Half-cell reaction (anode)	Half-cell reaction (cathode)	Predicted cell voltage	Observed cell voltage

C. Zinc Dry Cell

Cell voltage______________________ Flashlight cell voltage______________________

Which electrode is negative?______________________ Half-cell reaction (anode) ______________________

Half-cell reaction (cathode)______________________ How could you change this cell to get a higher voltage?

____

____

Discuss possible practical deficiences of this cell:

D. Lead Storage Cell

Cell voltage before "charging" ____________ Which electrode gained the PbO_2 coating? ____________

Cell voltage after charging ____________ Equation for formation of PbO_2: ____________

Overall reactions in lead storage cell (indicate charge and discharge reactions). ____________

____

EXPERIMENT 45

Analysis of Some Commercial Bleaches

Purpose: To test the validity of commercial claims by analyzing several bleach solutions by oxidation-reduction titration.

Materials: Two or three commercial bleaches available from the grocery or discount store, *standardized* approx. 0.13M $Na_2S_2O_3$ solution, 10% KI solution, 3M H_2SO_4, freshly-prepared soluble starch solution.

Apparatus: 100-mL volumetric flask and stopper, 10-ml and 25-mL pipets, pipet bulbs, 50-mL buret, 50- or 100-mL graduated cylinder, 250-mL Erlenmeyer flask.

Safety Precautions: Wear your safety glasses at all times during this experiment. Bleach, H_2SO_4, and HCl can cause serious chemical burns and can ruin clothes. Wash thoroughly with soap and water if skin contact occurs.

Hazardous Waste Disposal: All solutions may be rinsed down the drain with plenty of water.

INTRODUCTION

Hopefully, study of science will provide you with a healthy skepticism of various commercial claims. Listen more carefully to the next commercial you hear and pay special attention to the wording.

Most of the time a company's advertising claim takes the form of "nobody's product is superior to ours," when in fact most of the competing products will be *identical* in quality to the one being advertised. For a wide variety of products, the government has established standards and all reputable brands will adhere to those guidelines.

Consider the common pain reliever aspirin. A regular strength aspirin tablet contains 325 mg of acetylsalicylic acid and inert starch pressed to hold it all together. This is true if you buy Bayer, Norwich, Wal-Mart, Shopko, or generic.

Some of these manufacturers offer special reasons to purchase their product. For example, many companies coat their aspirin tablets to prevent them from dissolving on the tongue and leaving a sour taste. Some may be flavored to disguise the taste (children's chewables). A special shape, caplets, is offered and presented as being easier to swallow. (Capsules, those gelatin tubes with the medicine inside, have all but been eliminated due to several tragic tampering incidents.) In advertising their products with special features, the companies are avoiding admitting that medicinally, an aspirin is 325 mg of acetylsalicylic acid and that there is essentially no difference among brands.

Consider also the size of the product. "Extra-strength" aspirin is an aspirin tablet that is a 500-mg dose. There is nothing *stronger* about the aspirin itself—it is just a bigger dose! So when the advertisement claims "extra-strength (aspirin) contains more pain reliever than any other *regular-strength* aspirin," it's true only because 500 mg is more than 325 mg. Two tablets is the normal dosage of these products. Therefore three regular aspirins (3×325 mg) contain almost the same amount as two extra-strength (2×500 mg) tablets.

Also, read labels to discover what those "extra ingredients" are. One product says "...contains all the pain-fighter of regular-strength aspirin *plus an extra ingredient.*" Check the fine print. The extra ingredient is caffeine, in about the same amount found in a cup of coffee or a soda. Caffeine may be beneficial for the intended use, but be aware that you may be paying for something that's not so special, unique, or exotic as the ads imply.

Not all advertisements are fraudulent or irrelevant. The vast majority of commercial products are made by reputable companies and perform precisely as claimed. However, if you make purchases based on advertisements, be sure you really listen to any claims or guarantees. Or better, *read the label*! Compare it to the labels of competing products. Develop the habit of reading the label on every new product before you buy it. This applies to foods and medicines, as well as most household products.

In this experiment you will compare two different brands of household bleach. The government defines bleach as a particular weight % aqueous solution of sodium hypochlorite, NaClO.

You will react the hypochlorite with excess iodide ion to liberate iodine, then titrate the liberated iodine with sodium thiosulfate. The *unbalanced* reactions are

$$H^+ + OCl^- + I^- \longrightarrow I_2 + Cl^- + H_2O \qquad \text{Equation 1}$$

and

$$I_2 + S_2O_3^{-2} \longrightarrow I^- + S_4O_6^{-2} \qquad \text{Equation 2}$$

PROCEDURE

A. Titration of Bleach Solutions

1. Obtain a 100-mL volumetric flask with a stopper. Please be careful with this glassware.
2. Pipet 10 mL of the first bleach into the 100-mL volumetric flask. Make up to the mark with distilled water and mix well. Record the brand of bleach used. Also record the price paid per gallon of bleach. If the price is not visible on the bottle, your instructor will provide it.
3. Rinse and fill a buret with sodium thiosulfate solution. Record the molarity.
4. With a graduated cylinder measure out 10 mL of 10% KI solution and transfer to a 250-mL Erlenmeyer flask. To this add 7 mL of 3 M H_2SO_4. Mix well.
5. Pipet 25 mL of your *diluted* bleach solution into the Erlenmeyer with the KI and H_2SO_4. Note the formation of the deep red-brown color as iodine is formed as indicated in equation 1. If large amounts of SOLID iodine form, consult your TA or instructor.
6. Record the initial volume on the $S_2O_3^{-2}$ buret and titrate the solution to the *disappearance* of the iodine color. A good end point will make the solution go from yellow to colorless in 1 drop. Record the final volume. Discard the titrated solution down the sink with plenty of water.
7. Now you will repeat the titration with a variation—an indicator. The titration doesn't absolutely require one as you have seen, but an indicator can enhance the visibility of the end point. The indicator is soluble starch. Starch and iodine form a deep-blue complex.

   [Ironic—Chemists use starch to indicate the presence of iodine. Biologists use iodine to indicate the presence of starch.]

   Begin the titration as before, but 1 or 2 mL before the anticipated end point add 2 mL of starch. Titrate to the disappearance of the blue color. Record initial and final buret readings.
8. If the two titrations differ by more than 1 mL, perform a third titration with or without starch as you prefer.
9. Dump the diluted bleach solution down the sink and flush with plenty of water. Rinse the volumetric flask thoroughly with distilled water.
10. Choose a second brand of bleach and repeat steps 1–9 (you need not rinse the buret).
11. Wash all glassware thoroughly and return anything borrowed.

B. Density of Bleach

In order to perform some of the calculations, you will need to know the density of the original bleach solution. Pipet 25 mL of bleach into a tared-weight beaker (or determine mass by subtraction). Record the mass of 25 mL of the bleach. Calculate the density in grams per milliliter.

C. Statistics

Collect and record the % NaOCl values from five other students for two different brands. Alternatively, your instructor may collect all your answers and distribute them to the whole class, or have you write the values on the laboratory blackboard.

CALCULATIONS

1. Balance equations 1 and 2 and write on your report sheet.
2. Average titration volumes for each brand separately. Convert mL $S_2O_3^{-2}$ to liters, then liters to moles using the molarity.
3. Convert moles $S_2O_3^{-2}$ to moles I_2 using the balanced equation.
4. Convert moles I_2 to moles OCl^- using the balanced equation.
5. Moles OCl^- = moles NaOCl. Convert moles NaOCl to grams NaOCl.
6. The grams NaOCl is for the 25-mL aliquot of diluted solution. The entire solution was 100-mL, so the mass of NaOCl in the volumetric was grams NaOCl × 100 mL/25 mL = total grams NaOCl

   Therefore, just multiply by 4.
7. You also took a 10-mL sample of the original bleach. Use the provided density to change this from milliliters of bleach to grams of bleach.

   Note: Sodium hypochlorite is written as both NaOCl and NaClO
8. Calculate %NaOCl in each of the brands of bleach.

$$\%\text{NaOCl} = \frac{\text{total grams NaOCl}}{\text{grams of bleach}} \times 100\%$$

9. Calculate the average and standard deviation for the six data points (yours plus those of five other students) for the first brand of bleach. Repeat the calculation for the second brand of bleach.
10. Finally, consider the results of the experiment and comment.

Experiment 45

Analysis of Some Commercial Bleaches

Name______________________________________

Date_______________________________________

REVIEW QUESTIONS

1. Compare your results to the labels. How good are your results? Are the namebrand and the cheaper bleach equal in NaOCl content?

2. Why might a person who knows the results of your experiment still buy the higher priced bleach?

3. Calculate the molarity of the first bleach solution you titrated.

4. What is the oxidation state of Cl in OCl^- ? Is bleach an oxidizing agent or a reducing agent? Explain.

5. Why do some environmentalists suggest curtailing or eliminating the use of bleaches? Be specific.

6. Would you expect a bleach solution, NaOCl, to be acidic, basic, or neutral? Write a chemical equation to support your answer.

7. Which brand was more economical? Are there other considerations that this experiment did not test? Describe.

8. Knowing that its structure is very similar to that for sulfate, draw a Lewis Dot structure for thiosulfate ion.

Experiment 45

Analysis of Some Commercial Bleaches

Name____________________________

Date____________________________

REPORT SHEET

Titrant____________________ M $Na_2S_2O_3$

Bleach 10 mL diluted to 100 mL

Density of bleach ____________________ g/mL

A. Titrations Used 25 mL of diluted bleach solution

Brand 1 ____________		Brand 2 ____________	
Cost____________ /gallon		Cost____________ /gallon	
Trial 1	Trial 2	Trial 1	Trial 2
mL	mL	mL	mL
– mL	– mL	– mL	– mL
mL	mL	mL	mL

Balanced equations

Calculations show all calculations for %NaOCl

B. Density of Bleach

Brand 1 ______________________ Mass of 25 mL ______________________ g

Density ______________________ g/mL

Brand 2 ______________________ Mass of 25 mL ______________________ g

Density ______________________ g/mL

C. Statistics

	%NaOCl	
	Brand 1	*Brand 2*
1)		
2)		
3)		
4)		
5)		
Average		
Std. dev.		

Conclusions on the experiment:

EXPERIMENT 46

Spectrophotometry

Purpose: To practice the techniques of pipetting, make dilutions of a specific concentration, and learn the method of quantitative analysis by the use of a spectrophotometer.

Materials: Stock 0.0050 M $KMnO_4$ solution.

Apparatus: 100-mL volumetric flask and stopper, 1-mL pipet, 2-mL pipet, pipet bulb, large test tubes with rubber stoppers or corks, "Spec-20" or other similar spectrophotometer.

Safety Precautions: $KMnO_4$ is a strong oxidizer and must be handled carefully. It will stain clothes and skin, and it can cause chemical burns. Wash spills off immediately. Also, be aware the $KMnO_4$ will stain glassware! Rinse out pipets and beakers as soon as you are finished using them. Wear safety glasses at all times.

Hazardous Waste Disposal: The $KMnO_4$ solutions can be rinsed down the drain with plenty of water.

INTRODUCTION

A very wide variety of analytical techniques exist to identify and quantify chemical substances. Modern instrumentation can separate hundreds or thousands of components into individual species, identify each one, and calculate the exact amounts present. Paper chromatography, a technique you may have performed, is a simple variation on a separations method.

Qualitative analysis is a branch of chemistry that determines *what* chemicals are present. *Quantitative analysis* tells *how much*. Spectrophotometry is a quantitative method that is used to determine very low concentrations, usually in the 10^{-3} to 10^{-7} M ranges. As its name implies, spectro-photo-metry measures the light absorbed in a particular part of the spectrum (usually the visible spectrum from about 350 nm to 700 nm). The method is based on Beer's Law, which states that for a specific wavelength

$$A = \varepsilon bC$$

A = absorbance, ε = molar absorptivity constant, b = path length of light, C = concentration, mol/L

With a little reflection, experience will tell you that this is obvious. Consider two test tubes, X and Y, with different concentrations of a colored chemical species. Hold them both up to the light. Which one is darker? Logic tells you that the darker one is more concentrated, and if it's twice as dark, it must be twice as concentrated. But why does test tube X appear darker? When you hold it up to the light, the light passes through the test tube and is *absorbed* by a small amount every time a light particle encounters a light-absorbing (*colored*) chemical species. The more light-absorbing particles, the greater the concentration, and the more that the light intensity is diminished as it emerges on the other side of the test tube.

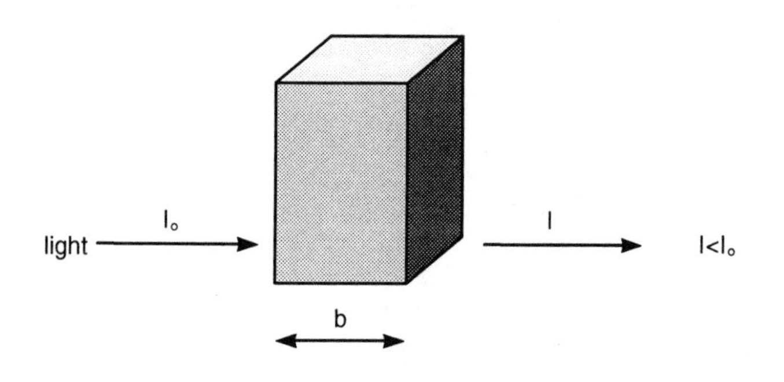

The transmittance is defined as $T=I/I_0$ The "lost" intensity has been absorbed and the spectrophotometer measures this absorbance. Notice that the effect is also dependent on how far the light must travel. The path length, b, is usually measured in centimeters, and the standard cuvette (most spectrophotometers don't use test tubes) is exactly 1 cm wide.

Spectrophotometry is probably more widely used than any other analytical technique and is the method of choice in the field of medicine. Chemical substances that are not colored can often be "tagged" by reacting them with a dyeing agent. For example, diabetics can monitor their blood sugar levels at home with a simple, portable spectrophotometer. They take a blood sample by finger pin-pick and touch the drop of blood to a strip of paper. Chemicals in the paper react with glucose. The end of the paper is placed in the device and the absorbance of the glucose-colored species is recorded. The device then displays a calculated blood sugar level. Thousands of laboratory tests are conducted in hospitals and clinics by this technique. Some blood tests can measure 24 or more different substances on one sample relatively quickly.

In this experiment you will prepare a *standard curve* for potassium permanganate $KMnO_4$ and use it to determine the concentration of an unknown.

PROCEDURE

A. Preparation of Solutions

Obtain a 100-mL volumetric flask and a stopper. This is expensive glassware. Be careful!

Obtain 60–75 mL of stock $KMnO_4$ solution in a beaker. Record the exact molarity from the bottle. Cover the beaker with a paper towel.

You will need a 1-mL and 2-mL pipet, but these will have to be shared. You will use a pipet bulb—NO MOUTH PIPETTING!!

To save on volumetrics, you will store your solutions after you make them. Thoroughly clean six test tubes. Label them 1-6.

Practice your pipetting technique with a beaker of water. **Note:** Avoid sucking solution up into the bulb. If you do, rinse the bulb and thoroughly blow the water out onto a paper towel. Volumetric glassware comes in two basic types: TC (to contain) and TD (to deliver). Check your pipet. It should be TD. This means that left to gravity, the pipet will deliver the specified volume. TD pipets are further divided into single-volume and graduated pipets. Single-volume pipets have one scored mark above the reservoir and are intended to deliver exactly the amount indicated on the label. To use this kind of pipet, draw liquid above the line and then hold with your thumb or forefinger. Gently rock or wiggle your finger and allow the excess solution to drain down to the mark. The bottom of the meniscus should just touch the mark. Transfer the pipet to the receiving container and let the liquid flow out. When it is nearly finished, touch the tip of the pipet to the inside of the container to draw off any last amount. A small amount of solution will be left inside the pipet. DO NOT BLOW IT OUT WITH THE PIPET BULB! The pipet has delivered what it should.

The graduated pipet is marked with many lines in the same way a graduated cylinder is. It is useful when a variety of different volumes are required; e.g., a 1-mL pipet marked off in 0.1 mL intervals. The difference here is that you must stop the pipet from completely draining. If you wanted 0.6 mL, you would start at the 0.0 mark and then stop the drainage at the 0.6 mark. However, for some graduated pipets, there is no bottom mark and the pipet is then allowed to drain as in a single-volume pipet. Examine yours carefully. If you are uncertain, ask a TA.

When you are proficient with a pipet, begin making your dilutions. You will take 1 mL, 2 mL, 4 mL, 6 mL, 8 mL, and then 10 mL, and dilute each to exactly 100 mL. Start with the most dilute solution first. The volumetric must be clean but not necessarily dry. First rinse the pipet with a small amount of stock $KMnO_4$. Then pipet 1 mL of the stock solution into the volumetric flask. Rinse the pipet immediately with distilled water. Dilute with distilled water until the meniscus bottom just touches the mark. Cap and mix *thoroughly*. Transfer the solution to the test tube marked "1," first rinsing and discarding about 10 mL, then filling the test tube nearly full. Stopper or cork and set aside. Discard the remainder of the solution down the sink and flush with water. Rinse the volumetric flask with distilled water.

Repeat the process for solutions 2–6 (2 mL, 4 mL, 6 mL, 8 mL, 10 mL).

B. The Standard Curve

Now you will use the Spec-20 spectrophotometer to measure the absorbance of your *standards* (solutions of known concentration). The wavelength is 526 nm.

The TA will help you with the cuvettes and instruct you in the use of the instrument. Adjust the dark current with no cuvette in the spectrophotometer. Then adjust to 100% T with distilled water. Start with the most dilute solution. Measure the absorbance of each solution and record the value on your report sheet.

Obtain an unknown and record its code. Your eyes are fairly sensitive photometers. Just using visual approximation, estimate the concentration of your unknown. Record your estimate. Before you leave the spectrophotometer, measure the absorbance of the unknown.

Rinse all glassware thoroughly. Discard all solutions down the sink with plenty of water. Return any glassware and equipment and your unknown test tube.

CALCULATIONS

First calculate the concentration of each of the standards by applying the appropriate dilution factor. For example, suppose the stock solution was 0.00531 M (5.31×10^{-3}) and you took a 4-mL aliquot and diluted it to 100 mL. The dilution factor is then 4 mL/100 mL and the concentration is reduced by that amount

$$0.00531 \text{ M} \times \frac{4 \text{ mL}}{100 \text{ mL}} = 2.124 \times 10^{-4} \text{ M}$$
$$= 2.12 \times 10^{-4} \text{ M}$$

Round off each to three significant figures. (You generally can't plot more than 3 figures.)

Now construct a *standard curve*. Plot absorbance (a unitless number) versus concentration (mol/L). Consult the information on graphing techniques in the Appendix if you need help. Draw the best-fitting straight line that probably should, but need not, go through the origin. If any one point seems "bad," ignore it.

Finally, use your standard curve to "read" the concentration of your unknown. To do this, find the position on the absorbance scale that corresponds to the absorbance reading for your unknown. Scan over horizontally until you reach the straight line you drew. Scan down to read the corresponding concentration to the nearest three significant figures. Record this concentration for your unknown on your report sheet. Turn in your graph with your report sheet. A sample standard curve sketch follows. (How good was your visual estimate?)

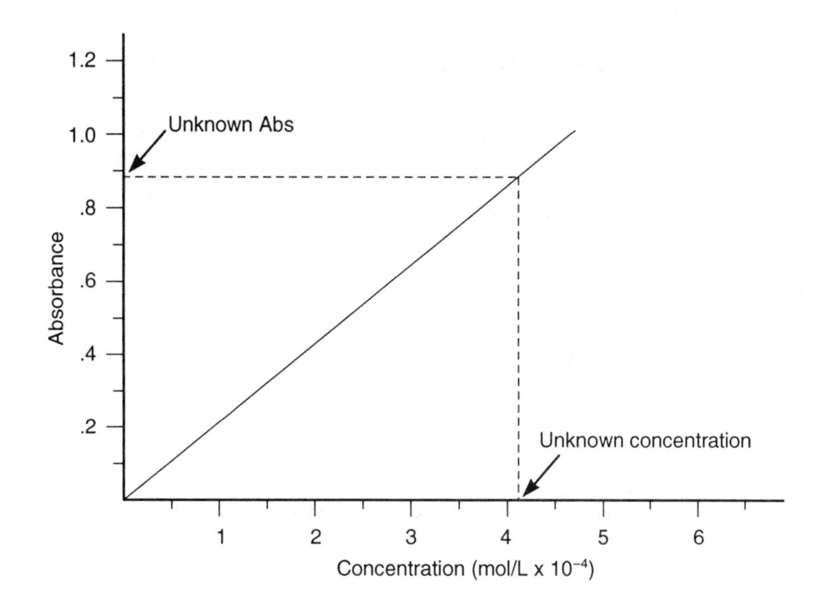

Experiment 46

Spectrophotometry

Name________________________________

Date_________________________________

REVIEW QUESTIONS

1. $K_2Cr_2O_7$ can oxidize alcohol (ethanol) to carboxylic acid. In the process the orange-yellow color of dichromate turns to the green of Cr(III) and the intensity of the green color is measured. Blood alcohol levels of suspected drunk drivers can be determined this way. A blood sample so treated gives an absorbance of 0.635 in a 1-cm cell. If the absorptivity constant (in terms of %) is 3.52/cm · %, is the driver legally drunk in your state? Show your work.

2. Why is $KMnO_4$ purple? (Hint: At what wavelength did you measure absorbance?)

3. What are the meanings of the terms of "near-ultraviolet" and "near-infrared?"

4. A major peak in the $KMnO_4$ spectrum is at 545 nm. What amount of energy in kJ/mol is absorbed for this transition?

5. Given that $A = -\log I/I_o$, what is the corresponding absorbance for a solution that has 60% transmittance (T = 0.60) at 500 nm?

6. Medical technologists who use spectrophotometry on a daily basis usually prepare a new standard curve every day. Why might this be necessary?

7. Assuming that the cuvettes you used were exactly 1 cm, what is the value of the molar absorptivity constant ε for $KMnO_4$?

Experiment 46
Spectrophotometry

Name______________________________

Date______________________________

REPORT SHEET

Concentration of stock $KMnO_4$ ____________________ M

Unknown code ____________________

Solution #	mL Stock	Absorbance	Concentration M
1	1		
2	2		
3	4		
4	6		
5	8		
6	10		
Unknown	X		

Estimate of unknown:

EXPERIMENT 47

The Mystery of the Thirteen Test Tubes

Purpose: This experiment is designed to give students who do not perform the standard qualitative analysis scheme a chance to investigate some simple descriptive chemistry. It avoids the toxicity concerns of mercury and cadmium, and can be performed in one laboratory period, assuming preparation by the student in reading the appropriate literature.

Materials: A uniquely-labeled set of thirteen unknowns in test tubes containing each of the following solutions, one set per student:

1M H_2SO_4	6M NH_3	1M K_2CrO_4	1M NaCl	1M $Fe(NO_3)_3$	1M $K_2C_2O_4$
0.5M Na_2S	1M $CuSO_4$	1M $NiSO_4$	0.1M $SnCl_2$	1M KNO_3	1M KSCN
1M $Ba(NO_3)_2$					

Apparatus: Test tubes, stirring rods, Bunsen burner, litmus paper or pH paper, spray atomizer or mister (such as for cosmetics or house plants) OR nichrome wire for flame tests.

Safety Precautions: You can't tell a harmful chemical from a harmless one just by looking. Follow all normal laboratory practices during this experiment. Avoid skin contact with the chemicals. Wear your safety glasses at all times.

Hazardous Waste Disposal: All the solutions may be washed down the drain with a large volume of water.

INTRODUCTION

One of the TAs (we don't know who) has filled thirteen test tubes with thirteen different solutions and labeled them with a secret code. The actual identity of each solution is a mystery.

You are the chemical detective. You must correctly identify each solution before time runs out (at the end of the laboratory period).

Now since the Chemical Abstracts Services (CAS) has registered over ten million compounds, this would seem to be an impossible task. In this case, however, you're in luck! All the other compounds have an alibi for the time in question and we know for certain what the thirteen solutions are.

The Suspects

H_2SO_4	$SnCl_2$
NH_3	KNO_3
K_2CrO_4	KSCN
NaCl	$Ba(NO_3)^2$
$Fe(NO_3)_3$	$K_2C_2O_4$
Na_2S	$Cu(NO_3)_2$
$NiSO_4$	

All you have to do is match these identities to the code on each test tube. This kind of process is called qualitative analysis. This means identifying *what* is present. The second semester of General Chemistry is entitled Qualitative Analysis at many colleges because of the focus on this topic in the second semester. In the last 10 years, however, there has been a shift in emphasis away from areas of descriptive chemistry; i.e., a physical description of what compounds look like.

The other part of Analytical Chemistry is *quantitative analysis*. After the identity of a compound is established, the next question is *how much*? Many of your other experiments were intended to address this question. You will recall that the answer was often a "number."

You are fortunate to have a limited list of suspects. It would be a much harder job to identify one out of ten million. But on the other hand, your investigation will be severely limited. No fancy electronic equipment, and you may not question any other chemicals or have them assist in your investigation. This means that all your chemical tests will have to make use of these chemicals themselves. Once you identify one chemical, you may be able to use it to identify others.

You may submit each to a lie detector test in the form of litmus paper or a flame test.

You must be very careful to avoid contaminating one solution with another. Don't forget to clean and dry the stirring rod between solutions. Contamination of solutions is a gross violation of their chemical rights and shall result in dismissed cases (and no credit).

One set will be provided for each detective. If it is spilled, lost, dropped, or stolen NO MORE will be given. Every set is different, so don't bother consulting your neighbor. Do your own work.

Before you begin, do some homework. Find out what you can about your suspects. Make a plan for how you will proceed. Your textbook, *The CRC Handbook of Chemistry and Physics*, and other standard reference books might be helpful. The following may also be useful:

H_2SO_4 — Strong acid, most powerful industrial chemical in the world, may produce insoluble sulfates if metaled with.

NH_3 — Ammonia, alias ammonium hydroxide (NH_4OH) has done important work in homes, last known employment as fertilizer, can turn ugly on any nosey detective.

K_2CrO_4 — Best known for its bright disposition, potassium is almost inert, but the chromate may drop out if faced with silver, lead, or barium. Remains bright even when it lays low.

NaCl — Nothing but a common salt, almost impossible to recognize in a crowd, but shows quite a yellow streak when the real heat is on.

$Fe(NO_3)_3$ — Ferric nitrate, alias "iron three"; ferric is more reactive than younger brother ferrous; may be recognized by color if not confused with other species, can be definitively identified by "bloody" encounter with greatest rival thiocyanate.

$K_2C_2O_4$ — Potassium oxalate, actions not well known but moderate toxicity noted, handle with care, believed to have had a "falling out" with barium.

Na_2S — Sodium sulfide, alias "Le Pew," a real loner, possible messy confrontations with copper, ferric, nickel or tin; tends to linger on the skin if touched (not recommended).

$Cu(NO_3)_2$ — Cupric nitrate, first name officially changed to "copper II"; leading chemical citizen, many business ventures include electrical wire manufacturing and production of alloys, notably brass; in solution easily recognized by "melancholy" disposition. Once suspected of conspiring with ammonia to impersonate ink.

$NiSO_4$ — Nickel sulfate, once very valuable, now net worth greatly reduced; "Nick" is easily recognized by his "envious" nature.

$SnCl_2$ — Stannous chloride, a.k.a. tin chloride, a hard worker, known since ancient times, currently employed in food packaging industry, recyclable; fluoride form prevents tooth decay; unfortunate confrontation with "Le Pew" (Note: the preparation of aqueous solutions often uses large amounts of HCl which may make the solution appear more acidic than normal.)

KNO_3 — Potassium nitrate, *the* most boring substance known outside of the noble gasses, chronically unemployed, does not participate in chemical reaction but often "watches," can be distinguished from the other "common salt" by its pale violet response to any "trial by fire."

KSCN — Potassium thiocyanate, poisonous little creature, approach with caution, long-standing "blood" feud with the iron brothers.

$Ba(NO_3)_2$ — Barium nitrate, little known on this one, chance encounters with sulfuric acid have often "precipitated" pale consequences.

An anonymous tipster has left this message: "The nitrates are a red herring."

PROCEDURE

Round up your container of suspects. Record the code for your set on your report sheet. Take care of your suspects. You cannot get replacements.

You may (and should) perform litmus tests.

Flame tests will help you with a couple of suspects. Ask the TA for help if you are uncertain how to perform or interpret these tests. You may use a nichrome wire, or aspirate the solution gently into the flame with an atomizer. Take care, however, that you do not wastefully exhaust all of your solution.

Do not collaborate with any other detectives.

Collect solids and dispose in the trash. Solutions may be flushed down the drain.

You may bring any notes, books, or lists with you that you think you might need.

Record any useful observations on your report sheet.

Experiment 47

The Mystery of the Thirteen Test Tubes

Name________________________________

Date________________________________

REVIEW QUESTIONS

1. Using the K_{sp}, calculate the approximate solubility of $BaSO_4$ in grams per liter.

2. Why are chromate, ferric, cupric, and nickelous ions all colored?

3. What is "hard" water?

4. Write a net ionic equation for the reaction between barium nitrate and potassium chromate.

5. Why are aqueous solutions of ammonia sometimes called ammonium hydroxide?

6. Explain the difference between the two main branches of analytical chemistry.

7. Give an example of a chemical reagent, not on the list of suspects, that would have made the identification of NaCl easier.

8. A favorite trick of graduate assistants is to give a student a test tube of distilled water. If you suspected that one of your unknowns was distilled water, how might you test that theory? You may suggest any common laboratory method or technique.

Experiment 47

The Mystery of the Thirteen Test Tubes

Name ____________________

Date ____________________

REPORT SHEET

Code for your set ____________________

1. ____________________
2. ____________________
3. ____________________
4. ____________________
5. ____________________
6. ____________________
7. ____________________
8. ____________________
9. ____________________
10. ____________________
11. ____________________
12. ____________________
13. ____________________

Notes and observations

Experiment 47 Name

The Mystery of the Thirteen Test Tubes Date

REPORT SHEET

EXPERIMENT 48

Qualitative Analysis of Group I Cations

Purpose: To investigate the chemistry of the Group I cations and develop a method for selectively separating and identifying each. Further, the techniques used in this laboratory will be used in the Qualitative Analysis experiments that follow.

Materials: Solutions in dropping bottles:

0.2 M $AgNO_3$	0.2 M $Hg_2(NO_3)_2$	0.2 M $Pb(NO_3)_2$	15 M NH_3
3 M HCl	0.25 M K_2CrO_4	6 M HNO_3	

Apparatus: 10–15 small centrifuge-size test tubes; stirring rods; centrifuge; test tube holder; pH or litmus paper; hot water bath

Safety Precautions: Solutions of mercury and lead are extremely toxic. Avoid skin contact with these solutions. Silver solutions will stain skin and clothing dark-brown or black. Wash hands thoroughly before leaving the laboratory. Wear your safety glasses at all times during this experiment.

Hazardous Waste Disposal: Dispose of all solutions in properly labeled waste receptacles.

INTRODUCTION

Qualitative analysis is the process of separating and identifying ions—in this experiment cations—based on their unique chemical reactions. In theory it is possible to separate ions, one at a time, from a mixture containing twenty or more different cations. In practice, the procedure is simplified by analyzing smaller numbers of ions grouped by a common characteristic.

The method of qualitative analysis employs precipitation followed by separation of the solid portion from the remaining solution. Subsequent precipitation and separations eventually yield the single ion as either a solid or in solution. Generally the presence of the ion is then determined unequivocally by the use of a confirming test. (Chemists are very meticulous!)

The qualitative analysis scheme is traditionally broken down into five groups. They are:

Groups	Characteristics	Members					
I	insoluble Cl^-	Pb^{+2}	Hg_2^{+2}	Ag^+			
II	insoluble S^{-2} (basic)	Hg^{+2}	Cu^{+2}	Cd^{+2}	Bi^{+3}	Pb^{+2}	
III	insoluble S^{-2} (acidic)	Al^{+3}	Ni^{+2}	Fe^{+3}	Cr^{+3}	Co^{+2}	Zn^{+2}
IV	insoluble PO_4^{-3}	Ca^{+2}	Ba^{+2}	Sr^{+2}	Mg^{+2}		
V	soluble ions	NH_4^+	K^+	Na^+			

Several ions have been omitted due to time constraints and because their separations pose particular difficulties.

You will notice that Pb^{+2} is found in both Group I and Group II. This points out the greatest difficulty associated with the selective precipitation method—sometimes the reactions do not "go to completion." The K_{sp} of $PbCl_2$ is approximately 1.7×10^{-5}. That's relatively high for an "insoluble" chloride, and it means that some of the Pb^{+2} will be precipitated as $PbCl_2$ while some will remain in solution as Pb^{+2} and be carried over into the Group II ions.

Notice also that the mercurous ion $Hg_2{}^{+2}$ is found in Group I because it forms Hg_2Cl_2, but the mercuric ion Hg^{+2} is not because $HgCl_2$ is soluble.

You will begin by making several chemical tests on individual ions and recording your observations. Using this information and the qualitative scheme developed, you will apply the separation scheme to an unknown and a known mixture of all the ions in the group for comparison. The unknown in Group I may contain one, two, or all three ions.

This qualitative analysis scheme has a history that goes back more than one hundred years. Before modern instrumentation was routinely available, chemists were forced to rely solely on observation and the interpretation of chemical equilibria. As you prepare for this lab, you may wish to review the material in your textbook on acid-base, precipitation and solubility, oxidation-reduction, and compleximetric equilibria.

Several standard techniques and concepts that will be used in these experiments are described in the following sections. You may wish to refer back to them during your work.

Acid/base: When the instructions are to make a solution acidic or basic, use the specified reagent in small amounts and check the pH with litmus paper or universal indicator paper with each addition. To test the pH, dip the end of a clean, dry glass rod into the solution and touch the droplet that clings to it to the piece of pH paper. Never drop or dip the pH paper into the solution. Remember that acid turns blue litmus paper red and base turns red litmus paper blue. Most pH paper, such as Universal Indicator Paper, comes with a color key to estimate pH.

Amphoteric: A substance that reacts with both H^+ and OH^- is said to be amphoteric. Particularly applicable to qualitative analysis are the amphoteric hydroxides; e.g., of Al^{+3}. Starting with an acidic solution of aluminum ion, addition of OH^- begins precipitation of the trihydroxide:

$$Al^{+3} + 3OH^- \longrightarrow Al(OH)_{3\,(s)}$$

Upon further addition of OH^-, the solid dissolves to give the tetrahydroxoaluminate complex ion:

$$Al(OH)_{3\,(s)} + OH^- \longrightarrow Al(OH)_4{}^-$$

Centrifuge: A device for separating solids from liquids. Select a test tube for your work that easily fits into the cylinders. Before using a centrifuge, make certain it is well anchored (usually with rubber feet) and that there is no debris in the test tube cylinders. (Test tubes sometime shatter under the pressure of centrifugation. Check that no small slivers of glass remain.) To use a centrifuge place the test tube with the mixture to be separated into one of the cylinders. Place another test tube with a nearly equal amount of water on the opposite side for balance. Never operate the centrifuge unbalanced. Most centrifuges will accept two, four, six, or eight test tubes at a time. Place the cover on the centrifuge to protect people from flying glass in the event of breakage. For most mixtures, 1 to 2 minutes is all that is necessary for separation. Longer centrifugation may compact the solid and make it difficult to remove. Allow the centrifuge to stop completely before removing the cover. If the mixture is incompletely separated, repeat. Small amounts of flocculent (fluffy) solids that float on the surface can be removed with a glass rod or twist of paper towel.

Decant: To pour off. Usually after centrifugation, the solution on the top, the *supernatant* solution, or simply the supernate, is removed by gently pouring it into another test tube. Care is taken to avoid contaminating the solution with small particles of solid.

Flame test: A method of qualitative analysis based on the appearance of a substance when burned in a flame. Adjust the Bunsen burner to give a blue nonluminous flame. Dip the end of a nichrome or platinum wire into 3 M HCl and heat to incandescence in the flame. Repeat several times to remove any contaminants on the wire. Dip the cooled wire into the acid and then into a fresh sample of unknown solution. The HCl tends to produce the more volatile chlorides of the cations. Hold the wire with the droplet of unknown in the nonluminous flame and observe the appearance as it burns. It may be necessary to view the flame through a small piece of blue cobalt glass. This will filter out certain colors that may mask others. Do not use a glass rod to perform a flame test. Sodium ion in glass will interfere. As an alternative, if sufficient sample exists, the solution can be aspirated directly into the flame. To do this, a pump or trigger sprayer (such as that used in cleaning or cosmetic products) is first cleaned by drawing a large amount of distilled water through it. Then the solution itself is sprayed into the flame to form a very fine mist. A perfume atomizer produces the finest mists and uses very little solution. This procedure gives a much clearer flame appearance but sacrifices 4–6 mL of solution. Always use a *fresh* sample of solution for an unknown test. Reagents added to the qualitative scheme may contaminate the solution, especially with NH_4^+, K^+, or Na^+.

Heat: Often a test reagent must be heated with the solution to yield complete reaction. Do not heat a test tube with solution directly in a Bunsen burner's flame. Since these are aqueous solutions, the maximum temperature is only about 100°C, not the 300°–400°C produced in the flame. Instead prepare a water bath by filling a 400-mL beaker about half-full with water and placing it on a ring stand with wire gauze and an ion ring. Heat the water gently to a low boil. Keep the hot water bath available throughout the lab time by occasionally replenishing the water supply. When heat is called for, simply place the test tube in the hot water bath. Use a test tube holder to remove.

Precipitate: The solid part of a reaction mixture.

Wash: You will sometimes be asked to wash a precipitate. To do so, centrifuge the mixture and decant off the solution. (You may or may not need to save the solution depending on whether or not there may be potential ions remaining.) Add a small amount of distilled water (1–2 mL) to the solid and stir with a clean stirring rod. The procedure is intended to remove contaminating ions from the surface of the solid, so a thorough mixing is needed. Take care, however, not to break the stirring rod or the bottom of the test tube. Re-centrifuge and decant the wash. If you saved the supernatant solution, you should combine the wash to the saved solution. Usually, a precipitate is washed at least twice.

PROCEDURE

A. Standard Tests

1. For each of the three approximately 0.2 M ions, obtain about 10 drops of solution and place in labeled test tubes. A marking pen that writes on glass and is removed with ethanol or acetone is very useful for identifying the many test tubes you will use.
2. To each of the Pb^{+2}, Hg_2^{+2}, and Ag^+ solutions, add 6 drops of 3 M HCl and record your observations. Add about 2 mL of water to each and heat the solutions for a few minutes in the hot water bath. What happens? If the result is unclear, add 1 mL more water and stir well.
3. Using about 5 drops of the cation and about 5 drops of the reagent, perform the tests for 15 M NH_3 (also may be labeled NH_4OH) and 0.25 M K_2CrO_4 and record your observations.
4. Complete and balance the equations based on your results.

B. The Qualitative Analysis Scheme

Using what you have learned about the behavior of the three ions, a plan can be developed to separate and identify each. A partially completed scheme is presented on the report sheet. Fill in the missing ions, formulas, or appearances. This will be your guide for the known and unknown. When a line diverges into two boxes, it implies centrifugation. One side must be a solid or solids, the other an ion or ions in solution.

C. The Known and Unknown

1. Prepare a known containing 8–10 drops each of Ag^+, Hg_2^{+2}, and Pb^{+2} reagent. Each is now diluted by one-third so the reactions may be slightly more difficult to observe. Label the test tube.

   Obtain an unknown from your TA or instructor. Record the unknown's code on your report form. Your unknown may contain one, two, or all three ions.
2. Beginning with the 3 M HCl, follow the qualitative scheme with the known and unknown simultaneously. Label test tubes carefully to keep track of the many solutions.
3. Indicate on the qualitative scheme which tests are positive for your *unknown.* To do this make a large check mark in colored pen, write "POSITIVE," or otherwise clearly indicate your observations for the chemical species detected.

Experiment 48
Qualitative Analysis Group I Cations

Name______________________________

Date______________________________

REVIEW QUESTIONS

1. a. In the known solution, the concentration of HCl after it is diluted is about 0.88 M. At what concentration of $[Pb^{+2}]$ will lead chloride first appear?

   b. Under the same conditions, at what concentration of $[Ag^{+}]$ will silver chloride first appear?

2. Explain why the mercurous ion is written Hg_2^{+2} and not simply Hg^{+}.

3. What causes the silver finish on platters, jewelry, teapots, or other products to tarnish? How can the tarnish be removed?

4. A chelating agent, EDTA, is often used to treat persons with lead poisoning. How does this work?

5. Explain why lead is found in both Groups I and II of the qualitative analysis scheme.

6. Besides chromate, what other confirmatory test could be used for lead ion? Write a balanced equation for the reaction and describe the product.

Experiment 48

Qualitative Analysis Group I Cations

Name____________________________________

Date_____________________________________

REPORT SHEET

A.

	Ag^{+}	Hg^{+2}	Pb^{+2}
HCl			
Heat			
NH_3			
K_2CrO_4			

Equations

$Pb^{+2} + Cl^{-} \longrightarrow$

$Hg_2^{+2} + Cl^{-} \longrightarrow$

$Ag^{+} + Cl^{-} \longrightarrow$

$Pb^{+2} + CrO_4^{-2} \longrightarrow$

$PbCl_2 + H_2O + \text{heat} \longrightarrow$

$AgCl + NH_3 \longrightarrow$

$[Ag(NH_3)_2]^{+} + Cl^{-} + H^{+} \longrightarrow$

B. Group I Qualitative Analysis Scheme

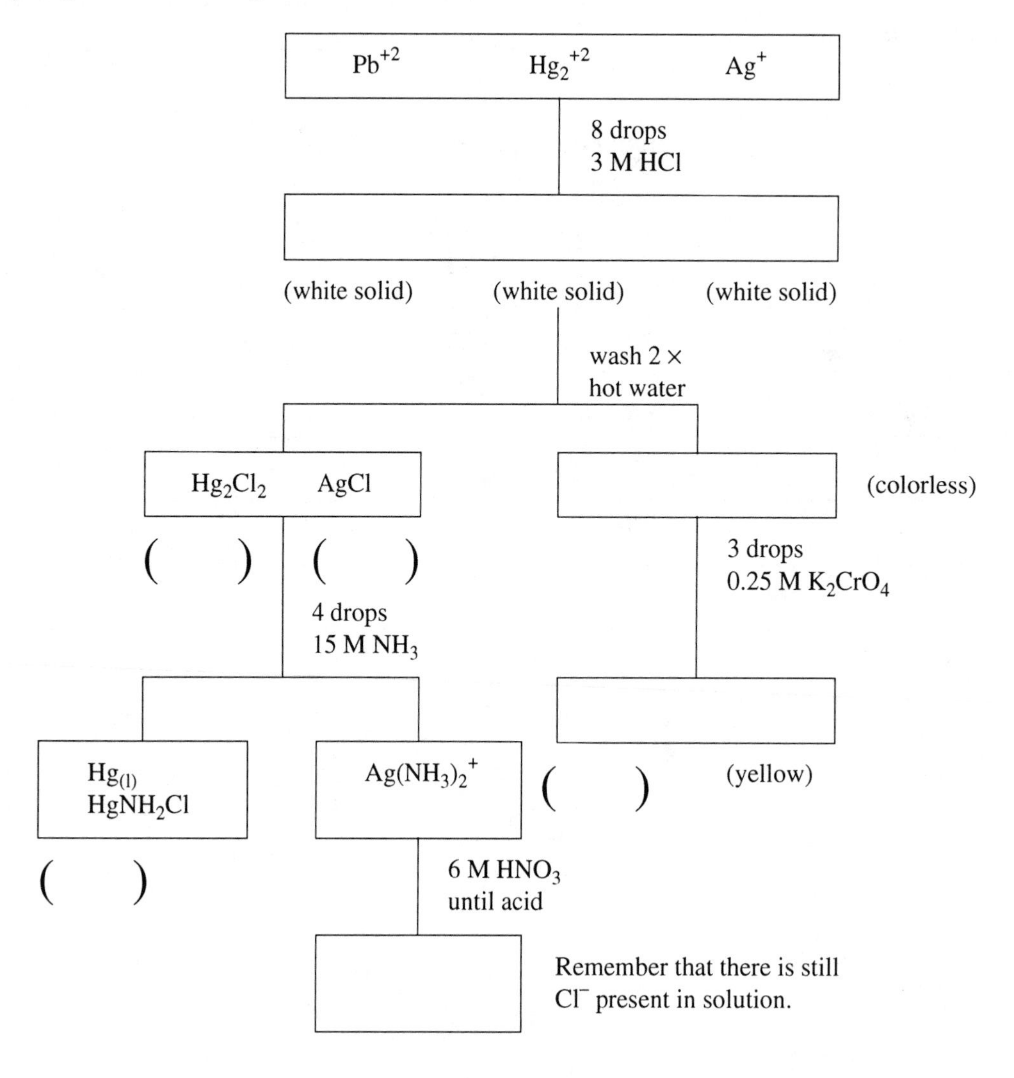

C. Unknown code ______________________ **Ion(s) present** ______________________

EXPERIMENT 49

Qualitative Analysis of Group II Cations

Purpose: To separate and selectively identify the cations of Group II; to investigate some of the corresponding chemistry of the Group II cations.

This experiment is designed to take two laboratory periods. Part A can be done in the first laboratory period, and Parts B and C can be done in the second.

Materials: dropping bottles of:

0.2 M	$Hg(NO_3)_2$	0.2 M	$Pb(NO_3)_2$	0.2 M	$Cu(NO_3)_2$
0.2 M	$Cd(NO_3)_2$	0.2 M	$Bi(NO_3)_2$	3 M	HCl
6 M	HCl	0.25 M	$SnCl_2$	15 M	NH_3 (or NH_4OH)
sat'd	$(NH_4)_2SO_4$	1 M	$NH_4C_2H_3O_2$	0.25 M	K_2CrO_4
6 M	HNO_3	1.2 M	CH_3CSNH_2 (thioacetamide)	Solid	$Na_2S_2O_4$

Apparatus: Centrifuge-size test tubes, test tube rack, test tube brush, stirring rods, pH paper, hot water bath, test tube holder, 10-mL graduated cylinder, wash bottle, marking pen.

Safety Precautions: The Group II ions, Hg^{+2}, Pb^{+2}, Cu^{+2}, Cd^{+2}, and Bi^{+3} are all toxic (although Cu^{+2} is only moderately so). All of the reagents are also toxic and should be handled carefully. Carry out the thioacetamide-sulfide precipitation in a fume hood. Wash hands thoroughly before leaving the lab. Wear your safety glasses at all times during this experiment.

Hazardous Waste Disposal: Dispose of all solutions in the properly labeled waste containers. No solutions in this experiment should be poured down the sink.

INTRODUCTION

The Group II cations of qualitative analysis are a continuation of the overall qual scheme begun in the previous experiment. If you did not perform the qualitative analysis of Group I, you should read the introduction to the lab for the appropriate background.

The Group I cations (Ag^+, Hg_2^{+2}, Pb^{+2}) are the insoluble chlorides. The Group II and III cations are the insoluble sulfides. The separation of the two groups is accomplished by carefully controlling the concentration of $[S^{-2}]$. An examination of the K_{sp} values of the sulfides is instructive.

Group II		Group III	
HgS	1×10^{-52}	CoS	7×10^{-23}
PbS	3.4×10^{-28}	NiS	3×10^{-21}
CuS	8.5×10^{-45}	FeS	3.7×10^{-19}
CdS	3.6×10^{-29}	MnS	7×10^{-16}
Bi_2S_3	1.6×10^{-72}		

Note that the Group II sulfides are considerably more insoluble than those of Group III. By keeping $[S^{-2}]$ at an extremely small amount, the Group II sulfides can selectively be precipitated first. The control of $[S^{-2}]$ is accomplished by considering the H_2S equilibrium

$$H_2S \rightleftharpoons 2H^+ + S^{-2} \qquad K = 1 \times 10^{-20}$$

By keeping $[H^+]$ very high (strongly acidic conditions) Le Chatelier's Principle states that the equilibrium is shifted to the left and the $[S^{-2}]$ decreases. Later the Group III cations are precipitated by making the solution basic. This reverses the equilibrium causing $[S^{-2}]$ to increase to quantities necessary for precipitation of Group III to occur. (At the same time other ions come out as hydroxides.)

In the past H_2S was used directly as the source of sulfide ion. Under normal atmospheric conditions, H_2S is a gas—it is, in fact, the evil-smelling odor of rotten eggs. Moreover, H_2S is actually *more* toxic than HCN, cyanide gas. One might wonder why more chemists haven't succumbed to the toxic effects of H_2S. The answer is obvious: it smells so bad that it is nearly impossible to accidently inhale a lethal amount.

Improvements in the qualitative analysis scheme have caused the adoption of thioacetamide as the reagent of choice for the production of H_2S. There are many advantages: the solution of thioacetamide is much easier to handle than a gas, the amount of H_2S produced is small, and the amount is easily controlled. When aqueous solutions are heated in the presence of either acid or base the following reaction occurs:

$$CH_3CSNH_2 + H_2O \longrightarrow H_2S_{(g)} + CH_3CONH_2$$

In practice a few drops of thioacetamide are added to the solution and the mixture is heated in a water bath.

PROCEDURE

A. Standard Tests

Share dropping bottles of the cations and reagents.

1. It will not be necessary (or practical) to test every reagent with every cation. To help you develop a separation scheme, the following tests should be performed. Remember to label all test tubes carefully.
2. Precipitation of sulfide—place 5–10 drops of Hg^{+2}, Pb^{+2}, Cu^{+2}, Cd^{+2}, and Bi^{+3} in each of the five labeled test tubes. Add 3–4 drops of 3 M HCl and verify that each is acidic. Add 5 drops of 1.2 M thioacetamide to each and place them in a boiling water bath for at least 5 minutes. CARRY THIS PROCEDURE OUT IN A FUME HOOD. Record your observations on your Report Sheet.

   Centrifuge, decant, and discard the supernatant. Wash each precipitate 2× in warm H_2O and discard the wash. Add 15 drops of 6 M HNO_3 and heat in a hot water bath. If part of the solid (but not all) dissolves, add 5 more drops 6 M HNO_3 and continue heating. The solid should either dissolve completely or not at all. Record your observations.
3. To the one precipitate that did not dissolve, add 5 more drops 6 M HNO_3 and 10 drops 6 M HCl and heat. Record the results. Do not add HCl to the others that have already dissolved. If all precipitate does not dissolve, centrifuge and discard any solid. To the clear supernatant add 2 mL of 1 M $SnCl_2$. The formation of a white precipitate of Hg_2Cl_2 is a confirmatory test for mercury. Record on your report sheet under "Confirming Tests." If no solid forms, add 2 more mL of $SnCl_2$ dropwise until the solid can be seen. The solid may darken over time due to the presence of $Hg_{(l)}$, further evidence of mercury.
4. To each of the other four solutions add 1 drop 15 M NH_3 and 3 drops saturated $(NH_4)_2SO_4$. Record any observations. If a precipitate forms, centrifuge, decant, and discard the supernatant. Add 1 mL of 1 M ammonium acetate, $NH_4C_2H_3O_2$

and stir. You may need to warm the test tube briefly to get the solid to dissolve. When dissolved, add 2 drops 0.25 M K_2CrO_4 and record any observations.

5. To the remaining three test tubes add 15 M NH_3 until the solutions are basic. Record observations for both changes.

   Centrifuge the test tube with the solid, decant, and discard the supernatant solution. Wash the solid twice with H_2O and discard the washes. In a separate test tube combine 1 mL $SnCl_2$ with enough 6 M NaOH to make it strongly basic. A precipitate will form initially, which subsequently dissolves upon further addition of OH^-. (What is this behavior called?) Add the $SnCl_2/OH^-$ mixture all at once to the washed solid and record any observations.

6. Copper ion must be suspected whenever a blue solution is encountered. The addition of concentrated ammonia to form the deep-blue complex ion $Cu(NH_3)_4^{+2}$ is a confirmation of copper. In a mixture, cadmium is harder to detect. The dark color of the CuS would completely obscure the brighter color of CdS. To separate Cd^{+2} from Cu^{+2}, we take advantage of the fact that Cu^{+2} is more easily reduced.

   Pour one-half of the copper solution *and* one-half of the cadmium solution into one test tube. Label it as "Cu/Cd."

   Prepare a solution of sodium dithionite, $Na_2S_2O_4$, by measuring 3 mL of 15 M NH_3 into a graduated cylinder. Add H_2O to just under the 10 mL mark (about 9 mL). Add 0.4 g $Na_2S_2O_4$ to the solution and stir gently. When the solid has dissolved, dilute it to 10 mL exactly and mix well. Sodium dithionite is unstable over time and very unstable in acidic solution. It must be prepared fresh just before use.

   Using a dropper add the $Na_2S_2O_4$ solution until the blue color is discharged in the Cu^{+2} solution and the Cu/Cd mixture. Add 5 drops to the pure Cd^{+2} for comparison. Place all three in a hot/boiling water bath for 2–3 minutes. Record the results. The dark-red solid formed is elemental copper. Centrifuge the pure Cu^{+2} solution and save the supernatant. Centrifuge the Cu/Cd mixture and save the supernatant.

   How would you then determine whether or not Cd^{+2} is present? Since the interfering Cu^{+2} has been removed, the sulfide of cadmium should be detectable. To the supernatant of the pure Cu^{+2}, the supernatant of the Cu/Cd mixture, and the Cd^{+2} mixture, add 5 drops of thioacetamide and heat. The appearance of the bright CdS is confirmatory of cadmium. Notice that small traces of copper left over in the Cu/Cd mixture give the sulfide a green appearance. Any appearance of yellowish-green color is taken as positive for cadmium.

B. Devising the Qualitative Analysis Scheme

Using what you have learned about the reactions of the five Group II cations, a plan can be developed to separate and identify each. A partially completed qualitative analysis scheme is presented on the report sheet. Fill in the missing descriptions, ions, or formulas. When a line diverges into two boxes, it implies centrifugation to separate solid(s) from solution. Have your TA or instructor approve your scheme before you proceed. Use this scheme as your guide for your unknown.

C. The Known and Unknown

Obtain an unknown from your TA or instructor and record its code on your report sheet. Your unknown may contain one, two, three, four, or five of the possible cations.

Prepare a known mixture of all five cations by placing 8 drops of each in a single test tube. You will carry out the tests on the unknown simultaneously with the known for comparison. Label each carefully.

Use your completed qual scheme to separate and identify the ions present. Place a check mark, or otherwise indicate those tests that are "POSITIVE" for your unknown. Report your findings.

Experiment 49
Qualitative Analysis Group II Cations

Name______________________________
Date______________________________

REVIEW QUESTIONS

1. Why would ingesting Hg_2^{+2} solutions be less harmful (but not harmless!) than ingesting Hg^{+2} solutions?

2. Could sulfuric acid be used to adjust the pH instead of hydrochloric acid in the first step? Explain.

3. What causes copper solutions to be colored, while cadmium solutions are not?

4. If $[H^+] = 0.3$ M and $[H_2S] = 0.10$ M, show whether 0.1 M solutions of Hg^{+2} and Fe^{+2} will precipitate. The K's you need are in the introduction.

5. If $[H^+]$ is subsequently lowered to 1×10^{-9} M, will the Fe^{+2} solution above precipitate? Explain.

6. Why does the addition of "ammonia" to a solution of Bi^{+3} cause the precipitation of bismuth *hydroxide*?

7. Why must copper ion be removed before cadmium can be confirmed as the sulfide?

Experiment 49

Qualitative Analysis Group II Cations

Name____

Date____

REPORT SHEET

	S^{-2}, Acidic	HNO_3, Heat	HNO_3, HCl, Heat
Hg^{+2}			
Pb^{+2}			
Cu^{+2}			
Cd^{+2}			
Bi^{+3}			

CONFIRMING TESTS

Ion	Reagent	Observation	Formula of Product
	$SnCl_2$		
	SO_4^{-2}		
	CrO_4^{-2}		
	NH_3		
	NH_3		
	$SnCl_2/OH^-$		
	$S_2O_4^{-2}$		
	S^{-2}		

C. Unknown code___________________________ **Ion(s) present**________________________

B. Qualitative Analysis Scheme Group II

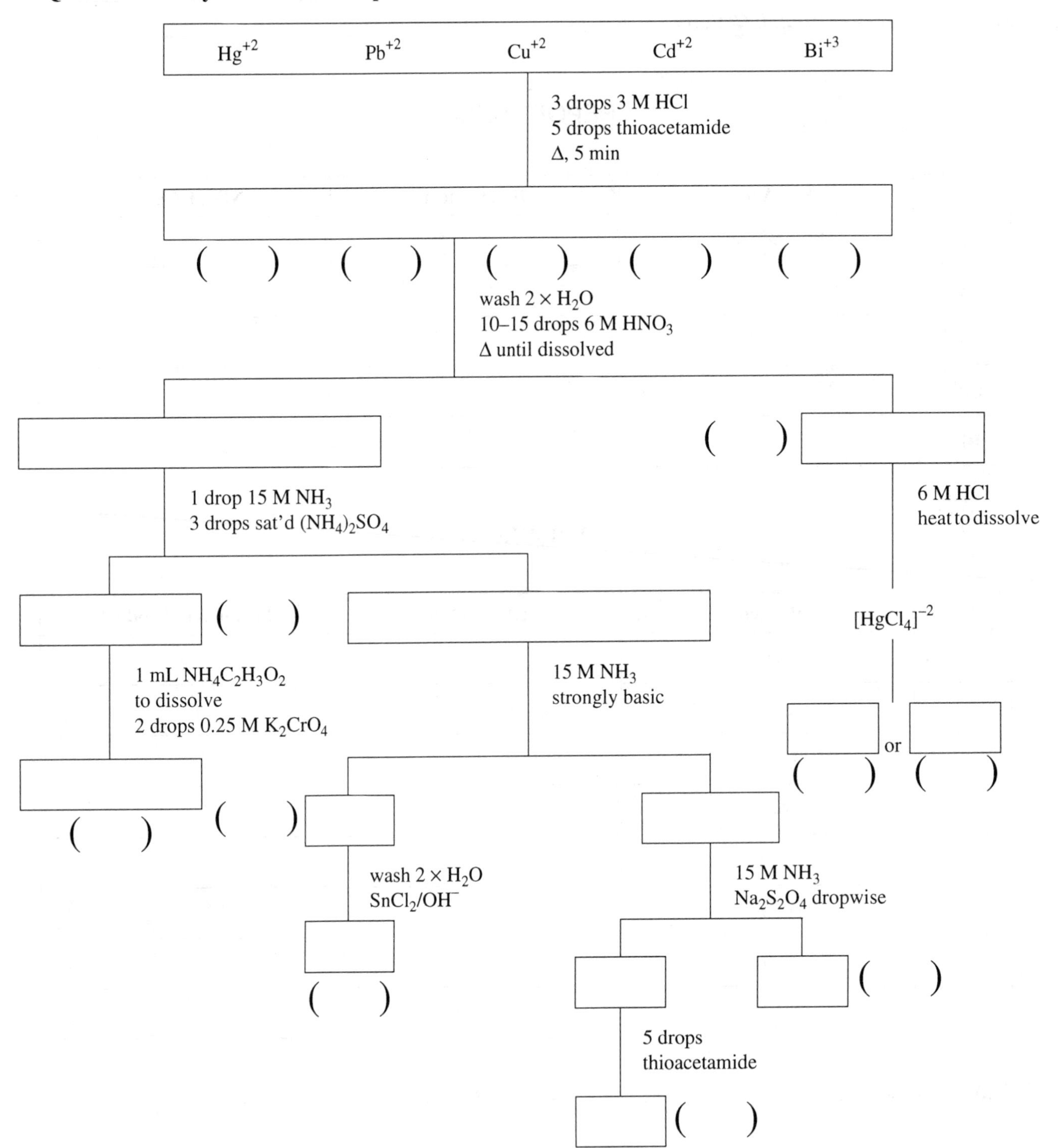

EXPERIMENT 50

Qualitative Analysis of Group III Cations

Purpose: To selectively isolate and identify cations of the Group III Qualitative Analysis Scheme. In addition, the student will investigate specific chemical reactions of the ions and the concepts of acid-base, redox, and compleximetric equilibria will be reinforced.

This experiment is designed for two laboratory periods. Parts A and B can be done in the first period and Parts C and D in the second.

Materials: Dropping bottles of:

1 M $Co(NO_3)_2$	1 M $Al(NO_3)_3$	1 M $Cr(NO_3)_3$
1 M $Zn(NO_3)_2$	1 M $Ni(NO_3)_2$	1 M $Fe(NO_3)_3$
1 M $Mn(NO_3)_2$	6 M HCl	6 M HNO_3
15 M NH_3	dimethylglyoxime (0.3g/100mL EtOH)	
3% H_2O_2	1 M KSCN	1 M $BaCl_2$
1.2 M thioacetamide	6 M NaOH	aluminon (1g/L)
3 M HCl	3 M NH_3	
$NaBiO_3$ (solid)	KSCN (solid)	acetone

Apparatus: 10–15 centrifuge-size test tubes, test tube rack, test tube holders, microspatula, stirring rods, pH paper, boiling water bath in fume hood.

Safety Precautions: Consider all the solutions as toxic and avoid skin contact. Carry out the sulfide precipitation in a fume hood, as the H_2S produced is toxic. Wash hands thoroughly before leaving the laboratory. Wear your safety glasses at all times during this experiment.

Hazardous Waste Disposal: Dispose of all solutions in the appropriately labeled waste container.

INTRODUCTION

If you have not performed the qualitative analysis of the Group I and II cations, you should read through the two previous experiments.

The Group I insoluble chlorides have been removed in the first step and the Group II acid-insoluble sulfides were removed second. The separation of the Group III cations is accomplished in a twofold fashion, relying on the insolubility of the sulfides and the equilibrium of H_2S. By making the solutions strongly basic, the H_2S equilibrium is shifted to favor higher concentrations of sulfide ion.

$$H_2S \rightleftharpoons 2H^+ + S^{-2}$$

Addition of base removes H^+ thus favoring the right-hand side of the equation. Recall that the Group II sulfides were precipitated under acidic conditions to keep $[S^{-2}]$ very low to *avoid* precipitating Group III at the same time.

There are seven ions in the Group III category and the number itself presents a formidable task. After the ions are precipitated, the major separations are achieved by carefully controlling the pH. Special advantage is taken of the amphoteric ions. These are soluble in both strong acid and strong base. For example, the hydroxide of aluminum

$$Al(OH)_3 + 3H^+ \longrightarrow Al^{+3} + 3H_2O$$
$$Al(OH)_3 + OH^- \longrightarrow Al(OH)_4^-$$

forms the Al^{+3} ion (actually $Al(H_2O)_6^{+3}$) under acidic conditions and the complex $Al(OH)_4^-$ under basic conditions. The solid hydroxide $Al(OH)_3$ can be precipitated only at near neutral to slightly basic pH.

Ammonia (15 M NH_3 or NH_4OH) is used to make solutions slightly basic. Sodium hydroxide is used to make solutions very basic.

PROCEDURE

A. Preliminary Tests

1. Place 10–15 drops of each of the seven cations into centrifuge-size test tubes. Make each basic with 15 M NH_3, add 5 drops of 1.2 M thioacetamide, and place in a boiling water bath for 4 to 5 minutes. Record the appearances of any precipitates. The Al^{+3} and Cr^{+3} form only hydroxides, not sulfides. The process reduces Fe^{+3} to Fe^{+2} to form FeS.
2. While the sulfide precipitation is ongoing, investigate the acid-base nature of each of the ions. Take 10 drops of the first cation in a test tube. (This is best done one ion at a time.) Check the pH. If not already acidic, add 1 or 2 drops of 6 M HNO_3 to make it so. Record the color of the acidic solution.

   Now begin adding 15 M NH_3 dropwise to the solution. Stir after each addition. Check the pH occasionally. If a precipitate forms that persists with stirring, record its appearance under "Slightly Basic." At this point the pH should be around 9. Whether or not a precipitate forms, proceed with the next test. Begin adding 6 M NaOH dropwise and with stirring. Note whether any precipitate dissolves under "Strongly Basic."

   Repeat for the other six ions.

B. Confirmatory Tests

Centrifuge the NiS and CoS samples and decant off the supernatant solutions. Add 8–10 drops of 6 M HCl to each and stir well. Is there any reaction? Next, to each add 5 drops more of 6 M HCl and 15 drops of 6 M HNO_3. Stir each thoroughly (you may need to warm them for about 1 minute). Record the results. The mixture of HCl and HNO_3 is called *aqua regia* and it is sufficiently acidic and oxidizing to dissolve most solids. (NiS and CoS will dissolve slowly in HCl, especially if warm.)

Combine the Ni^{+2} and Co^{+2} solutions and mix well. This will simulate the solution appearance when both ions are present in an unknown.

Take 8–10 drops of the mixture, place in a clean test tube and make basic with 15 M NH_3. Add about 5 drops (or more if you wish) of dimethylglyoxime solution. What happens? This formation of a "lake" or gelatinous colloid with dimethylyglyoxime is an extremely sensitive test for nickel.

Test the other portion of the solution for cobalt. Prepare the test reagent by dissolving 0.1 g to 0.2 g of KSCN in approximately 3 mL of acetone. (Alternatively, this reagent may be prepared for you.) Layer the KSCN/acetone solution over the Co^{+2}/Ni^{+2} solution. Record the results. Briefly verify in a separate test that Co^{+2} solution and *aqueous* KSCN solution do not give the same results.

The test for chromium is more difficult, although the presence of chromium is often suspected from the beginning because of its highly colored appearance. Chromium is confirmed by oxidation of Cr^{+3} to Cr^{+6} in the CrO_4^{-2} anion, followed by precipitation of $BaCrO_4$.

Take 10 drops of Cr^{+3} and make basic with 6 M NaOH. Add 10 drops of 3% H_2O_2, mix, and heat. Record any changes. If the dark hydroxide does not change to the bright color of CrO_4^{-2}, add a few more drops of H_2O_2 and continue heating. After heating for 3–4 minutes, stir the solution in the hot water bath and observe the evolution of bubbles of excess H_2O_2. Continue until most of the bubbling ceases. Remove from the warm bath, add 10 drops of 1 M $BaCl_2$, mix, and wait for the appearance of a cloudy, yellow-white precipitate of $BaCrO_4$. Centrifuge, discard the supernatant, and dissolve the solid in a minimum amount of 6 M HNO_3 (it will remain slightly cloudy). If the amount of solid is small and Cr is still uncertain, squirt a small amount of 3% H_2O_2 rapidly into the test tube. The appearance of a blue color that may disappear rapidly confirms chromium.

Aluminum is particularly easy to miss because the pH conditions necessary to precipitate $Al(OH)_3$ are somewhat fragile. Place 10 drops of Al^{+3} solution in a test tube and 1 mL of aluminon solution. Add 3 M NH_3 dropwise until a flocculent, white solid appears. Check the pH and add 1 drop more to ensure a pH of about 9. Record any changes. A true, positive test has the orange color of the aluminon solution completely discharged, while the white solid becomes pinkish-red. Centrifuge to observe this change more closely. The aluminon dye has been adsorbed onto the surface of the $Al(OH)_3$ particles.

Prepare a mixture of Fe^{+3} and Mn^{+2} (10 drops each). Withdraw a small portion into a clean test tube and add a few drops 6 M HNO_3. Add 1 drop of aqueous KSCN solution, and then a few more. This test for iron is extremely sensitive. Record the results. Since both Co^{+2} and Fe^{+3} are detected by SCN^-, you should be aware that small contaminants of Fe can obscure the Co test.

To the remainder of the solution add several drops 6 M HNO_3. Obtain a small amount of solid $NaBiO_3$, sodium bismuthate, on the tip of a microspatula. Drop the $NaBiO_3$ into the Fe^{+3}/Mn^{+2} solution all at once, without stirring. Record your observations. The intense color of the MnO_4^- ion is confirmatory for Mn. (Caution! $NaBiO_3$ is a powerful oxidizing agent; do not place the solid on paper as it could start a fire.) The dark solid that develops in the test tube is elemental bismuth.

Zinc is only confirmed by the precipitation of its sulfide, which is pure white. Unfortunately, the presence of even small amounts of the ions that form black sulfides (e.g., NiS, CoS) can obscure the ZnS. Multiple cycles of precipitation and washing usually are necessary.

Prepare a mixture of 10 drops of Zn^{+2} and 1 drop of Co^{+2}. Carry out the precipitation of the sulfides as before: 15 M NH_3 + 5 drops of thioacetamide, heat for 5 minutes. What color is the solid? Centrifuge and discard the solution. Dissolve the material in 3 M HCl. Ideally, the CoS should *not* dissolve, but some will. Centrifuge and discard the solid. Make the solution basic with 15 M NH_3. If any solid forms, centrifuge and discard it; zinc should remain in solution as $Zn(NH_3)_4^{+2}$. Carry out the sulfide precipitation a second time. Is the ZnS appreciably lighter in appearance? Multiple (4 or 5 times) cycles of precipitations followed by acid and base washes may be necessary to confirm zinc. Any solid that persists and is at least medium-beige in color is probably zinc.

C. The Qualitative Analysis Scheme

Using what you have learned from your preliminary and confirming tests, complete the qual scheme provided. Most of the items have been done for you. Complete the appearances by filling in the ( ). Special instructions have been included as footnotes marked with a*.

D. The Unknown

Because of the extreme difficulty, you will not employ a known with all seven ions. Obtain an unknown and record its code. Have your TA or instructor approve your qualitative analysis scheme and use it to test your unknown. Indicate on the scheme any positive results. Your unknown should contain no more than three or four of the possible ions. List the ions that you detect in your unknown.

Experiment 50
Qualitative Analysis Group III Cations

Name____

Date____

REVIEW QUESTIONS

1. Write the names of the following complex ions or compounds:
   (a) $[Zn(NH_3)_4]^{+2}$

   (b) $[Al(OH)_4]^-$

   (c) KSCN

   (d) $[Co(SCN)_4]^{-2}$

2. You may have noticed the evolution of white smoke when an acid was added to a basic solution, or a base to an acidic solution. What is the most likely cause of this?

3. Write equations to show the amphoteric nature of zinc.

4. Write a balanced equation to show $Cr(OH)_3$ being oxidized to CrO_4^{-2} by H_2O_2 in basic solution. The hydrogen peroxide is reduced to water.

5. A pink-colored Group III ion in solution turned blue upon addition of concentrated HCl. Write a chemical equation to indicate the ion and explain its behavior.

6. Why must the Co^{+2} test be carried out with KSCN dissolved in acetone rather than in aqueous solution?

7. Explain how you would separate
   (a) Ni^{+2} and Cd^{+2}

   (b) Co^{+2} and Zn^{+2}

Experiment 50
Qualitative Analysis Group III Cations

Name________________________________

Date________________________________

REPORT SHEET

A1. Basic Sulfide Precipitation

$Al(OH)_3$ $Cr(OH)_3$ ZnS NiS CoS MnS FeS

2.

Cation	Acidic	Slightly Basic	Strongly Basic
Al^{+3}			
Cr^{+3}			
Zn^{+2}			
Ni^{+2}			
Co^{+2}			
Mn^{+2}			
Fe^{+3}			

B. Appearances and Observations

	HCl	**HCl/HNO_3**
NiS		
CoS		

Ni^{+2} with dimethylglyoxime: ________________________________

Co^{+2} with KSCN/acetone: ________________________________

Cr^{+3} ________________________________

CrO_4^{-2} (by H_2O_2) ________________________________

$BaCrO_4$ (by xs Ba^{+2}) ________________________________

additional H_2O_2 ________________________________

$Al(OH)_3$ + aluminon ________________________________

Fe^{+3} + SCN^- ______________________________

Mn^{+2} + $NaBiO_3$ ______________________________

ZnS + CoS ______________________________

ZnS (2nd ppt.) ______________________________

C. Qualitative Analysis Group III

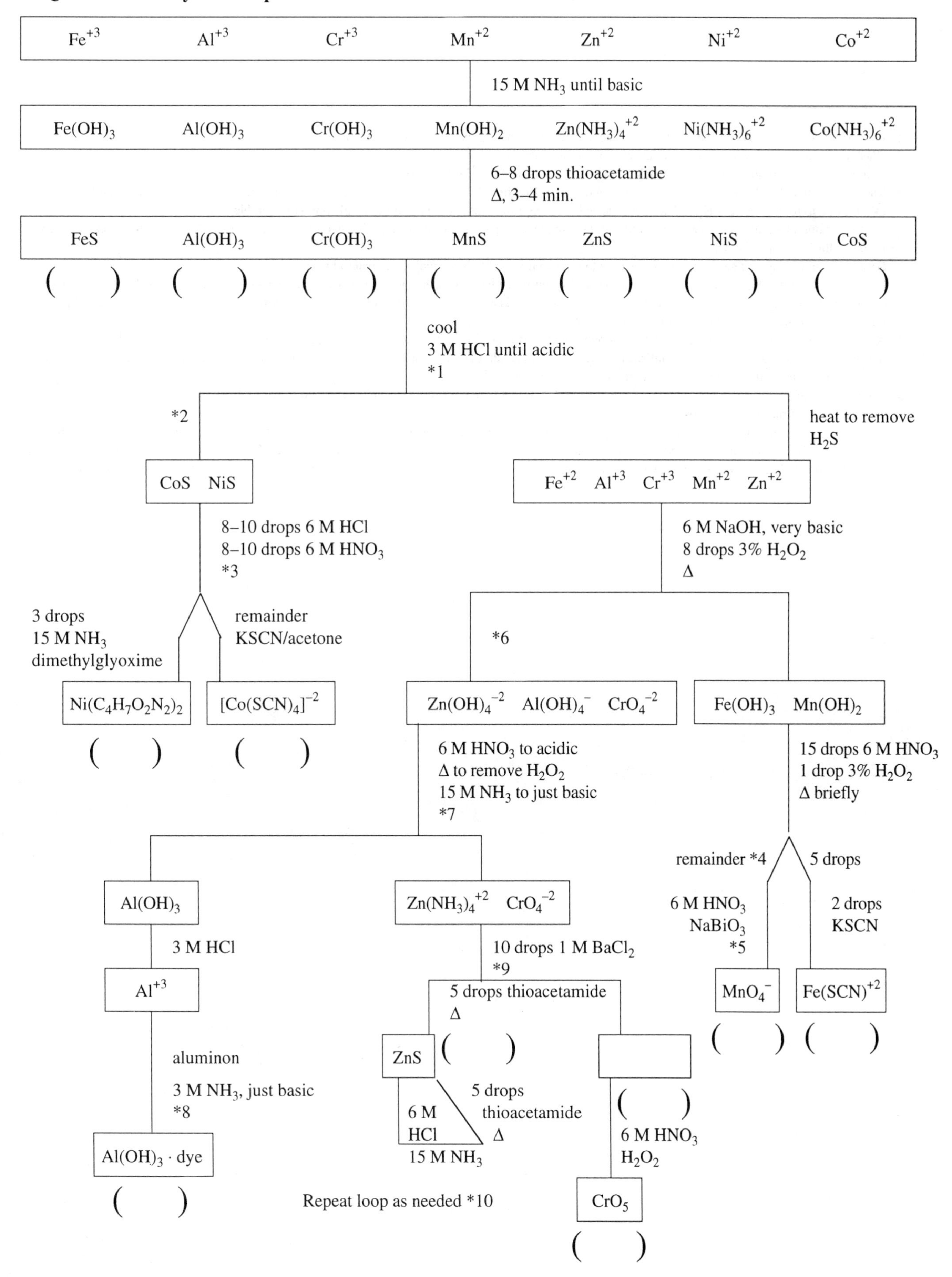

D. Unknown code ______________________

Ions present ______________________

---

1. CoS and NiS will dissolve slowly in HCl. The rate of dissolution increases with temperature. Be sure the solution is cool before adding HCl.
2. Wash the CoS and NiS with water once. Discard the wash.
3. Heat gently, 1–2 minutes. Any large particles that do not dissolve should be fished out.
4. The solution may be cloudy as all the solid will not dissolve. Remove about 5 drops for the Fe test and return the rest to the hot water bath for evaporation to remove excess H_2O_2. This can be speeded by vigorously stirring the solution while hot. H_2O_2 bubbles out rapidly.
5. $NaBiO_3$, sodium bismuthate, is a powerful oxidant. Do not place on paper! The amount on the tip of a microspatula is sufficient. An immediate deep-purple color is positive for Mn.
6. Split the sample in half. The aluminum test is easy to miss so it is best to have additional material in reserve.
7. It is essential not to add too much base. Add dropwise NH_3 until just basic. Check pH as you proceed. If the solution becomes cloudy, this is indicative of Al. If the solution becomes thoroughly basic and no precipitate is noted, *back-titrate* with 6 M HNO_3 dropwise to nearly neutral. $Al(OH)_3$, if present, may begin to form at that point, and only forms slowly.
8. Add the aluminon dye first and then 3 M NH_3 dropwise until just basic (check with pH paper). The presence of a white flocculent solid strongly indicates Al, but the aluminon test is confirmatory. In a positive result, the red color is *transferred* from the solution to the solid. Centrifuge if you are unsure of the results. If Al is present, the solution will be colorless.
9. The presence of a yellow solution indicates Cr as CrO_4^{-2}. The $BaCrO_4$ may form slowly and be difficult to see. A fine white precipitate *together with* the decolorization of the solution is positive for Cr.
10. The ZnS may appear gray or off-white due to the presence of small amounts of contaminants. If the first precipitate is indeterminate for Zn, centrifuge and then dissolve in a few drops of 6 M HCl. Centrifuge and discard any solid that does not dissolve. Add 15 M NH_3 to the solution until basic, 3 drops of thioacetamide, and heat. If present, the ZnS should be nearly white. It may be necessary to repeat this cycle of acid/base/ppt washings to be certain of Zn.

EXPERIMENT 51

Qualitative Analysis of Groups IV and V Cations

Purpose: To perform tests on the ions of Groups IV and V to prepare a qualitative analysis scheme for selectively separating and identifying each. To test an unknown for the presence of these ions.

This experiment is designed for two laboratory periods. Parts A and B can be done in the first period and Parts C and D in the second.

Materials: Dropping bottles of

0.2 M $Ca(NO_3)_2$	0.2 M $Mg(NO_3)_2$	0.2 M $(NH_4)_2HPO_4$
0.2 M $Sr(NO_3)_2$	0.2 M $Ba(NO_3)_2$	6 M NH_3
15 M NH_3	0.25 M $(NH_4)_2C_2O_4$	6 M HCl
conc. HNO_3	6 M $NH_4C_2H_3O_2$	3 M $(NH_4)_2SO_4$
6 M NaOH	1 M NH_4NO_3	1 M $NaNO_3$
1 M KCl or KNO_3	acetone	"magneson" reagent

("magneson" is *p*-nitro-benzeneazoresorcinol, 0.5 g in 100 mL of 0.25 M NaOH)

Apparatus: Test tubes, test tube rack, test tube holder, stirring rods, boiling water bath, ice water bath, platinum or nichrome wire OR mist sprayer or atomizer, cobalt glass, pH paper

Safety Precautions: The ions of Groups IV and V are relatively safe, but be cautious with Sr^{+2} and Ba^{+2} solutions. Take special care to avoid skin contact with concentrated nitric acid, HNO_3. It is corrosive. Solutions of oxalate ion, $C_2O_4^{-2}$, are moderately toxic. Avoid getting acetone on your skin; it is a powerful dehydrating agent. KEEP ACETONE AWAY FROM FLAMES OR OTHER HEAT SOURCES.

Hazardous Waste Disposal: All the solutions may be washed down the drain with a large volume of water unless your instructor specifies otherwise.

INTRODUCTION

The Groups IV and V ions are the last part of the total qualitative analysis scheme. If you did not perform the experiments for the first three groups, you should read those experiments before this one.

The cations have been removed from a theoretical unknown containing all ions in the following fashion:

Group I	excess Cl^-	Ag^+ Hg_2^{+2} Pb^{+2}
Group II	acidic S^{-2}	Hg^{+2} Pb^{+2} Bi^{+3} Cu^{+2} Cd^{+2}
Group III	basic S^{-2}	Cr^{+3} Al^{+3} Fe^{+3} Mn^{+2} Zn^{+2} Ni^{+2} Co^{+2}
Group IV	excess HPO_4^{-2}	Mg^{+2} Ca^{+2} Sr^{+2} Ba^{+2}
Group V	soluble	Na^+ K^+ NH_4^+

In some schemes, the Group IV ions are precipitated as the carbonates. However, $MgCO_3$ is partially soluble and would be carried over into Group V. The use of $(NH_4)_2HPO_4$ under basic conditions precipitates the Mg^{+2} as $MgNH_4PO_4$ and the others as the common phosphates. Notice that the Group IV ions comprise the portion of the Periodic Table known as the alkaline earth metals.

The Group V ions are sometimes known as the "soluble cations." Nearly all of their compounds are soluble and all of their complexes are colorless. This makes chemical analysis difficult.

Identification of Na^+ and K^+ relies on flame tests. The flame coloration is caused by movement of electrons in the atom or ion to higher energy levels. This is followed rapidly by cascade of the electrons back to lower levels with emission of characteristic wavelengths of light.

The ammonium ion does not give a flame test but can be detected chemically after conversion to ammonia.

Some other ions give flame tests and you may wish to try them (get permission from TA or instructor first). Some interesting ions are Ba^{+2}, Cu^{+2}, Sr^{+2}, and Li^+.

PROCEDURE

A. Preliminary and Confirmatory Tests of Group IV

1. Place 10 drops each of the 0.2 M solutions of $Mg(NO_3)_2$, $Ca(NO_3)_2$, $Sr(NO_3)_2$, and $Ba(NO_3)_2$ in each of four test tubes. In a fifth test tube prepare a combined unknown made of 5 drops of each of the cations together. Label each test tube.

   Make all five solutions basic with 3 drops of 6 M NH_3 in each of the single cation solutions, and 6 drops of 6 M NH_3 in the combined known. Add 2 mL of 1 M $(NH_4)_2HPO_4$ to each and heat in a boiling water bath for 4 minutes, stirring occasionally.

   Remove all the test tubes, cool for 10 minutes, centrifuge each and discard the supernatant solution. Wash with 1 mL of cool water and discard. Record the appearances of the solids on your report sheet.

   Wash each solid in 2 mL of acetone (FLAMMABLE!). Centrifuge and discard the supernatant solution. Repeat. This step helps to remove water from the solid. The cations are separated based on their solubility in cold, concentrated nitric acid, HNO_3. The presence of any water greatly increases the solubility of the nitrates, so acetone is used to dry the precipitates. After the second wash with acetone, turn the test tubes upside down so that the last drops of the acetone and water drain out, but the solid remains inside. Drain for 3–4 minutes. The test tube and solid must be completely dry.

   To each dry test tube add 1 mL of concentrated nitric acid, HNO_3. (CAUTION! Nitric acid is strongly corrosive. If any skin contact occurs, wash immediately with a large volume of water.) Stir each and quickly place in an ice water bath for 5 minutes. Be careful not to splash any water into the test tubes. Record the solubility of the precipitates in cold, concentrated nitric acid.

2. In the known mixture of all cations, two dissolved and two did not (the results you should have obtained for the single ions). Centrifuge the two solids from the individual ions (they are now *nitrates* of the cations) and cautiously decant the nitric acid. Centrifuge the known mixture of the four ions, but this time save the supernatant solution in a *large*, clean test tube. Add 1 mL of cold, concentrated HNO_3 to the remaining solid in the known mixture, stir briefly and return to the ice bath for 5 minutes. Centrifuge and combine the supernatant with the first portion saved. Set aside this solution and the two individual ions solutions that dissolved.

   To the two solids that did not dissolve and the solid portion from the mixture, proceed as follows. Add 5 drops of water and 1 mL of 6 M ammonium acetate, $NH_4C_2H_3O_2$. Warm to dissolve. Add 1 mL of 1 M K_2CrO_4, stir, and heat 3 minutes in a hot water bath. You should observe a precipitate forming immediately; the heating "ages" the precipitate, permitting a cleaner separation by centrifugation. Write a balanced chemical equation for the formation of the precipitate and briefly note its appearance.

   Wash the barium chromate solid from the individual cation solution in 1 mL of water to remove excess chromate ion and reveal the color of the pure solid.

   Centrifuge the known mixture and save the supernatant solution. It should contain only Sr^{+2} and excess chromate. To the Sr^{+2} from the known mixture add 1 mL of 3 M $(NH_4)_2SO_4$, stir, and heat 1–2 minutes. Cool under running water, centrifuge, and discard the supernatant. Wash twice with 2 mL of water + 2 drops of 6 M HCl. Write a balanced chemical equation for the formation of the precipitate and briefly note its appearance. For comparison, quickly combine 1 mL of 0.2 M $Sr(NO_3)_2$ and 1 mL of 3 M $(NH_4)_2SO_4$. Are the precipitates the same? Why?

3. Return to the Ca^{+2}, Mg^{+2}, and combined Ca^{+2}/Mg^{+2} solutions, add 1 mL 1 M $(NH_4)_2HPO_4$ to each and place the three test tubes in an ice bath. Next you must neutralize the concentrated nitric acid solutions. VERY CAREFULLY, 3 to 5 drops at a time and with stirring, add 15 M NH_3 until the solution is basic. It will take 4–6 mL of NH_3 to make the solutions

basic, so do not stop to check the pH until a precipitate forms and persists with stirring. Add a few drops of the 15 M NH_3 in excess. Centrifuge and wash the solids with 1 mL of water.

Redissolve the solids by adding 1 mL of 6 M acetic acid, CH_3COOH, and 1 mL of water. Warm, if necessary, to dissolve. If any solid does not dissolve, centrifuge and discard the solid.

To the Ca^{+2} solution, the Mg^{+2} solution, and the Ca^{+2}/Mg^{+2} mixture, add 1 mL of 0.25 M $(NH_4)_2C_2O_4$, ammonium oxalate. Heat in a boiling water bath for 2 minutes.

Centrifuge both test tubes that contain precipitates. Save the supernatant solution from the Ca^{+2}/Mg^{+2} mixture. Write a balanced chemical equation for the oxalate precipitation and briefly note its appearance on your report sheet.

To the remaining solution, Mg^{+2} and the Mg^{+2} from the mixture, add 1 mL 1 M $(NH_4)_2HPO_4$ and then 15 M NH_3 until basic. Stir and cool under running water. Centrifuge and discard the supernatant. Dissolve the precipitates in 6 drops of 6 M HCl. Add 2 drops of "magneson" reagent and 1 to 2 mL of 6 M NaOH. Stir and heat for 1 to 2 minutes. The formation of an intensely colored "lake" is positive for magnesium. The complex is the hydroxide of magnesium coordinated to the "magneson" dye. Write an equation for the formation of the magnesium hydroxide and note its color in the presence of the "magneson" dye.

B. Flame Tests for Group V

Always test for Group V ions on the ORIGINAL SOLUTION.

1. A flame test is the easiest way to determine Na^+ and K^+. Flame tests have traditionally been carried out using nichrome or platinum wires embedded in a glass holder. To perform the test, dip the wire in 3 M HCl (or 6 M) in a side beaker. Hold the wire in the hottest part of the nonluminous flame of a Bunsen burner. Repeat several times to burn off any impurities. Dip the clean wire into the solution to be tested (to which a few drops of HCl have been added). Hold the wire with the droplet in the hottest portion of the flame and observe.

   Alternatively, if enough solution is available, the appearance of the flame may be easier to see if the solution is *aspirated* into the flame. Using a mist sprayer (such as for house plants) or atomizer (such as for perfumes or cosmetics), a fine mist containing ions is introduced into the flame.

   First, clean the atomizer thoroughly and *pump a large volume of distilled or deionized water through it.* (Careful! Soaps and certain detergents used to clean glassware will often give false sodium tests.)

   Obtain the solutions to be tested with the wire loops or by aspiration. While 1–2 mL of sample is all that is needed for the wires, at least 8 mL of each solution is needed for aspirated samples. (This includes a portion to "prime" the pump.)

   Test a 1 M KCl or KNO_3 solution to which a few drops of HCl have been added. Look closely, the color is faint. It is best to work with lighting subdued and at your back. Now try viewing the flame through cobalt glass. The blue glass filters out yellows and oranges. Record your observations on your report sheet.

   Make a mixture of about two-thirds K^+ and one-third Na^+ and test it. The color of the sodium flame totally obscures the potassium. Next try viewing the flame of the mixture through the cobalt glass. You should be able to see the potassium flame this way.
2. Ammonium ion, since it is a complex ion, not a metallic cation, does not give a flame test. However, if present in sufficient quantity, its conversion to ammonia can be detected by odor or with pH paper.

   Place 1 mL of 1 M NH_4NO_3 in a test tube, add 7 drops of 6 M NaOH, and stir. Can you detect the smell of ammonia? Do not inhale deeply, but pull the fumes toward your nose. Check that the pH is strongly basic and warm the test tube in a hot water bath. Place a piece of pH paper over the mouth of the test tube, taking care not to get any NaOH solution on it. If ammonia is given off, the pH paper will show basic. Record your observations.

C. The Qualitative Analysis Scheme

Complete the qualitative analysis scheme for Group IV by supplying formulas or appearances in the ( ). Have your scheme checked by your TA or instructor before you proceed.

D. The Unknowns

Obtain Group IV and Group V unknowns. Alternatively, your instructor may provide *one* unknown with both groups in it. Record the code(s). There may be one or all of the ions in your unknown. Use your qual scheme to test for Group IV ions, and flame tests or the ammonia test for Group V. Be sure to use *original* solution for these tests. The qual scheme often introduces Group V ions. Mark those tests that are POSITIVE on your qual scheme. Report your findings. The details for carrying out the scheme are in the Procedure, Part A.

Experiment 51

Qualitative Analysis Groups IV and V Cations

Name________________________________

Date_________________________________

REVIEW QUESTIONS

1. Why would you expect the alkaline earth metals to behave similarly?

2. Barium chromate and lead chromate are both insoluble yellow precipitates. Explain how you might distinguish between them.

3. Magnesium hydroxide is partially soluble. A suspension of the solid is sold as "milk of magnesia." How is the partial solubility an advantage and what chemical reaction is desired?

4. What is the approximate chemical composition of glass? Which ion often leaches from glass to contaminate an unknown?

5. Why is HCl often added before flame tests?

6. Metallic ions are often added to fireworks although they do not actually explode. Why?

7. Why is Ba^{+2} separated as the chromate rather than the much less soluble sulfate?

8. Devise a simple qualitative analysis scheme for separating and identifying a solution that may contain only these ions: Ag^{+}, Cd^{+2}, Fe^{+3}, Mg^{+2} .

Experiment 51
Qualitative Analysis Group IV and V Cations

Name_______________________________________

Date_______________________________________

REPORT SHEET

A. Group IV

Precipitate	Appearance	Solubility in conc. HNO_3
$MgNH_4PO_4$		
$Ca_3(PO_4)_2$		
$Sr_3(PO_4)_2$		
$Ba_3(PO_4)_2$		

Precipitate with CrO_4^{-2}:

Precipitate with SO_4^{-2}:

Precipitate with $C_2O_4^{-2}$:

Precipitate with OH^-:

B. Group V

Ion	Flame	Flame with Cobalt Glass
K^+		
Na^+		
K^+/Na^+		

Test for NH_4^+

$NH_4^+ + OH^- \longrightarrow$

Observations:

C. Qualitative Analysis Group IV

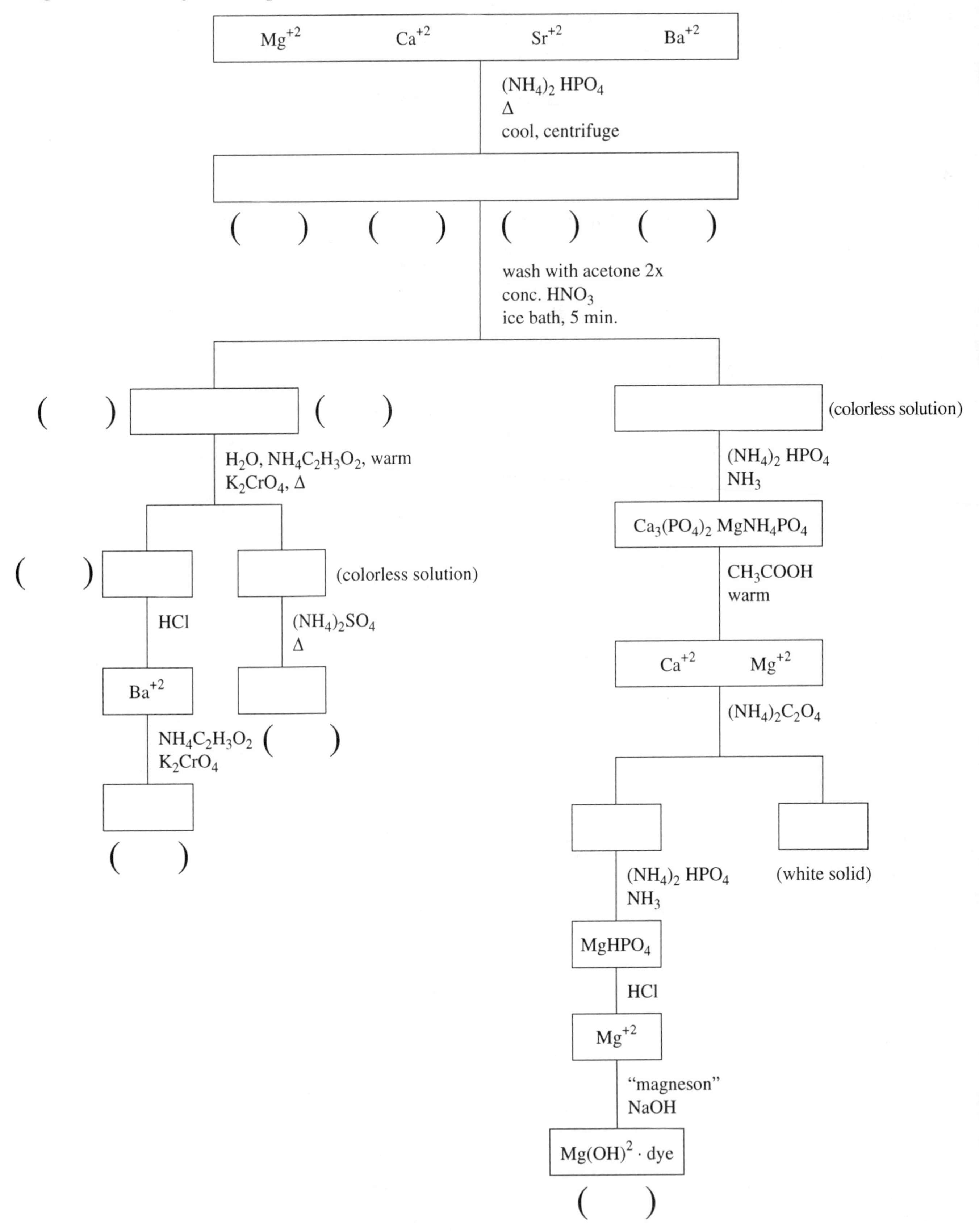

D. Unknowns

Code________________________ Ion(s)________________________

Code ________________________ Ion(s) ________________________

EXPERIMENT 52

Qualitative Analysis for Anions

Purpose: To investigate spot tests for the common anions; to be able to identify an anion in an unknown.

Materials: Dropping bottles of:

1 M $NaNO_3$
1 M Na_3PO_4
1 M Na_2SO_4
0.5 M Na_2S
6 M HCl
1 M $BaCl_2$
sat'd KNO_2
hexanes, copper wool or fine wire

1M NaCl
1 M NaBr
1 M NaI
conc. H_2SO_4
0.1 M $Fe(NO_3)_3$
0.2 M $FeSO_4$

1 M NaSCN or KSCN
1 M $NaC_2H_3O_2$
1 M Na_2CO_3
6 M H_2SO_4
6 M HNO_3
bleach (5.25% NaOCl)

Apparatus: Test tubes, test tube rack, test tube holder, stirring rods, pH paper, Bunsen burner, lead acetate paper, centrifuge, 10-mL graduated cylinder, ring stand and clamp, boiling water bath, ice water bath.

Safety Precautions: Handle all the chemicals in this experiment with care. Avoid getting Ag^+ on skin or clothing; it will stain dark-brown or black. Be especially careful when handling concentrated sulfuric acid, H_2SO_4. It is one of the strongest concentrated acids. If any skin contact occurs, wash with copious amounts of water. If instructed to sample odors, do not inhale deeply, gently waft the fumes toward your nose. Wear your safety glasses at all times.

Hazardous Waste Disposal: To avoid smelling up the lab, the sulfide solution should be discarded in a waste solution container in a hood. All of the other aqueous solutions may be washed down the drain with a large volume of water. The hexanes or other organic solvents should be collected in a container for organic wastes.

INTRODUCTION

The qualitative analysis, or identification, of the common anions is markedly simpler than the analysis of the cations. One reason is that there are many fewer possibilities for the anions. Another is that analysis of anions usually relies on *spot tests* of the anions rather than separations followed by confirming tests. For these reasons, the study of qualitative analysis often begins with the anions.

The common anions you will test for are carbonate, phosphate, sulfate, bromide, chloride, iodide, sulfide, acetate, thiocyanate, and nitrate. The anions that are oxymetallic anions, such as chromate and permanganante, are considered in the analysis of the chromium and manganese cations.

Before beginning this experiment, you may wish to review the chemistry of these ions. The ten anions can be categorized into four groups.

The Acid-Volatile Group includes the carbonate and sulfide ions. Upon addition of strong acid, these anions form gases that are readily evolved from solution. For carbonate:

$$CO_3{}^{-2}{}_{(aq)} + 2\,H^+{}_{(aq)} \longrightarrow H_2CO_{3(aq)}$$

Carbonic acid, H_2CO_3, is unstable and is rapidly decomposed to carbon dioxide and water.

$$H_2CO_{3(aq)} \longrightarrow CO_{2(g)} + H_2O_{(l)}$$

Sulfide ion, when acidified, produces the foul-smelling hydrogen sulfide gas:

$$S^{-2}{}_{(aq)} + 2\,H^+{}_{(aq)} \longrightarrow H_2S_{(aq)}$$

The H_2S is usually unavoidably detected by the odor of rotten eggs, but since the gas is toxic, you should not inhale it. Sulfide ion is confirmed by holding moistened lead acetate paper over the mouth of the test tube.

The Barium Precipitate Group includes sulfate and phosphate ions. These are the only ions on our list that form precipitates upon the addition of excess Ba^{+2} ion. Sulfate can be differentiated from phosphate in that barium phosphate is soluble in HCl, while barium sulfate is not.

The Silver Precipitate Group includes the halides iodide, bromide, and chloride, and thiocyanate, which is often called a pseudohalide. All of these form light-colored precipitates with excess Ag^+ ion. The precipitates vary slightly in appearance, which helps to distinguish them. Thiocyanate ion is readily confirmed by the blood-red complex it forms with Fe^{+3}. The halides can be oxidized to the halogens, then extracted into an organic layer and identified by color.

The Soluble Group is made up of the last two of the anions you will encounter, nitrate and acetate. Nitrate ion is identified by the very specific brown ring test, or by the production of Cu^{+2} in the presence of elemental copper. Acetate ion is identified by the vinegar odor of acetic acid.

PROCEDURE

A. Spot Tests on Known Anions

1. *Test for* $CO_3{}^{-2}$
   Place 10 drops of 1 M Na_2CO_3 into a small test tube. Dilute with distilled water to about double the volume and mix. Add 6 M H_2SO_4, 1 drop at a time, and record the results. Continue until the effervescence ceases. The bubbling is only barely detectable under dilute conditions.
2. *Test for* S^{-2}
   Place 10 drops of 0.5 M Na_2S in a small test tube. (Caution! Do not get the sulfide ion on the skin. It has a foul odor that tends to persist, even with washing.) Dilute with distilled water to about double the volume and mix. Note the slight yellow appearance of the solution (before dilution), characteristic of sulfide. Dry the top and just inside the mouth of the test tube with a twist of paper towel. This is necessary to remove traces of the sulfide solution. Lay a strip of lead acetate paper over the mouth of the test tube and slightly to one side. Through the opening add 6 M H_2SO_4 dropwise.
   After about 10 drops of acid, or when the solution stops changing, check the lead acetate paper. Darkening or formation of dark streaks is indicative of H_2S. Do not deliberately inhale the H_2S, but at this point, the odor should be unmistakable. Record your observations. Discard the solution in the appropriately labeled container in the hood.
3. *Test for* $SO_4{}^{-2}$
   Place 10 drops of 1 M Na_2SO_4 solution in a small test tube and dilute slightly. Add 4 drops of 1 M $BaCl_2$ and mix well. The precipitate may form slowly, especially if the sulfate is very dilute. Centrifuge and decant the supernatant. Add 6 drops 6 M HCl and stir. Does the precipitate dissolve? If you are uncertain, centrifuge again. Record all observations.
4. *Test for* $PO_4{}^{-3}$
   Place 10 drops of 1 M Na_3PO_4 solution in a small test tube and dilute slightly. Add 6 drops of 1 M $BaCl_2$ and mix well. Centrifuge and decant the supernatant. Add 7 drops of 6 M HCl and stir. Is the solid soluble in HCl? Record all observations and write equations for the formation of the precipitates in steps 3 and 4.
5. *Test for* SCN^-
   Thiocyanate produces a precipitate in the presence of Ag^+ in a manner similar to the halides. In a small test tube mix 5 drops of 1 M KSCN (or NaSCN) with 2 drops $AgNO_3$. Record the results. Save the precipitate for later comparison to your unknown.

Thiocyanate is readily confirmed by reaction with ferric ion. In fact, thiocyanate itself is the reagent used to identify iron in the qualitative analysis of the cations.

Place 5 drops of 1 M KSCN (or NaSCN) in a test tube and dilute just slightly. Add 2 drops of 0.1 M $Fe(NO_3)_3$ and record the results. Dilute the complex with water until the test tube is nearly full. Notice that the color is still detectable, even at very dilute concentrations. The reaction is

$$Fe^{+3}_{(aq)} + SCN^{-}_{(aq)} \longrightarrow Fe(SCN)^{+2}_{(aq)}$$

6. *Test for I^-*

Place 6 drops of 1 M NaI solution in a test tube and add 4 drops of 0.2 M $AgNO_3$. Note the appearance of the precipitate and set it aside. In another test tube place 5 drops of 1 M NaI, add 5 drops of 6 M acetic acid, CH_3COOH, and 1 drop of saturated KNO_2 solution. The evolution of brown gas is confirmatory for I_2. Dilute the solution with distilled water slightly and add 1 mL of hexanes gently down the side of the test tube. The nonpolar I_2 will dissolve readily in the (upper) organic layer. Shake or agitate briefly. Record its appearance, stopper, and set aside.

Starch Test For I^-. Take 5 drops of 1 M NaI into a test tube and add 1 drop of 6 M HNO_3. Dilute slightly with water. Spot a piece of filter paper with 1 drop of freshly prepared, iodide-free starch solution. Add 1 drop of the acidified, diluted NaI solution on top of the starch spot. Apply 1 drop of saturated KNO_2 solution on top of that. Formation of a blue-black starch complex is confirmatory for I.

7. *Test for Br^-*

Place 6 drops of 1 M NaBr solution in a test tube and add 4 drops of 0.2 M $AgNO_3$. Note the appearance of the precipitate, and set it aside for later comparison.

WORK IN THE HOOD FOR THIS TEST! Place 6 drops of 1 M NaBr in a small test tube and acidify with 4 drops of 6 M HCl. Dilute slightly with water. Check with pH paper that the solution is strongly acidic. Add 1 mL of hexanes gently down the side of the test tube. (It is important that the organic layer is present *before* the evolution of the Br_2.)

In a 10-mL graduated cylinder, prepare a solution of chlorine water. Measure 4 mL of water into the graduated cylinder and add 5 drops of commercial bleach solution. Mix well. (Commercial bleach solution is 5.25% NaOCl, sodium hypochlorite.)

Add 2 drops of chlorine water and note the yellow appearance as Cl_2 is absorbed into the upper, organic layer. Shake the test tube gently. Continue adding the chlorine water dropwise and stir vigorously or agitate. A darker color will develop in the hexane layer as the reaction continues. The reaction is a redox

$$Br^{-}_{(aq)} + Cl_{2(aq)} \longrightarrow Br_{2(hex)} + Cl^{-}_{(aq)}$$

Record the results, stopper the test tube, and set it aside.

8. *Test for Cl^-*

Place 6 drops of 1 M NaCl in a test tube and add 4 drops of 0.2 M $AgNO_3$. Note the appearance of the precipitate and set it aside.

In another test tube place 6 drops of 1 M NaCl and acidify with 4 drops of 6 M HCl. Dilute slightly with water. Prepare chlorine water as in part 7 (4 mL water + 5 drops bleach). Add 1 mL hexanes gently over the aqueous solution and begin adding the chlorine water dropwise. Shake vigorously. Add nearly all of the chlorine water. The color will be extremely pale, if it is visible at all. You are not actually producing chlorine here, only trapping what free Cl_2 is present in the chlorine water. Stopper the test tube and set it aside for later comparison.

9. *Test for NO_3^-*

The qualitative test for nitrate has traditionally been the "brown ring" test. The brown color is caused by the formation of $Fe(NO)^{+2}$ in the presence of NO_3^- and excess Fe^{+2} in a two-step reaction:

$$3\ Fe^{+3}_{(aq)} + NO_3^{-}{}_{(aq)} + 4\ H^{+}_{(aq)} \longrightarrow 3\ Fe^{+2}_{(aq)} + NO_{(aq)} + 2\ H_2O$$

$$NO_{(aq)} + Fe+^{2}_{(aq)} \longrightarrow Fe(NO)^{+2}_{(aq)}\ \text{(brown)}$$

The H^+ is provided by concentrated sulfuric acid. Because of its density, H_2SO_4 will form a lower layer when added to an aqueous solution. The solutions are layered rather than mixed because the heat of dilution of sulfuric acid is enough to destroy the brown Fe complex. The "brown ring" forms at the interface between the two layers.

BE PARTICULARLY CAREFUL WITH CONCENTRATED SULFURIC ACID, H_2SO_4! If you get any on yourself, wash immediately with lots of water.

Place 10 drops of 1 M $NaNO_3$ in a test tube and add 5 drops of 0.2 M ferrous sulfate, $FeSO_4$. Mix well. Place the test tube in an ice water bath. Transfer 2 mL of concentrated H_2SO_4 in a separate test tube and chill it in the ice water also, 3 to 5 minutes.

Clamp the test tube with the NO_3^-/Fe^{+2} solutions to a ring stand at an angle of about 30°. Pour 1 to 1.5 mL of the chilled sulfuric acid down the side of the test tube gently, to avoid mixing the layers. Let the mixture stand undisturbed for at least 5 minutes. Describe the appearance of the solution.

Alternative Test for NO_3^-

It can be difficult to get good results from the brown ring test for nitrate. The nitrate ion must be present in high enough concentrations, the solution of $FeSO_4$ must be freshly prepared (Fe^{+2} in solution is rapidly oxidized by air to Fe^{+3}), and the two solutions must be chilled.

An alternative test uses concentrated H_2SO_4 (again) to convert the NO_3^- to HNO_3, which is detected by the oxidation of elemental copper.

Obtain a small amount of copper wool (like "steel wool" but made of copper; the kind used to pack fractionating columns works well). Twist it into a loose wad and place a pea-sized mass (about 0.05 g) at the bottom of a small test tube. Use a stirring rod to place the copper wool. If copper wool is not available, fine copper wire works, but be certain the wire is not coated in a plastic sheath.

Add 10 drops of 1 M $NaNO_3$ to the test tube and dilute with 10 drops of distilled water. Prepare a blank test tube for comparison with 20 drops of distilled water on a piece of copper wool. WORK IN THE HOOD FOR THE REMAINDER OF THIS TEST!

To both test tubes add about 1 mL of concentrated H_2SO_4. CAUTION! It will get hot and may spatter!! Use a test tube holder or rack. Shake the test tube moderately to mix. Avoid breathing the gas that evolves. It is toxic, acrid NO. The color of the solution develops as Cu is oxidized to Cu^{+2}. Record the results of the test.

The concentration of NO_3^- in this test is 0.5 M before H_2SO_4 is added. The nitrate ion is detectable down to about 0.04 M, and the amount of nitrate ion can be estimated by the intensity of color formed in solution.

Bromide and iodide ions interfere with this test (as they do for the brown ring test) and their absence must be established before the test is carried out. The presence of chloride ion alters the solution color to yellow.

10. *Test for CH_3COO^-*

Like nitrate, most compounds of acetate are soluble. Although the test is sometimes inconclusive, the simplest test is the conversion of acetate to acetic acid, which can be confirmed by smell.

Place 12 drops of 1 M $NaC_2H_3O_2$, sodium acetate solution, in a small test tube and add 4 drops of concentrated sulfuric acid, H_2SO_4. Heat in a hot water bath for 1 to 2 minutes. Sample the vapors from the test tube by gently wafting them to your nose. Does it smell like vinegar? If not, add 2 more drops of sulfuric acid and heat a little longer.

If the test for acetate is inconclusive, add 10 drops of ethanol to the mixture and heat for 2 minutes in a boiling water bath. Remove and sample the odor of this preparation, ethyl acetate, which has a sweet, fruity smell. Your instructor may provide a sample of pure ethyl acetate for your comparison.

B. Unknowns

Your TA or instructor will provide several unknowns. Each will have only one anion per test tube. While it is possible to remove any interfering ions from a mixture, this experiment does not deal with any mixtures.

Based on the observations you have made, prepare a plan, flowchart, or algorithm to test any given unknown for the ten anions. Show your plan to your TA or instructor before you proceed.

Make your preliminary tests first (acid-volatile, Ba^{+2} ppt., Ag^+ ppt.) on *small* amounts of your unknowns. These "spot tests" need not consume more than a few drops. Then narrow down the identity of each unknown by elimination—first by group, then specific anion. Report your findings.

Experiment 52
Qualitative Analysis for Anions

Name____

Date____

REVIEW QUESTIONS

1. Why do the halogens, I_2, Br_2, Cl_2, dissolve so easily in hexane?

2. Your phosphate standard solution is prepared from Na_3PO_4. Would you expect an aqueous solution of Na_3PO_4 to be acidic, basic, or neutral? Write a chemical equation to support your answer.

3. Why is table salt usually sold commercially as "iodized salt"? Be specific.

4. I_2 dissolves very slightly in water, but dissolves readily in a solution of sodium iodide. Explain this observation.

5. Draw a Lewis Dot Structure for carbonate ion and for nitrate ion. What electronic characteristic do these two share? (Hint: How many structures can you draw for each?)

6. Write a structural equation for the formation of ethyl acetate from acetic acid and ethanol.

7. For the three halide ions, X^-: place them in order of ease by which they are oxidized (easiest to hardest).

8. Is it possible to oxidize I^- to an oxidation state higher than that in I_2? What anion is formed and what is the oxidation number of I in it?

Experiment 52
Qualitative Analysis for Anions

Name________________________________

Date________________________________

REPORT SHEET

A1. $CO_3^{-2} + H_2SO_4$

Observations:

2. $S^{-2} + H_2SO_4$

Observations:

3. $SO_4^{-2} + Ba^{+2}$

Observations:

Equation:

4. $PO_4^{-3} + Ba^{+2}$

Observations:

Equation:

5. $SCN^- + Ag^+$

Observations:

$SCN^- + Fe^{+3}$

Observations:

	Ag^+	Hexane Layer
6. I^-		
7. Br^-		
8. Cl^-		

9. $NO_3^- + H^+ + Fe^{+2}$

Observations:

$NO_3^- + H^+ + Cu$

Observations:

10. $CH_3COO^- + H^+$

Observations:

B. Plan for testing unknowns:

Unknown code____________________ Anion ____________________

Unknown code____________________ Anion ____________________

Unknown code____________________ Anion ____________________

Unknown code____________________ Anion ____________________

APPENDIX A: TECHNIQUES FOR GRAPHING

In many cases experimental results will rely on some form of graphical analysis to reach the desired conclusions. This is true in all branches of science, not just chemistry. You should, therefore, undertake this small portion of your education so that you will always be equipped to prepare and interpret a graph.

1. *A graph must be neat and carefully crafted.* It should not have to be said, but first and foremost, a graph must be drawn on graph paper. Do not try to scratch out a sketch with notebook paper and homemade lines. This is not accurate, and if you did not want accuracy, you would not be making a graph.

   Be neat! Throughout your career the quality of your work will be judged as much (sometimes more!) by its appearance as by its content. In practical terms, this often means making a rough draft of your plot, adjusting the fine points without regard to neatness, and then recopying the graph in its final form.
2. *Use a ruler to draw in the two main axes.* Use a pen, not pencil, and trace the axes over lines on the graph paper. (Allow a small margin from the edge of the paper.)
3. *Labeling.* Label both axes with the variables they represent and include the units in parentheses. Examples: Temperature (°C), Pressure (torr), Time (minutes), and Acceleration (ft/s^2).
4. *The range and intervals.* Next you must choose what values to assign to each mark. Select the most appropriate ranges and employ convenient intervals.

   NOT EVERY SCALE SHOULD START AT ZERO!! In fact, unless you are planning to extrapolate back to an axial intercept, there is no reason that it *must* do so. However, if the numbers are small enough and zero can be included without sacrificing too much space, it is often somehow psychologically satisfying.

   Ideally, a graph should be space-filling. This means that you choose ranges that will scatter the data points over the field. It is not useful to draw a large graph and then display the data points all bunched up in one place. The rest of the space is just wasted, and more importantly, it is harder to read the value of the points because the range is too wide.

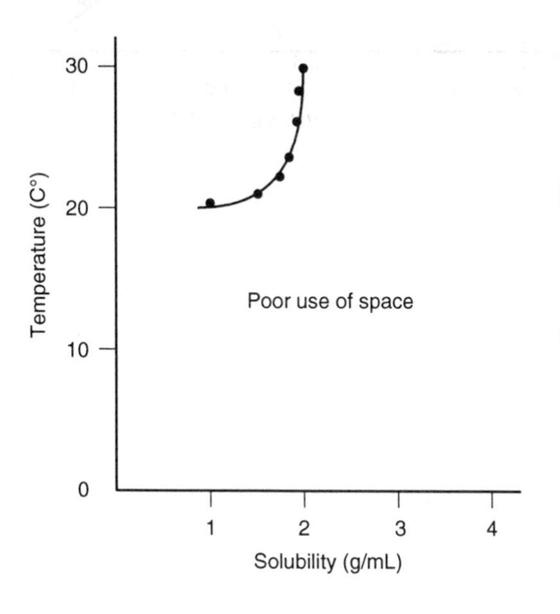

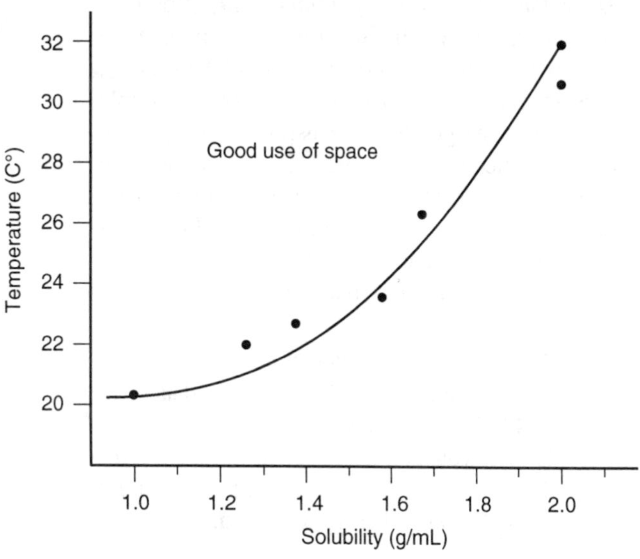

How do you choose a range and a scale (the value between each major mark)? For each axis subtract the lowest value from the highest. Round this up to a convenient number. Now count the number of boxes (or lines) along the axis, leaving a small margin. This is the number of intervals. Divide the range by the intervals and this gives you the approximate scale. Then choose the most logical, convenient increments. (Best choices are: 1, 2, 5, 10, 20, and multiples thereof.)

Consider the example for this data set:

Pressure (mm Hg)	710	740	785	793	812
Volume (L)	0.00149	0.00130	0.00117	0.00101	0.00093

Range Pressure: 812 – 710 = 102

Range Volume: 0.00149 – 0.00093 = 0.00056

Interval Pressure: 30

Interval Volume: 44

(Of course, by turning the graph paper by 90°, you can plot it in intervals of 44 and 30. Your choice.) Then

Pressure: 102 ÷ 30 = 3.4, round to 4
Volume: 0.00056 ÷ 44 = 0.000013, round to 0.00002

Therefore, a convenient scale would be: Pressure from 700 to 820 by 4s; and Volume from 0.00090 to 0.00152 by 2s, or by 0.00002s. This allows a margin, uses space adequately, and employs a simple-to-use scale. Note that the volume does not completely fill the lower axis, but it is too large to double and using an interval of 0.000013 would have been too awkward. The final graph then looks like:

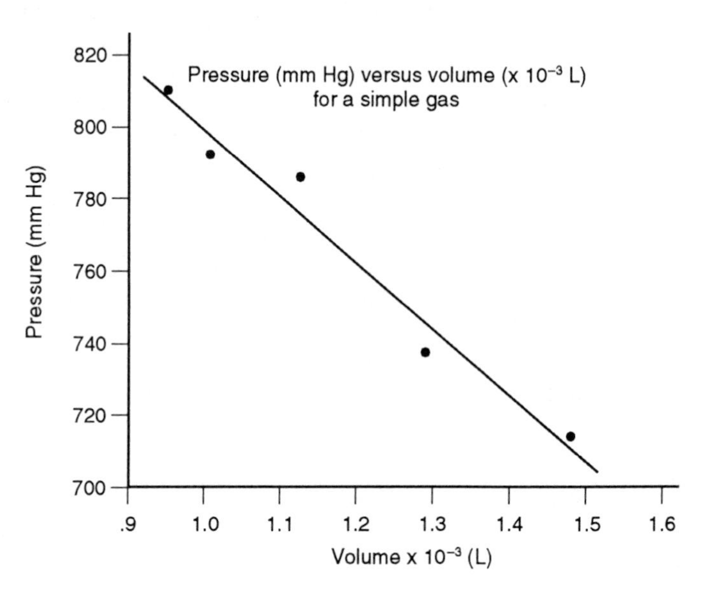

You could try to turn the graph by 90° to see if the space works out better.

It is a simple trick to handle very small numbers by ignoring the leading zeroes and thinking from 90 to 152 by 2s. Don't forget to write the zeroes in on your final graph, however.

5. *Draw the data points in permanent ink.* You may want to start by writing them in pencil, then going back over them in ink. For the "points" themselves, dots are acceptable but x's are also good. Multiple data sets on one graph can be distinguished by using dots for one, x's, open circles, squares, etc. The overriding concern is to keep them small so that a reader can accurately determine the value. In its fanciest version, data points are drawn with *error bars*. These visually display the *absolute uncertainty* of each data point.

The error in the horizontal direction is larger than the error in the vertical direction. One data point might be 42.77±0.05 grams, 2.715±0.002 magnetos.

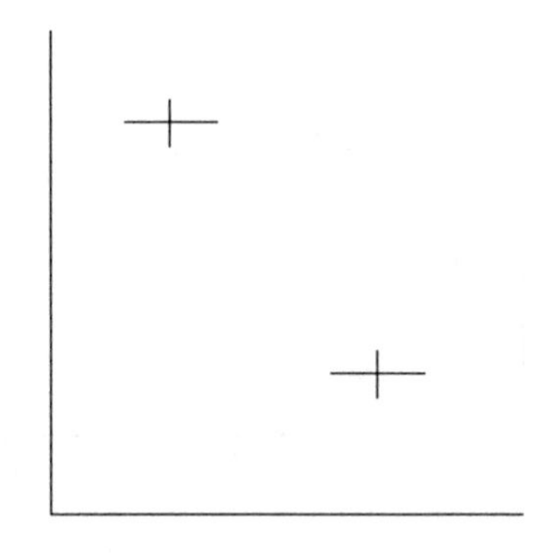

Unless you are instructed to employ error bars, do not use them. Nevertheless, you should know what they are when you see them.

6. *Curve fitting*. Sometimes a graph is for illustration and sometimes it is used to arrive at a number (usually the slope of a line). In either case, you should draw the "best-fitting, smooth curve." (Never make a jagged graph by simply connecting the dots!) Remember that geometrically speaking, a straight line is also a curve. If the data form a curved curve, draw it in (permanent ink) as smoothly as you can. A device called a flexible curve is useful for this, but it is just as well done freehand. If the data form a straight line, *use a ruler* to draw in the best line. Estimate visually so that the average number and distance of data points above the line are the same as those below the line. (Fancy calculators can calculate the slope and intercept for you.)
7. *Calculating the slope*. If you draw in the best line, you probably will also need to calculate the slope. To do this, you will need to select two points *from the line* (NOT data points!) Your line-drawing is a technique similar to averaging data; therefore, the "average" line is superior to any individual measurement. To make it easier, try to select points that cross major "intersections." Also, try to choose values a good distance apart; this will minimize reading errors.

   You can perform the calculation on the graph if you wish. Circle or otherwise indicate the chosen points. Remember that the equation for slope is:

$$\text{slope} = \frac{\Delta y}{\Delta x} = \frac{y_2 - y_1}{x_2 - x_1}$$

   for the two points (x_1, y_1) and (x_2, y_2). Be careful — x is the horizontal axis and y is the vertical axis.
8. Graphs are always referred to in terms of "y versus x"; that is, in terms of the vertical versus the horizontal, in that order.
9. *Title*. Every graph must have a title! This tells the reader at a glance what the subject is. In a book or journal, it would appear as a figure caption. On your graph you may write the title at the top (most common), at the bottom, or on the side—anywhere as long as it is neat and out of the way. The title should be direct and descriptive. Whimsical or abstract titles ("Ralph," "Blue Number 3 with Red Circles") are inappropriate. The simplest titles are in the "y versus x" mode, as in "Time versus Temperature for Iodate Solutions" or "pH versus mL NaOH."
10. Finally for obvious reasons, it is a good idea to put your name on your graph.

    SPECIAL NOTE: Many personal computers will do graphs for you, as well as make useful calculations. This is perfectly acceptable. It is preferable to print the graph on a laser printer or at least letter-quality printer. (Dot-matrix printers with exaggerated pixeling fail on the "neatness" criterion.) Nevertheless, you must acquire the skill of hand-crafting graphs just in case you ever need to draw them yourself.

APPENDIX B: METRIC SYSTEM

LENGTH

1 kilometer (km) = 10 hectometers = 0.62 miles
1 hectometer (hm) = 10 dekameters
1 dekameter (dkm) = 10 meters
1 meter (m) = 10 decimeters = 39.27 inches
1 decimeter (dm) = 10 centimeters
1 centimeter (cm) = 10 millimeters (mm) (1 inch = 2.54 cm)

MASS

1 kilogram (kg) = 10 hectograms = 2.2 lb
1 hectogram (hg) = 10 dekagrams
1 dekagram (dkg) = 10 grams
1 gram (g) = 10 decigrams (453.6 g = 1 lb)
1 decigram (dg) = 10 centigrams
1 centigram (cg) = 10 milligrams (mg)

VOLUME

1 kiloliter (kL) = 10 hectoliters
1 hectoliter (hL) = 10 dekaliters
1 dekaliter (dkL) = 10 liters
1 liter (L) = 10 deciliters = 1.06 U.S. liquid quart
1 deciliter (dL) = 10 centiliters
1 centiliter (cL) = 10 milliliters (mL)

APPENDIX C: COMMON IONS

POSITIVE IONS

One Positive Charge

NH_4^+	Ammonium
Cu^+	Copper(I) (cuprous)
H^+	Hydrogen
K^+	Potassium
Ag^+	Silver
Na^+	Sodium

Two Positive Charges

Ba^{2+}	Barium
Ca^{2+}	Calcium
Co^{2+}	Cobalt
Cu^{2+}	Copper(II) (cupric)
Fe^{2+}	Iron(II) (ferrous)
Pb^{2+}	Lead
Mg^{2+}	Magnesium
Hg_2^{2+}	Mercury(I) (mercurous)
Hg^{2+}	Mercury(II) (mercuric)
Sn^{2+}	Tin(II) (stannous)
Zn^{2+}	Zinc

Three Positive Charges

Al^{3+}	Aluminum
As^{3+}	Arsenic(III) (arsenious)
Bi^{3+}	Bismuth
Cr^{3+}	Chromium(III) (chromic)
Fe^{3+}	Iron(III) (ferric)

Four Positive Charges

Sn^{4+}	Tin(IV) (stannic)

Five Positive Charges

As^{5+}	Arsenic(V) (arsenic)
Sb^{5+}	Antimony(V) (antimonic)

NEGATIVE IONS

One Negative Charge

$C_2H_3O_2^-$	Acetate
Br^-	Bromide
ClO_3^-	Chlorate
Cl^-	Chloride
CN^-	Cyanide
F^-	Fluoride
HCO_3^-	Hydrogen Carbonate (bicarbonate)
$H_2PO_4^-$	Dihydrogen Phosphate
OH^-	Hydroxide
I^-	Iodide
NO_3^-	Nitrate
NO_2^-	Nitrite
MnO_4^-	Permanganate

Two Negative Charges

CO_3^{2-}	Carbonate
CrO_4^{2-}	Chromate
$Cr_2O_7^{2-}$	Dichromate
HPO_4^{2-}	Hydrogen Phosphate
O^{2-}	Oxide
SO_4^{2-}	Sulfate
S^{2-}	Sulfide
SO_3^{2-}	Sulfite
$C_2O_4^{2-}$	Oxalate

Three Negative Charges

AsO_4^{3-}	Arsenate
AsO_3^{3-}	Arsenite
PO_4^{3-}	Phosphate
$Fe(CN)_6^{3-}$	Ferricyanide

APPENDIX D: GENERAL SOLUBILITY RULES

Few things are absolutely insoluble, but many are for most practical purposes. The following rules can serve as a rough guide.

The following are soluble in water:

1. All simple Na^+, K^+, and NH_4^+ compounds.
2. All nitrates, acetates [$Bi(NO_3)_3$, and most acetates hydrolyze to slightly soluble products].
3. All chlorides except Pb^{2+}, Ag^+, Hg_2^{2+}, and Cu^+. (Sb^{3+}, As^{3+}, Bi^{3+}, Sn^{2+}, and Sn^{4+} chlorides hydrolyze to insoluble products.)
4. All sulfates except Pb^{2+}, Ag^+, Hg_2^{2+}, Ba^{2+}, Ca^{2+}, and Sr^{2+}.

With a few exceptions all alkali metal salts and all other strong acids are soluble. Most weak acid salts, except those of the alakali metals, are slightly soluble or hydrolyze to slightly soluble products.

The following are very slightly soluble in water:

1. All metals except Na, K, Ba, and Sr, which react with H_2O.
2. All oxides and hydroxides except Na^+, K^+, NH_4^+.
   (Ba^{2+} , Ca^{2+} , Mg^{2+} , and Sr^{2+} oxides react with water to form slightly soluble hydroxides.)
3. All sulfides, oxalates, phosphates, carbonates, and silicates except those of Na^+, K^+, and NH_4^+. (Most acetates, the soluble silicates, and the sulfides of Ba^{2+} , Ca^{2+} , and Sr^{2+} hydrolyze to slightly soluble products.)

APPENDIX E: COMMON STRONG AND WEAK ACIDS AND BASES

Strong Acids	Weak Acids	Strong Bases	Weak Bases
HNO_3	CH_3COOH ($HC_2H_3O_2$)	NaOH	NH_4OH or NH_3
HCl	$H_2C_2O_4$	KOH	
H_2SO_4	H_2S		
	HCN		
	H_2CO_3		
	H_3PO_4		

APPENDIX F: USEFUL CONVERSIONS

LENGTH

1 Angstrom unit (Å) = 10^{-8} cm
1 micro meter (μm) = 10^{3} nanometers (nm) = 10^{4} Å
1 inch = 2.54 cm
1 foot = 30.48 cm
1 meter = 39.37 in
1 rod = 16.5 ft
1 kilometer = 0.62 mile   1 mile = 1.61 kilometer
1 light year = 5.88×10^{12} miles (speed of light = 186,000 miles/sec = 3×10^{10} cm/sec)

VOLUME

1 teaspoon = 5 mL (approx.)
1 tablespoon = 15 mL (approx.)
1 cu in = 16.4 mL
1 oz (fluid) = 29.6 mL
1 cup = 8 oz = 237 mL
1 quart (U.S. liq) = 0.946 liter = 946 mL
1 liter = 1.06 qt = 61.03 cu inches
1 gallon = 3.785 liters
1 cu ft = 7.48 gal = 28.30 liters = 28,300 mL
1 bushel (U.S.) = 35.24 liters

MASS

1 carat = 0.2 g
1 ounce (Avoir) = 28.35 g = 0.991 troy oz
1 pound = 453.6 g
1 kilogram = 2.201 lb
1 ton (short) = 2,000 lb = 907.2 kg
1 ton (metric) = 1,000 kg = 2,204.6
1 ton (long) = 2,240 lb

MISCELLANEOUS

1 acre U.S. = 43,560 sq ft = 0.405 hectares (square hectometers)
1 atmosphere = 760 mm Hg = 30 inches Hg = 14.7 lb/sq inch
1° Fahrenheit = 5/9° Celsius
1 British thermal unit (Btu) = 252 cal   1 cal/g = 1.8 Btu/lb
1 kilocalorie = 1,000 calories   1 joule = 10^{7} ergs = 0.239 cal
1 horsepower = 745 watts = 550 ft lbs/sec
1 kilowatt = 1,000 watts
1 faraday = 96,500 coulombs (ampere seconds)

APPENDIX G: VAPOR PRESSURE OF WATER

Temperature	Pressure	Temperature	Pressure
0° C	4.6 mm	23° C	21.1 mm
5°	6.5	24°	22.4
10°	9.2	25°	23.8
11°	9.8	26°	25.2
12°	10.5	27°	26.7
13°	11.2	28°	28.3
14°	12.0	29°	30.0
15°	12.0	30°	31.8
16°	13.6	31°	33.7
17°	14.5	32°	35.7
18°	15.5	33°	37.7
19°	16.5	34°	39.9
20°	17.5	35°	42.2
21°	18.6	———	———
22°	19.8	100°	760.0

APPENDIX H: ACTIVITY SERIES

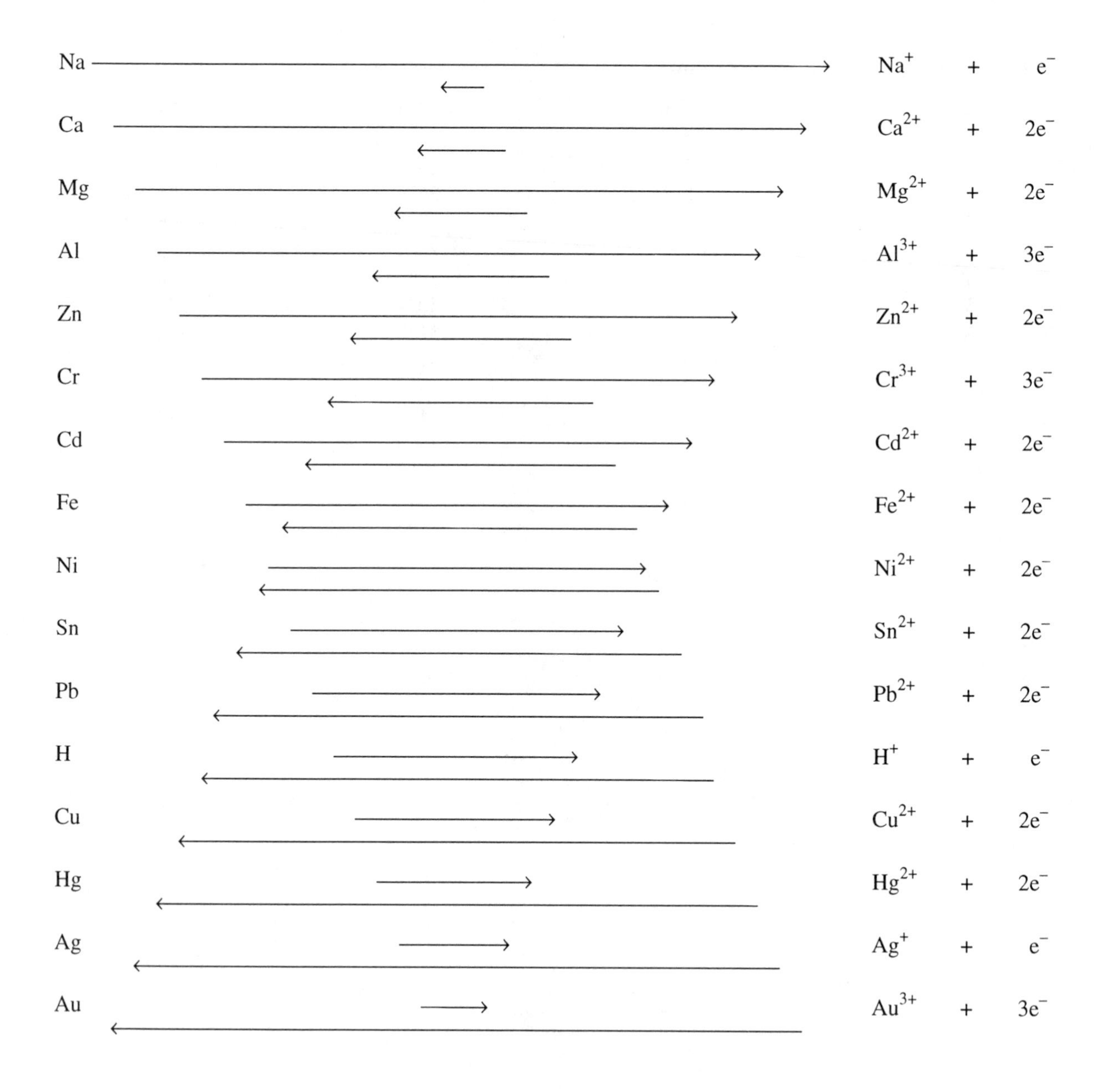

APPENDIX I: TEST REACTIONS FOR COMMON SUBSTANCES

1. Acid. Blue litmus turns red. Phenolphthalein turns colorless.
2. Base. Red litmus turns blue. Phenolphthalein turns red.
3. Ammonium salts (NH_4^+). Odor of ammonia when 6 N NaOH is added to dry salt. When mixture is warmed, moist red litmus paper held in the *vapor* turns blue.
4. Carbohydrates. Molisch test—1 drop 10% α-napthol in alcohol added to 5-ml test solution and layered over concentrated H_2SO_4. Violet ring appears at liquid boundary if carbohydrates are present.
5. Carbonate ion (CO_3^{2-}). Effervesces if dilute HCl is added to sample.
6. Carbon dioxide (CO_2). White precipitate of $BaCO_3$ appears if CO_2 is brought into contact with clear saturated $Ba(OH)_2$ solution.
7. Chlorine (Cl_2). Moist starch-KI paper turns blue (test is positive for most oxidizing agents).
8. Chloride ion (Cl^-). White precipitate of AgCl if $AgNO_3$ slightly acidified with HNO_3 is added.
9. Collodial sols. Transparent sols show Tyndall effect with light beam.
10. Hydrogen (H_2). Whistling explosion when H_2-air mixture is ignited in test tube. (Other gases form explosive mixtures with air, but usually in narrow composition range.)
11. Iron, Iron III (Fe^{3+}). Solutions give deep-red color with KSCN. Iron II (Fe^{2+}). Solutions give dark-blue precipitate with $K_3Fe(CN)_6$.
12. Oxygen (O_2). Glowing splint bursts into flame.
13. Proteins. Biuret test—test solution made alkaline with NaOH and 1 drop $CuSO_4$ solution added. Violet color indicates proteins.
14. Sulfate ion (SO_4^{2-}). White precipitate of $BaSO_4$ when $BaCl_2$ solution slightly acidified with HCl is added. Precipitate may form slowly.
15. Nitrate ion (NO_3^-). To 1 ml of test solution add 1 ml of freshly made 0.5 M $FeSO_4$ solution. Mix thoroughly and carefully underlay with concentrated H_2SO_4. A brown ring appears at the interface if NO_3^- is present.
16. Sugars. Fehling's test for reducing sugars only—add 1 ml each of Fehling's solutions A and B to 5 ml test solution and warm. Orange-red precipitate forms if reducing sugars are present.
17. Flame tests. Dip platinum wire into concentrated HCl, then into sample, and hold in edge of Bunsen flame.
    a. Barium ion (Ba^{2+})—Green flame, very persistent
    b. Calcium ion (Ca^{2+})—Brick-red flame, soon disappears
    c. Copper ion (Cu^{2+})—Green flame, soon disappears
    d. Potassium ion (K^+)—Violet flash when viewed through cobalt glass.
    e. Strontium (Sr^{2+})—Crimson flame, very persistent.

APPENDIX J: STANDARD REDUCTION POTENTIALS (E^0 VALUES AT 25°C)

Half-reaction	E^0 (V)
$Li^+ + e^- \longrightarrow Li^0$	–3.045
$Rb^+ + e^- \longrightarrow Rb^0$	–2.925
$K^+ + e^- \longrightarrow K^0$	–2.925
$Cs^+ + e^- \longrightarrow Cs^0$	–2.92
$Ba^{2+} + 2e^- \longrightarrow Ba^0$	–2.90
$Sr^{2+} + 2e^- \longrightarrow Sr^0$	–2.89
$Ca^{2+} + 2e^- \longrightarrow Ca^0$	–2.87
$Na^+ + e^- \longrightarrow Na^0$	–2.714
$Mg^{2+} + 2e^- \longrightarrow Mg^0$	–2.37
$H_2 + 2e^- \longrightarrow 2\,H^-$	–2.25
$Be^{2+} + 2e^- \longrightarrow Be^0$	–1.85
$Al^{3+} + 3e^- \longrightarrow Al^0$	–1.66
$Mn^{2+} + 2e^- \longrightarrow Mn^0$	–1.18
$V^{2+} + 2e^- \longrightarrow V^0$	–1.18
$Se + 2e^- \longrightarrow Se^{2-}$	–0.92
$Cr^{2+} + 2e^- \longrightarrow Cr^0$	–0.91
$Zn^{2+} + 2e^- \longrightarrow Zn^0$	–0.763
$Cr^{3+} + 3e^- \longrightarrow Cr^0$	–0.74
$S^0 + 2e^- \longrightarrow S^{2-}$	–0.48
$Fe^{2+} + 2e^- \longrightarrow Fe^0$	–0.440
$Cr^{3+} + e^- \longrightarrow Cr^{++2+}$	–0.41
$Tl^+ + e^- \longrightarrow Tl^0$	–0.3363
$CO^{2+} + 2e^- \longrightarrow Co^0$	–0.277
$V^{3+} + e^- \longrightarrow V^{2+}$	–0.255
$Ni^{2+} + 2e^- \longrightarrow Ni^0$	–0.250
$2\,SO_4^{2-} + 4\,H_3O^+ + 2e^- \longrightarrow S_2O_6^{2-} + 6\,H_2O$	–0.22
$CO_2 + 2\,H_3O^+ + 2e^- \longrightarrow HCOOH + 2\,H_2O$	–0.196
$Sn^{2+} + 2e^- \longrightarrow Sn^0$	–0.136
$Pb^{2+} + 2e^- \longrightarrow Pb^0$	–0.126
$2\,H_3O^+ + 2e^- \longrightarrow H_2^0 + 2\,H_2O$ (1 M H_3O^+)	0.00
$NO_3^- + H_2O + 2e^- \longrightarrow NO_2^- + 2\,OH^-$	+0.01
$HCOOH(aq) + 2\,H_3O^+ + 2e^- \longrightarrow 3\,H_2O + HCHO(aq)$	+0.056
$S_4O_6^{2-} + 2e^- \longrightarrow NO_2^- + 2\,OH^-$	+0.01
$Sn^{4+} + 2e^- \longrightarrow Sn^{2+}$	+0.15
$Cu^{2+} + e^- \longrightarrow Cu^+$	+0.153
$Cu^{2+} + 2e^- \longrightarrow Cu^0$	+0.337
$2\,H_2SO_3 + 2\,H_3O^+ + 4\,e^- \longrightarrow S_2O_3^{2-} + 5\,H_2O$	+0.40

Half-reaction	E^0 (V)
$Cu^+ + e^- \longrightarrow Cu^0$	+0.521
$I_2 + 2e^- \longrightarrow 2I^-$	+0.5355
$I_3^- + 2e^- \longrightarrow 3I^-$	+0.536
$MnO_4^- + e^- \longrightarrow MnO_4^{2-}$	+0.564
$MnO_4^{2-} + 2\,H_2O + 2e^- \longrightarrow MnO_2 + 4\,OH^-$	+0.60
$O_2 + 2\,H_3O^+ + 2e^- \longrightarrow H_2O_2 + 2\,H_2O^-$	+0.682
$Fe^{3+} + e^- \longrightarrow Fe^{2+}$	+0.771
$Hg_2^{2+} + 2e^- \longrightarrow 2\,Hg^0$	+0.789
$Ag^+ + e^- \longrightarrow Ag^0$	+0.7991
$O_2 + 4\,H_3O^+ + 4e^- \longrightarrow 6\,H_2O$ $(10^{-7}\ H_3O^+,\ 10^{-7}\ OH^-)$	+0.82
$NO_3^- + 10\,H_3O^+ + 8\,e^- \longrightarrow NH_4^+ + 13\,H_2O$	+0.84
$Hg^{2+} + 2e^- \longrightarrow Hg^0$	+0.854
$2\,Hg^{2+} + 2e^- \longrightarrow Hg_2^{2+}$	+0.920
$NO_3^- + 3\,H_3O^+ + 2e^- \longrightarrow HNO_2 + 4\,H_2O$	+0.94
$NO_3^- + 4\,H_3O^+ + 3e^- \longrightarrow NO + 6\,H_2O$	+0.96
$HNO_2 + H_3O^+ + e^- \longrightarrow NO + 2\,H_2O$	+1.00
$Br_2 + 2e^- \longrightarrow 2\,Br^-$	+1.0652
$ClO_4^- + 2\,H_3O^+ + 2e^- \longrightarrow ClO_3^- + 3\,H_2O$	+1.19
$IO_3^- + 6\,H_3O^+ + 5e^- \longrightarrow 1/2\,I_2 + 9\,H_2O$	+1.195
$ClO_3^- + 3\,H_3O^+ + 2e^- \longrightarrow 4\,H_2O + HClO_2$	+1.21
$O_2 + 4\,H_3O^+ + 4e^- \longrightarrow 6\,H_2O$ $(1\,M\ H_3O^+)$	+1.229
$MnO_2 + 4\,H_3O^+ + 2e^- \longrightarrow 6\,H_2O + Mn^{2+}$	+1.23
$Cr_2O_7^{2-} + 14\,H_3O^+ + 6e^- \longrightarrow 2\,Cr^{3+} + 21\,H_2O$	+1.33
$Cl_2 + 2e^- \longrightarrow 2\,Cl^-$	+1.3595
$HIO + H_3O^+ + e^- \longrightarrow 1/2\ I_2 + 2\,H_2O$	+1.45
$2\,ClO_3^- + 12\,H_3O^+ + 10e^- \longrightarrow Cl_2 + 18\,H_2O$	+1.47
$Au^{3+} + 3e^- \longrightarrow Au^0$	+1.50
$Mn^{3+} + e^- \longrightarrow Mn^{2+}$	+1.51
$MnO_4^- + 8\,H_3O^+ + 5e^- \longrightarrow Mn^{2+} + 12\,H_2O$	+1.51
$BrO_3^- + 6\,H_3O^+ + 5e^- \longrightarrow 1/2\,Br_2 + 9\,H_2O$	+1.52
$HBrO + H_3O^+ + e^- \longrightarrow 1/2\,Br_2 + 2\,H_2O$	+1.59
$Ce^{4+} + e^- \longrightarrow Ce^{3+}$	+1.61
$HClO + H_3O^+ + e \longrightarrow 1/2\ Cl_2 + 2\,H_2O$	+1.63
$MnO_4^- + 4\,H_3O^+ + 3e^- \longrightarrow MnO_2 + 6\,H_2O$	+1.695
$H_2O_2 + 2\,H_3O^+ + 2e^- \longrightarrow 4\,H_2O$	+1.77
$Co^{3+} + e^- \longrightarrow Co^{2+}$	+1.82
$O_3 + 2\,H_3O^+ + 2e^- \longrightarrow O_2 + 3\,H_2O$	+2.07
$F_2 + 2e^- \longrightarrow 2\,F^-$	+2.65